MEASUREMENT IN SCIENCE AND CIVIL ENGINEERING

Measurement in Science and Civil Engineering

Editor:

Nelson Bolívar, Ph.D

ArclerPress
www.arclerpress.com

Measurement in Science and Civil Engineering
Editor: Nelson Bolívar, Ph.D

708 3rd Avenue, 6th Floor
New York
NY 10017
United States of America
www.arclerpress.com

ISBN: 978-1-68094-372-6

Library of Congress Control Number: 2016950948

About the Editor

Nelson Bolívar, Ph.D

Nelson Bolivar has a PhD. in physics from the University of Lorraine in France finished in 2014. His expertise is in quantum systems and condensed matter. His interest includes spintronic devices and correspondences between general relativity and condensed matter. He is currently an associate professor at the Central University of Venezuela

Contents

List of Contributors

Yulin Zhou
School of Mechanical and Electrical Engineering, Wuyi University, Jiangmen, China

Xin Quan
School of Mechanical and Electrical Engineering, Wuyi University, Jiangmen, China

Tieniu Yang
School of Mechanical and Electrical Engineering, Wuyi University, Jiangmen, China

Shumpei Funatani
Department of Mechanical Systems Engineering, University of Yamanashi, Kofu, Japan

Koji Toriyama
Department of Mechanical Systems Engineering, University of Yamanashi, Kofu, Japan

Tetsuaki Takeda
Department of Mechanical Systems Engineering, University of Yamanashi, Kofu, Japan

David B. Sharp
Department of Engineering and Innovation, Open University, Milton Keynes, UK

Alistair Shawcross
School of Physics and Astronomy, University of Edinburgh, Edinburgh, UK

Clive A. Greated
School of Physics and Astronomy, University of Edinburgh, Edinburgh, UK

Zhenzhen Wang
State Key Laboratory of Multiphase Flow in Power Engineering, Xi'an Jiaotong University, Xi'an, China
Graduate School of Advanced Technology and Science, The University of Tokushima, Tokushima, Japan

Yoshihiro Deguchi
Graduate School of Advanced Technology and Science, The University of Tokushima, Tokushima, Japan

Hiroaki Watanabe
Energy Engineering Research Laboratory, The Central Research Institute of Electric Power Industry, Kanagawa, Japan

Ryoichi Kurose
Department of Mechanical Engineering and Science, Advanced Research Institute of Fluid Science and Engineering, Kyoto University, Kyoto, Japan

Junjie Yan
State Key Laboratory of Multiphase Flow in Power Engineering, Xi'an Jiaotong University, Xi'an, China

Jiping Liu
State Key Laboratory of Multiphase Flow in Power Engineering, Xi'an Jiaotong University, Xi'an, China

Ken McGill
Department of Chemistry, Physics, and Astronomy, Georgia College and State University, Milledgeville, GA, USA

Kathryn Ham
Department of Chemistry, Physics, and Astronomy, Georgia College and State University, Milledgeville, GA, USA

Kris Schock
Department of Chemistry, Physics, and Astronomy, Georgia College and State University, Milledgeville, GA, USA

Endang Suprastiwi
Department of Conservative Dentistry, Faculty of Dentistry, University of Indonesia, Jakarta, Indonesia

Ratna Meidyawati
Department of Conservative Dentistry, Faculty of Dentistry, University of Indonesia, Jakarta, Indonesia

Md. Tawhidul Islam Khan
Department of Advanced Technology Fusion, Saga University, Saga, Japan

Maeda Kazuhiko
Faculty of Science and Engineering, Saga University, Saga, Japan

Kenbu Teramoto
Department of Advanced Technology Fusion, Saga University, Saga, Japan

Md. Mahbub Hasan
Graduate School of Science and Engineering, Saga University, Saga, Japan

Fei Duan
School of Mechanical and Aerospace Engineering, Nanyang Technological University, 50 Nanyang Avenue, 639798, Singapore

Ting Foong Wong
School of Mechanical and Aerospace Engineering, Nanyang Technological University, 50 Nanyang Avenue, 639798, Singapore

Alexandru Crivoi
School of Mechanical and Aerospace Engineering, Nanyang Technological University, 50 Nanyang Avenue, 639798, Singapore

Joe-Ming Chang
Institute of NanoEngineering and MicroSystems, Hsinchu 30013, Taiwan.

Wei-Yu Chang
Engineering and System Science Department, National Tsing Hua University, Hsinchu 30013, Taiwan

Fu-Rong Chen
Engineering and System Science Department, National Tsing Hua University, Hsinchu 30013, Taiwan

Fan-Gang Tseng
Institute of NanoEngineering and MicroSystems, Hsinchu 30013, Taiwan.
Engineering and System Science Department, National Tsing Hua University, Hsinchu 30013, Taiwan
Applied Science Research Center, Academia Sinica, Taipei 11529, Taiwan

Clement Kleinstreuer
Department of Mechanical and Aerospace Engineering, NC State University, Raleigh, NC 27695-7910, USA

Yu Feng
Department of Mechanical and Aerospace Engineering, NC State University, Raleigh, NC 27695-7910, USA

Jaroslav Holis
Department of Electromagnetic Field, Czech Technical University in Prague, Technicka 2 Street, 166 27 Praha 6, Czech Republic

Pavel Pechac
Department of Electromagnetic Field, Czech Technical University in Prague, Technicka 2 Street, 166 27 Praha 6, Czech Republic

K. De Wilder
Department of Civil Engineering, Building Materials and Building Technology Section, KU Leuven, 3001 Heverlee, Belgium

G. De Roeck
Department of Civil Engineering, Strucutral Mechanics Section, KU Leuven, Kasteelpark Arenberg 40, Box 2448, 3001 Heverlee, Belgium.

L. Vandewalle
Department of Civil Engineering, Building Materials and Building Technology Section, KU Leuven, 3001 Heverlee, Belgium

Florian Moser
Institute of Engineering Geodesy and Measurement Systems, Graz University of Technology, Steyrergasse 30, 8010 Graz, Austria

Werner Lienhart
Institute of Engineering Geodesy and Measurement Systems, Graz University of Technology, Steyrergasse 30, 8010 Graz, Austria

Helmut Woschitz
Institute of Engineering Geodesy and Measurement Systems, Graz University of Technology, Steyrergasse 30, 8010 Graz, Austria

Hartmut Schuller
INSITU Geotechnik ZT GmbH, Gartengasse 19, 8010 Graz, Austria

Ping Wang
Key Laboratory of High-speed Railway Engineering, Ministry of Education, Southwest Jiaotong University, Chengdu 610031, China

School of Civil Engineering, Southwest Jiaotong University, Chengdu 610031, China

Li Wang
Key Laboratory of High-speed Railway Engineering, Ministry of Education, Southwest Jiaotong University, Chengdu 610031, China
School of Civil Engineering, Southwest Jiaotong University, Chengdu 610031, China

Rong Chen
Key Laboratory of High-speed Railway Engineering, Ministry of Education, Southwest Jiaotong University, Chengdu 610031, China
School of Civil Engineering, Southwest Jiaotong University, Chengdu 610031, China

Jinhui Xu
Key Laboratory of High-speed Railway Engineering, Ministry of Education, Southwest Jiaotong University, Chengdu 610031, China
School of Civil Engineering, Southwest Jiaotong University, Chengdu 610031, China

Jinmang Xu
Key Laboratory of High-speed Railway Engineering, Ministry of Education, Southwest Jiaotong University, Chengdu 610031, China
School of Civil Engineering, Southwest Jiaotong University, Chengdu 610031, China

Mingyuan Gao
Key Laboratory of High-speed Railway Engineering, Ministry of Education, Southwest Jiaotong University, Chengdu 610031, China
School of Civil Engineering, Southwest Jiaotong University, Chengdu 610031, China

S. Papini
Department of Energy Engineering "S. Stecco", University of Florence, Florence, Italy

L. Pugi
Department of Energy Engineering "S. Stecco", University of Florence, Florence, Italy

A. Rindi
Department of Energy Engineering "S. Stecco", University of Florence, Florence, Italy

E. Meli
Department of Energy Engineering "S. Stecco", University of Florence, Florence, Italy

I. W. Nam
Department of Civil and Environmental Engineering, Korea Advanced Institute of Science and Technology, Daejeon 34141, South Korea

H. K. Lee
Department of Civil and Environmental Engineering, Korea Advanced Institute of Science and Technology, Daejeon 34141, South Korea

Laura Chiacchiari
Dipartimento di Ingegneria Civile e AmbientaleSapienza, Università di Roma

Giuseppe Loprencipe
Dipartimento di Ingegneria Civile e AmbientaleSapienza, Università di Roma

Gun Kim
School of Civil and Environmental Engineering, Georgia Institute of Technology, Atlanta, GA 30332-0355, USA

Jin-Yeon Kim
School of Civil and Environmental Engineering, Georgia Institute of Technology, Atlanta, GA 30332-0355, USA

Kimberly E. Kurtis
School of Civil and Environmental Engineering, Georgia Institute of Technology, Atlanta, GA 30332-0355, USA

Laurence J. Jacobs
School of Civil and Environmental Engineering, Georgia Institute of Technology, Atlanta, GA 30332-0355, USA
G. W. Woodruff School of Mechanical Engineering, Georgia Institute of Technology, Atlanta, GA 30332-0355, USA

Yann Le Pape
Oak Ridge National Laboratory, Oak Ridge, TN 37831-6148, USA

Maria Guimaraes
Electric Power Research Institute (EPRI), Charlotte, NC 28262-8550, USA

Preface

In order to develop, research, model, or construct it is require to perform measurements on those systems we are interested. The type of measurement will be given by the nature of the object, or the subject under investigation.

In science the measurements often pursue to understand very complicated phenomena, inaccessible sometimes to the currents measurement devices or that may need some innovation in the protocols used. Of course it is in science and engineering where most of the innovations manifest primary and then are exported to other fields, for instance the measure devise in high energy experiments carry a lot of new techniques and protocols to improve the characterization of a variety of systems. In a laboratory the conditions to perform a measurement can be controlled to a certain degree, but facing measurement in outside scenarios involves more complex situations. Typically, any construction needs the proper elaboration of it constituents, saying concrete, steel of any other product of pre-manufacturing origin, that requires in every step of its fabrication a control and monitoring, even beyond the factories, the in-situ measure of certain terrains characteristics, tension on rail roads, porosity in the ground and in the structures, all of these depends of meticulous and appropriate measurements using a variety of techniques, very often originated in pure sciences. Students, civil engineers, and materials scientists will find this book to be a good comprehensive resource for learning about the fundamental methods used in those fields to achieve trustworthy measures, focusing on specific examples that are also top research in the area, suitable for the interested reader.

Editor

Nelson Bolívar, Ph.D

INTRODUCTION

Measurements for civil engineering works are mainly categorized for their nature, shape and size and for making payments to the contractor. The principle of units of measurements normally consists the following:

a Single units work like doors, windows, trusses etc., is expressed in numbers.

b Works consists linear measurements involve length like cornice, fencing, hand rail, bands of specified width etc., are expressed in running meters (RM)

c Works consists areal surface measurements involve area like plastering, white washing, partitions of specified thickness etc., and are expressed in square meters (m^2)

d Works consists cubical contents which involve volume like earth work, cement concrete; Masonry etc are expressed in Cubic meters.

The rules for measurement of each item are invariably described in IS – 1200. However some of the general rules are:

1. Measurement shall be made for finished item of work and description of each item shall include materials, transport, labor, fabrication tools and plant and all types of overheads for finished the work in required shape, size and specification.

2. In booking, the order shall be in sequence of length, breadth and height or thickness.

3. All works shall be measured subject to the following tolerances.

    - Linear measurement shall be measured to the nearest 0.01m.
    - Areas shall be measured to the nearest 0.01 SQM
    - Cubic contents shall be worked out to the nearest 0.01 cum.

4. Same type of work under different conditions and nature shall be measured separately under separate items.

5. The bill of quantities shall fully describe the materials, proportions, workmanships and accurately represent the work to be executed.

6. In case of masonry (stone or brick) or structural concrete, the categories shall be measured separately and the heights shall be described as:
    - From foundation to plinth level
    - From plinth to first floor level
    - From first floor to second floor level and so on.

Civil engineering is one of the oldest engineering disciplines, since civil engineers of one form or another have been around ever since humans started building major public works such as roads, bridges, tunnels, and large public buildings. It is also an incredibly broad discipline, spanning treatment of environmental issues, transportation, power generation, and major structures. To become a civil engineer, a person must typically study engineering at a university and then participate in field work for practical training. Many nations also require that students pass a competency exam to ensure that they will be able to design and build safe, stable structures.

There are many branches of civil engineering, and a wide range of specialties. Some engineers focus on conception and initial design of a project, analyzing the site, the needs, and the resources to come up with a workable project plan. Others specialize in contracting, physically building the structure, managing the site crew, and handling materials and supply. In other cases, civil engineers focus on maintenance of the project after it is completed, to make sure that it is safe and useful.

Most people pick a focus while they are receiving an education. Engineers who focus on transportation, for example, might choose to specialize in building bridges, tunnels, and roads. Others might lean towards power generation facilities, water treatment, waste management, construction of light railways and subways, or other disciplines. In all cases, extensive training is undertaken so that the prospective civil engineer understands his or her chosen field in depth. Behind every major public works is a team of civil engineers.

One of the primary concerns of civil engineering is public safety and health. A value is also placed on building structures that are functional, efficient, and also aesthetically pleasing. Structural soundness, conformity with local codes, and functionality are all issues which are faced in the discipline. Some civil engineers work directly for the public in the form of government agencies, while others find employment with public firms.

Education does not end with a degree and a course of fieldwork with trained and experienced civil engineers. Continuing education is also an important part of this discipline. As advances are made in the field, engineers are expected to keep pace with them, especially when the advances improve safety for workers and the public. There are many trade journals and annual conferences in the field to keep engineers updated.

Basic Weights, Measures and Conversion Factors in Civil Engineering

Metric System

A system of weights and measures which was first adopted by France and is now almost universally used in the sciences and arts. The greater part of the population of the world works on the metric system. The metric is so called because it uses the “meter” as the primary unit.

SI System of Units

The International System of units abbreviated as “SI”. From French the SI is accepted in United States and throughout the world. It is a modern version of Metric System of units by the international agreement.

U.S Customary Units

The U.S customary or British System of Units also called the Foot Pound Second System (FPS) System has been the common system in English speaking countries, although the system within time be replaced by the SI Units (Best for many Years).

Basic Quantities and Units

Quantity	Dimensions	SI Units	FPS Units
Mass	M	Kg	Slug
Length	L	M	ft.
Time	T	sec.	sec.
Force	F	Newton (N)	Pound (lb)

Basic Standards

The standard relationship between English measures and Metric measures is as follows;

1 inch = 25.4 millimeters = 2.54cm

1 meter = 39.37 inches =1.09 yards

1 liter = 0.22 galls (imp.)

1 gallon (imp.) = 4.546 liters

1 gallon (US) = 3.785 liters

1 Kilogram (kg) = 2.2046 pounds (lb)

Metric Units of Weights/Mass

1 tonne = 1,000,000 grams = 1000 kilograms

1 quintal = 100,000 grams = 100 kilograms

1 Slug = 14.606 kg

1 Slug = 32.2 lb

Measures of Length:

Following are the basic measures of length in British Units;

12 inches = 1 foot

3 feet = 1 yard

220 yards = 1 furlong

8 fulongs = 1 mile

Metric Units of Liquid Measures

1 liter = 1000ml

1 liter = 1kg

1 kiloliter (1000 liters) = 1 cu. Meter

Conversion Factors

1 cu. ft. = 28.317 liters

1 cu. ft. = 0.028 cu. meters

1 cu. ft. = 6.24 galls (imperial)

1 cu. ft. = 7.48 galls (US)

1 imp. gall = 1.20 galls (US), liquid

1 imp. gall = 1.03 galls (US), dry

Sub-disciplines of Civil Engineering

Civil engineering is a multiple science encompassing numerous sub-disciplines that are closely linked with each other. The various sub-disciplines of civil engineering are mentioned below:

- *Structural Engineering*: This discipline involves the design of structures that should be safe for the users, be economical, and accomplish the desired functions. The design and analysis should initially identify the loads that act on the structures, stresses that are created due to loads, and then design the structure to withstand these loads. It includes steel structures, buildings, tunnels, highways, dams, and bridges.

- *Geotechnical Engineering*: Geotechnical engineering deals with soils, rocks, foundations of buildings and bridges, highways, sewers and underground water systems. Technical information obtained from the sciences of geology, material testing, and hydraulics is applied in the design of foundations and structures to ensure safety and economy of construction.

- *Water Resources Engineering*: This discipline of civil engineering concerns the management of quantity and quality of water in the underground and above ground water resources, such as rivers, lakes and streams. Geographical areas are analyzed to forecast the amount of water that will flow into and out of a water source. Fields of hydrology, geology, and environmental science are included in this discipline of civil engineering.

- *Environmental Engineering*: It is related to the science of waste management of all types, purification of water, cleaning of contaminated areas, reduction of pollution, and industrial ecology. Technical data obtained due to environmental engineering assists the policy makers in making decisions related to environmental issues.

- *Other Disciplines*: Some of the other disciplines included in civil engineering include coastal engineering, construction engineering, earthquake engineering, materials science, transportation engineering, and surveying.

Scope of Different Fields of Civil Engineering

Civil Engineering may be divided into the following fields:

(i) Surveying

(ii) Building Materials

(iii) Construction Technology

(iv) Structural Engineering

(v) Geotechnical Engineering

(vi) Hydraulics

(vii) Water Resources and Irrigation Engineering

(viii) Transportation Engineering

(ix) Environmental Engineering and

(x) Architecture and Town planning

Scope of each one of these is discussed below.

Surveying

Surveying is the science of map makng. To start any development activity in an area the relative positions of various objects in the horizontal and vertical directions are required. This is approved by surveying the area. Earlier, the conventional instruments like chain, tape and levelling instruments were used. In this electronic era, modern equipments like distance meters and total stations are used to get more accurate results easily. The modern technologies like photogrammetry and remote sensing have made surveying easier.

Building Materials

Shelter is the basic need of civilization. To get good shelter continuous efforts are going on right from the beginning of civilization. Stones, bricks, timber, lime, cement, sand, jellies and tiles are the traditional building materials. Use of steel, aluminium, glass, glazed tiles, plaster of Paris, paints and varnishes have improved the quality of buildings. The appropriate mixture of binding materials like lime and cement with sand is known as mortar. The mixture of cement, sand and jelly (crushed stones) with water is known as concrete. The use of concrete with steel bars placed in appropriate position has helped in building strong and durable tall structures. The composite material of concrete and steel is called reinforced cement concrete which is popularly known as R C C. A civil engineer must know the properties of all the building materials so that they can be used appropriately. Improved versions of many building materials appear in the market A good civil engineer will make use of them at the earliest.

Construction Technology

Construction is the major activity of civil engineering which is continuously improving. As land cost is going up there is demand for tall structures in urban areas while in rural areas need is for low cost constructions. One has to develop technology using locally available materials. In India, contribution of Central Building Research Institute (CBRI)

- Roorkee and Gaziabad, several educational institutions throughout the country and Nirmithi Kendras in the technology development are noteworthy.

Structural Engineering

Load acting on a structure is ultimately transferred to ground. In doing so, various components of the structure are subjected to internal stresses. For example, in a building, load acting on a slab is transferred by slab to ground through beams, columns and footings. Assessing the internal stresses in the components of a structure is known as Structural Analysis and finding the suitable size of the structural component is known as design of structure. The structure to be analysed and designed may be of masonry, R. C C. or steel. Upto midsixties considerable improvements were seen in classical analysis. With the advent of computers numerical methods emerged and analysis and design packages are becoming popular. Matrix Method of analysis and Finite Elements Analysis have helped in the analysis of complex structures. A civil engineer has not only to give a safe structure but he has to give economical sections. To get economical section mathematical optimization techniques are used. Frequent earthquakes in the recent years have brought, importance of analysis of the structure for earthquake forces. Designing earthquake resistant structures is attracting lot of researches. All these aspects fall under structural engineering field.

Geotechnical Engineering

Soil property changes from place to place. Even in the same place it may not be uniform at various depths. The soil property may vary from season to season due to variation in moisture content. The load from the structure is to be safely transferred to soil. For this, safe bearing capacity of the soil is to be properly assessed. This branch of study in Civil Engineering is called as Geotechnical Engineering. Apart from finding safe bearing capacity for foundation of buildings, geotechnical engineering involves various studies required for the design of pavements, tunnels, earthen dams, canals and earth retaining structures. It involves study of ground improvement techniques also.

Hydraulics

Water is an important need for all living beings. Study of mechanics of water and its flow characteristics is another important field in Civil Engineering and it is known as hydraulics.

Water Resources and Irrigation Engineering

Water is to be supplied to agriculture field and for drinking purposes. Hence suitable water resources are to be identified and water is to be stored. Identifying, planning and building water retaining structures like tanks and dams and carrying stored water to fields is known as water resources and irrigation engineering.

Transportation Engineering

Transportation facility is another important need. Providing good and economical roads is an important duty of civil engineers. It involves design of base courses, suitable surface finishes, cross drainage works, road intersections, culverts, bridges, tunnels etc. Railway is another important long-way transport facility. Design, construction and maintenance of railway lines, signal system are part of transportation engineering. There is need for airports and harbours. For proper planning of these transportation facility, traffic survey is to be carried out Carrying out traffic survey, design, construction and maintenance of roads, bridges, railway, harbour and airports is known as transportation engineering.

Environmental Engineering

Proper distribution of water to rural areas, towns and cities and disposal of waste water and solid waste are another field of civil engineering. Industrialisation and increase in vehicular traffic are creating air pollution problems. Environmental engineering while tackling all these problems provides healthy environment to public.

Architecture and Town Planning

Aesthetically good structures are required. Towns and cities are to be planned properly. This field of engineering has grown considerably and has become a course separate from Civil Engineering.

Types of Infrastructure

Infrastructure facilities involve various civil engineering amenities, electricity, telephone, internet facility, educational and healthcare facilities.

Civil engineering amenities in the infrastructure developments are listed below:

- A good town planning and developing sites
- Providing suitable roads and network of roads
- Railway connection to important places
- Airports of national and international standards
- Assured water supply to towns, cities and rural areas
- A good drainage and waste disposal system
- Pollution free environment

Effect of Infrastructure Facilities

Connecting producing center to marketing places minimizes exploitation from middlemen. Both producer and consumers are benefitted. Imports and exports become easy as a result of which whole world becomes a village. The infrastructure development generates scope for lots of industries. Manpower is utilized for the benefit of mankind. Antisocial activities come under control. Improved education and healthcare give rise to skilled and healthy work force. Quality of life of the people is improved. In case of natural calamities assistance can be extended easily and misery of affected people is reduced. Infrastructure facility improves defense system and peace exists in the country. Improved economical power of the country brings a respectable status in the world. The world has realized that a government should not involve itself in production and distribution but should develop infrastructure to create an atmosphere for economical development.

Role of Civil Engineers in the Infrastructure Development

A civil engineer has to conceive, plan, estimate, get approval, create and maintain all civil engineering infrastructure activities. He has to carry out research and training programmes to improve the technology. Civil engineer has a very important role in the development of the following infrastructures:

(i) Town and city planning

(ii) Build suitable structures for the rural and urban areas for various utilities.

(iii) Build tanks, dams to exploit water resources.

(iv) Purify the water and supply water to needy areas like houses, schools, offices, and agriculture field.

(v) Provide good drainage system and purification plants.

(vi) Provide and maintain communication systems like roads, railways, harbours and airports.

(vii) Monitor land, water and air pollution and take measures to control them.

Contributions of Civil Engineers

They build the world, literally speaking. They are responsible for the analysis, design, construction, upgrading, and maintenance of all forms of infrastructure for better quality of life and sustainable economy. The scope of civil engineering is multifaceted and versatile. Civil engineers contribute in the following ways:

1. They contribute in the construction of the following:

- Houses and apartments
- Office buildings, hotels, and shopping malls
- Factories
- Planning and design of transportation facilities such as highways and railways
- Ports and harbours, railway stations, bus and truck terminals, and airports and helipads

- f Canals and drainage systems
- Reclaimed land and underground caverns
- Offshore platforms, oil refineries, and power stations
- Stadiums and amusement parks
- Hospitals, schools, army camps, bomb shelters, dams, desalination plants, etc

2. They also contribute to the water that we drink and to the clean air that we breathe in.
3. They also help to preserve our environment by assisting in the cleaning up of existing pollution and planning ways to reduce the future pollution of our air, land, and water.
4. They help in surveying and in the preparation of estimates.

Civil Engineering Materials

The basic materials used in civil engineering applications or in construction projects are given as follows:

Wood: It is derived from trees and can be put to use directly, as pieces of lumber cut from a log or as a raw material in the manufacture of various wood products or manufactured components. Plywood, glue-laminated timber, and oriented strand board are some of the wood products most commonly found in the construction of buildings and bridges.

Cement and concrete: Concrete is one of the most common construction materials, in which Portland cement is the essential ingredient. Portland cement (and other types of hydraulic cement) is also a key ingredient in the manufacture of many other cementitious products, such as masonry blocks, soil—cement bricks, and plaster. In combination with other materials, such as reinforcing bars, polypropylene fibres, and high-strength strands or wires, different types of concrete are produced, such as reinforced, fibre, and prestressed concrete.

Bitumen and bituminous materials: Bitumen, which comes in a variety of forms, is mixed with other raw materials for the construction of pavements, roof shingles, waterproofing com-pounds, and many other materials.

Structural clay and concrete units: They are the principal elements in the construction of masonry walls. They are commonly called bricks and blocks.

Reinforcing and structural steels: They are used in the construction of railroad ties, high-rise buildings, roof trusses, and many more structural elements. These basic materials or products are selected for their properties, performance, availability, aesthetics, and cost. The knowledge of all these aspects is essential in selecting a suitable material for a particular situation.

These are sometimes called structural materials. Added to these are plastics, soils, and aluminium. All these materials are employed in a variety of civil engineering structures such as dams, bridges, roads, foundations, liquid-retaining structures, waterfront construction, buildings, and retaining walls. The basic materials most common to highway construction are soils, aggregates, bituminous binders, lime, and cement. In addition to the materials mentioned above, there are a significant number of secondary construction materials common to engineering projects. Sealants, adhesives, floor and wall coverings, fasteners, and doors and windows fall into this category. Most of these, also called nonstructural materials, are chosen based on quality guidelines and aesthetic considerations.

1
CHAPTER

MEASUREMENT UNCERTAINTY AND ITS APPLICATION IN THE VACUUM PRESSURE MEASUREMENT

Yulin Zhou, Xin Quan, Tieniu Yang

Keywords: Capacitance Diaphragm Gauges, Pressure Measurement, Measurement Uncertainty

INTRODUCTION

The progress of science and technology depends on the development of microelectronics and semiconductor technology that the upgrading and renewing of equipment and technology of microelectronics and semiconductor consequently grow rapidly. This attracted some researchers to study on it. Regardless of from equipment research and development or technological parameters on the need to measure pressure in integrated circuit (IC) processing chamber.

Evaluation of uncertainty is widely used in the test, measurement and other fields of engineering research [1] [2] . Capacitance film gauges (CDGs) measuring the pressure have some advantages such as high accuracy, good linearity and measurement result has nothing to do with the gas composition and types so that it can be applied to measure pressure in IC processing chamber. Due to the difference in manufacture, using time and over-pressure environment that will cause the performance difference [3] . To study the pressure difference in vacuum chamber during dynamic gas flow when the inlet flow rate is low, differential pressure sometimes is only a fraction of Pa. Therefore, the uncertainty factors of capacitance film gauge are very important for the pressure measurement. This paper based on ASME PTC 19.2-2010 "Pressure Measurement Instruments and Apparatus Supplement" of China as the standard for studying the

Citation: Yulin Zhou, Xin Quan, Tieniu Yang (2015) Measurement Uncertainty and Its Application in the Vacuum Pressure Measurement. Open Journal of Applied Sciences,05,495-500. doi: 10.4236/ojapps.2015.58048.

uncertainty of pressure measurement. This experiment adopts three capacitance film gauges which come from INFICON Instruments Inc., USA and its range is 1333 Pa. Through the experiment, pressure difference between film gauges in measuring pressure is got. Various factors [4] which influence the uncertainty are analyzed and the results provide a base for measuring pressure or pressure differences.

EXPERIMENTAL APPARATUS

The experimental system mainly includes the gas intake system, pressure measurement system, extraction system. The height of the chamber is 320 mm while its inner diameter is 580 mm (capacity = 84.5 L) as shown in Figure 1. And the vacuum chamber made of 304 stainless steel that is pumped by a high vacuum pumping system is composed of a turbo molecular pump with pumping speed 300 $L \cdot s^{-1}$ for N_2 backed by a 82 $m^{3} \cdot s^{-1}$ roots pump. The chamber is extracted through a flapper valve of diameter 100 mm. At first, the vacuum chamber wall would suffer two hours' baking (about 100 degrees Celsius) by a heating jacket, then, the chamber would be extracted for about four hours. It could be achieved that the pressure of the chamber was less than 2 ´ 10^{-4} Pa. Based on the static pressure boosting method, the overall leak rate of the chamber was 8.84 ´ 10^{-6} $Pa \cdot m^{2} \cdot s^{-1}$ [5] .

For this purpose, three capacitance diaphragm gauges (CGG1, CGG2, CGG3, zero pressure less than 4 ´ 10^{-4} Pa) are flanged joint on the cavity wall as shown in Figure 2.

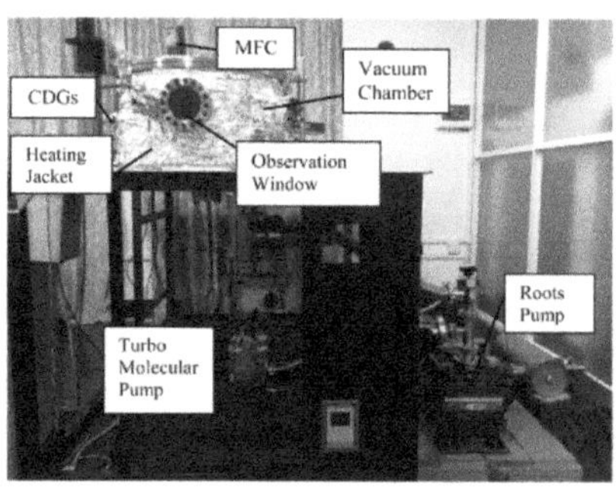

Figure 1. The basic composition of chamber.

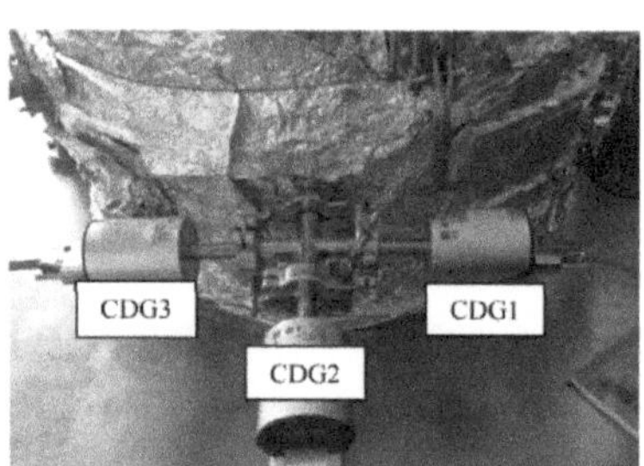

Figure 2. CDGs locations.

EXPERIMENTAL

Experimental Process

Experimental research on pressure in the range of 10 - 100 Pa, it is necessary to investigate pressure difference in static environment of the three film gauge. Setting the static pressure values such as 20 Pa, 30 Pa, 40 Pa, 50 Pa, 60 Pa, 70 Pa, 80 Pa, 90 Pa, 100 Pa and recording readings of the CDGs, to find out the difference between them. Specific process is as follows: firstly, increasing chamber pressure to over 100 Pa through the vent valve. Secondly, only starting the roots pump to make the pressure at setting value, stopping evacuating. Finally, Please wait until this date is stable and record data.

In the above experiment system, controlling strictly the influence of other experimental factors, such as, keeping the indoor temperature at 20°C ± 0.2°C and the humidity at 50% ± 2% RH, ensuring the Experiments occur in the condition that in absence of noise and vibration situation. In order to exclude specific situation, repeat the experiment 10 times.

Experimental Results

The experimental data are dealt with error processing, and the conclusions from these are compared. For example, in terms of 20 Pa, the average measured data, the residual error of the CDG can be obtained, respectively. The average of residual error in 10 times test to obtain average residual error of every CDG at 20 Pa shown in Table 1. The data processing steps of other pressure are as same as the condition of static pressure 20 Pa, the results as shown in Table 1.

Table 1 shows the average residual error difference of CDGs, for examples, 1 - 2 means difference between CDG1 and CDG2, others like them. The hydrostatic pressure difference of CDGs is shown in Table 2.

The static pressure of chamber should be equal. Table 2 shows that there are performance differences among CDGs, so the measurement results are corrected based on the pressure difference.

UNCERTAINTY CALCULATION

Mathematical Model Established

The pressure measurement mathematical model of CDG is denoted from the simple relation.

$$P = P_u + \delta P_s + \delta P_t \qquad (1)$$

where P_u (P_a) is the Instrument measurements; δP_s is the instrument error; δP_t is the environment error.

Uncertainty Evaluation

The experiment experience shows that the significant factors affecting the CDG accurate measurement such as measurement repeatability, apparatus, and temperature. Analysis of the uncertainty characteristics, the components (μ_1) is type A evaluation of standard uncertainty and components (μ_2, μ_3) are type B.

Table 1. The average residual error of CDG under different pressure.

CDG	20/Pa	30/Pa	40/Pa	50/Pa	60/Pa	70/Pa	80/Pa	90/Pa	100/Pa
1	0.08	0.18	0.14	0.11	0.06	0.18	0.16	0.06	0.12
2	−0.05	−0.06	−0.05	−0.03	−0.03	−0.08	−0.06	−0.02	−0.06
3	−0.03	−0.12	−0.09	−0.08	−0.03	−0.10	−0.10	−0.04	−0.06

Table 2. The differential pressure of the static pressure.

The differences of CDGs	20/Pa	30/Pa	40/Pa	50/Pa	60/Pa	70/Pa	80/Pa	90/Pa	100/Pa
1 - 2	0.13	0.24	0.19	0.13	0.9	0.26	0.22	0.08	0.18
1 - 3	0.11	0.30	0.23	0.19	0.9	0.28	0.26	0.10	0.18
2 - 3	−0.02	0.06	0.03	0.06	0.00	0.02	0.04	0.02	0.00

The Components (μ_1) of the Uncertainty Caused by the Repeatability

Using a Bessel method to calculate a single measure standard deviation and the results are shown in Table 3.

The standard deviation of CDG2 is the smallest of the three at same pressure that explaining its stability is best. On the contrary, the stability of the CDG1 is the worst.

The average standard deviation $\sigma_{\bar{P}} = \sigma/\sqrt{10}$, that is to say, The components of the uncertainty caused by the repeatability, $\mu_1 = \sigma_{\bar{P}}$. The results are as follows (Table 4).

The degrees of freedom $\upsilon_1 = n - 1 = 9$

The Components (μ_2) of the Uncertainty Caused by the Apparatus

According to the instrument specifications, CDGs' indication error (δ) is 0.2% readings. The indication error of setting pressure is shown in Table 5. According to uniformly distributed, the components of the uncertainty caused by the apparatus can be calculated as following.

Due to the stability of the instrument is reliable, the degrees of freedom $\upsilon_2 = \infty$.

The Components (μ_3) of the Uncertainty Caused by the Temperature

According to the instrument specifications, CDGs' error is 0.0050% F.S/°C. Due to the indoor temperature at 20°C ± 0.2°C, the error $\delta = 0.0050\% \times 1333$ Pa/°C × 0.4°C = 0.027 Pa. According to uniformly distributed, the components of the uncertainty caused by the temperature can be calculated from the simple relation.

$u_3 = 0.027/\sqrt{3} = 0.0156$ Pa

Due to the stability of the instrument is reliable, Degrees of freedom $\upsilon_2 = \infty$.

Combined Uncertainty

The uncertainty components such as u_1, u_2, u_3 are independent of each other, that is to say, $\rho_{ij} = 0$. The combined uncertainty $u = \sqrt{u_1^2 + u_2^2 + u_3^2}$, the results are as follows (Table 6).

The free degree of synthetic standard uncertainty was calculated from the following formula.

Table 3. Single measurement standard deviation.

CDG	20/Pa	30/Pa	40/Pa	50/Pa	60/Pa	70/Pa	80/Pa	90/Pa	100/Pa
1	0.016	0.014	0.029	0.012	0.016	0.010	0.011	0.009	0.011
2	0.009	0.013	0.013	0.011	0.016	0.009	0.009	0.009	0.007
3	0.017	0.010	0.018	0.010	0.009	0.013	0.010	0.013	0.012

Table 4. The components (μ_1) of the uncertainty caused by repeatability.

CDG	20/Pa	30/Pa	40/Pa	50/Pa	60/Pa	70/Pa	80/Pa	90/Pa	100/Pa
1	0.005	0.004	0.009	0.004	0.005	0.003	0.004	0.003	0.003
2	0.003	0.004	0.004	0.003	0.005	0.003	0.003	0.003	0.002
3	0.005	0.003	0.005	0.003	0.003	0.004	0.003	0.004	0.004

Table 5. The uncertainty caused by instrument error.

$$v = \frac{u^4}{\frac{u_1^4}{v_1} + \frac{u_2^4}{v_2} + \frac{u_3^4}{v_3}} \qquad (2)$$

The data in Table 7 are large number can explain the evaluation of uncertainty is better.

Expanded Uncertainty

Taking confidence probability P = 0.95, through the degrees of freedom in Table 7 to check the t distribution table [6] , $t_{0.95}(\upsilon) = 1.96$, that is to say, coverage factor k = 1.96. The expanded uncertainty of pressure measurement is shown in Table 8.

Discussion and Analysis

The results show the combined uncertainty and expanded uncertainty of CDGs are considered equal at same pressure. Relative to components μ_2 and μ_3, components μ_1 is much smaller. That is to say, the experiment error caused by the repeatability is minimum. The CDG adopted in the experiment is the most accurate to measure vacuum pressure in the market. Lacking of a more precise instrument for reference, it is difficult to determine which regulate has the highest accuracy through experiment result but its stability can be judged by the standard deviation. Learn from the hydrostatic pressure difference of the CDG in this experiment, Follow-up experiments can accurately measure the pressure difference between two points of interior chamber.

CONCLUSION

In a follow-up experiment, the measurement uncertainty was evaluated based on the results in Tables 6-8 when the CDG was used to measure the chamber pressure. The measurement results are corrected based on the hydrostatic pressure difference and measurement uncertainty when measuring pressure differences. The uncertainty evaluation process of measuring pressure can provide reference for vacuum measurement in the future and measurement pressure difference between two points of vacuum chamber can provide data support for designing process parameters.

Table 6. The value of synthetic standard uncertainty.

CDG	20/Pa	30/Pa	40/Pa	50/Pa	60/Pa	70/Pa	80/Pa	90/Pa	100/Pa
1	0.028	0.038	0.050	0.060	0.071	0.082	0.094	0.105	0.117
2	0.028	0.038	0.049	0.060	0.071	0.082	0.094	0.105	0.117
3	0.028	0.038	0.049	0.060	0.071	0.082	0.094	0.105	0.117

Table 7. The free degree of synthetic standard uncertainty.

CDG	20 Pa	30 Pa	40 Pa	50 Pa	60 Pa	70 Pa	80 Pa	90 Pa	100 Pa
1	9.2×10^3	5.7×10^4	7.6×10^3	5.6×10^5	3.2×10^5	4.0×10^6	4.7×10^6	1.6×10^7	1.4×10^7
2	9.0×10^4	8.3×10^4	1.7×10^5	9.8×10^5	3.7×10^5	6.8×10^6	1.3×10^7	2.1×10^7	7.2×10^7
3	7.4×10^3	1.6×10^5	4.6×10^4	1.1×10^6	3.3×10^6	1.5×10^6	8.6×10^6	3.5×10^6	9.0×10^6

Table 8. The value of expanded uncertainty.

CDG	20/Pa	30/Pa	40/Pa	50/Pa	60/Pa	70/Pa	80/Pa	90/Pa	100/Pa
1	0.056	0.075	0.097	0.117	0.140	0.162	0.184	0.206	0.230
2	0.055	0.075	0.096	0.117	0.140	0.162	0.184	0.206	0.230
3	0.056	0.075	0.096	0.117	0.140	0.162	0.184	0.206	0.230

ACKNOWLEDGEMENTS

The research work was supported by grant No.2011ZX02403-004 of the National

Key Technology Research and Development Program of the Ministry of Science and Technology of China. A simulate system and experimental platform of IC Equipment including Process Chamber, supported by Multidisciplinary Collaborative Designing.

REFERENCES

1. Wu, X.H., Wang, C.D. and Chen, F. (2015) Measurement Uncertainty Degree Analysis for the Actual Super Elevation of Calibrater for Track Inspection Instrument. Science Technology and Engineering, 15, 226-228, 238. (In Chinese)
2. Wang, Z.H.G., Ma, Y.T. and Lu, W. (2015) The Application of Uncertainty Analysis Theory in Thermal Efficiency Testing for Boiler. Proceedings of the CSEE, 25, 125-129. (In Chinese)
3. Li, Z.H. (1989) Several Questions Riveted Attention in the Application of Capacitance Diaphragm Gauge. Vacuum, 3, 51-53. (In Chinese)
4. Li, Z.H. (1998) Method for Evaluating the Uncertainty of Vacuum Measurement. Vacuum & Cryogenics, 4, 181-183. (In Chinese)
5. Zhou, Y.L. and Yang, T.N. (2015) Air Tightness Testing and Analysis of Vacuum Chambers of IC Equipment. Journal of Wuyi University, 29, 59-63, 78. (In Chinese)
6. Fei, Y.T. (2005) The Error Theory and Data Processing. 4th Edition, China Machine Press, Beijing. (In Chinese)

2

CHAPTER

TEMPERATURE MEASUREMENT OF AIR FLOW USING FLUORESCENT MISTS COMBINED WITH TWO-COLOR LIF

Shumpei Funatani, Koji Toriyama, Tetsuaki Takeda

Keywords: Flow Visualization, Two-Color LIF, Temperature Measurement, Spraying Mist

INTRODUCTION

In order to understand the transport phenomena of heat and fluid flow in various fields of engineering, it is essential to develop experimental methods for measuring the temperature field in the fluid flow.

For temperature field measurement of an airflow, the nitric oxide LIF (NO-LIF) method [1-3] or the acetone planar LIF (PLIF) method [4-5] have been employed. However, these methods have two drawbacks: the sensitivity of fluorescence intensity for temperature variation in these methods—NO-LIF: 0.05%/K [1] and PLIF: 0.4%/K [4]—is lower than that for the two-color laserinduced fluorescence (LIF) method for the temperature measurement of water flow—2.3%/K [6]—and a very expensive UV laser is required for the NO-LIF and PLIF methods.

The two-color LIF technique is suitable for performing temperature measurements of water flow. Doing so eliminates the influence of laser intensity fluctuations observed in the single-color LIF technique, which is a significant source of error in the temperature measurement [6]. But the two-color LIF technique has not been applied to airflows because while the fluorescent dyes could be dissolved and could generate water mist for visualizing the airflow, they vaporized easily and dissipated immediately and could thus not be visualized successfully.

Citation: S. Funatani, K. Toriyama and T. Takeda, "Temperature Measurement of Air Flow Using Fluorescent Mists Combined with Two-Color LIF," Journal of Flow Control, Measurement & Visualization, Vol. 1 No. 1, 2013, pp. 20-23. doi: 10.4236/jfcmv.2013.11004.

The present study aims to solve the abovementioned problems by developing a technique for the simultaneous measurement of the temperature field of an airflow by using the two-color LIF method.

MEASUREMENT TECHNIQUES

Improvement of Visualization Techniques

In this study, we used a single-color camera for temperature measurement using the two-color LIF technique [4]. Therefore, the fluid temperature could be measured if the relationship between the intensity ratio and the temperature was known. It should be noted that the R, G, and B filters of the color camera pass light having a wavelength of 575 - 640, 490 - 575, and 400 - 485 nm, respectively. Rhodamine B (emission wavelength: 575 nm) and Rhodamine 110 (emission wavelength: 520 nm) are selected as the temperature-sensitive and insensitive dyes, respectively. The R and G filters of the color camera serve to separate the excitation light and the emission light.

The temperature measurement of an airflow using two-color LIF was performed by spraying a mist of the fluorescent dye in the test vessel. A solution mist of fluorescent dyes could not be used for visualization because it vaporized and dissipated as soon as it was sprayed. To avoid this problem, we used propylene glycol, the vapor pressure of which (~10 Pa) is much lower than that of water (>2000 Pa), as the solvent of the fluorescent dyes. A supersonic moisture chamber was used as the atomizer and the Sauter mean diameter of the propylene glycol mist was ~3 μm, which was measured using a light-scattering airborne particle counter.

The fluorescence intensity from the mist is much smaller than that from the liquid solution, and it is difficult to visualize clearly. In the previous study, a 488 nm Arion laser was used as the light source. Such a process could not be applied when the fluorescence intensity becomes relatively too small compared to that by scattering using an incident laser beam. To solve this problem, we used a 445 nm diode laser (power: 1000 mW) as the light source, because the shorter wavelength can reduce the incidence of the laser beam to the green image of the color camera such that the fluorescence can be detected. Furthermore, we used a long-pass filter (>480 nm) to reduce the incidence of the 445 nm diode laser on the camera.

Experimental Setup

The temperature field of air was measured using the experimental apparatus shown in **Figure 1**. The test section's volume was 400 × 250 × 280 mm^3. The test section was visualized by using a 445 nm diode laser (power: 1000 mW). The laser light sheet was generated by scanning a laser beam by using a rotating mirror instead of a cylindrical lens, because the light sheet generated by a cylindrical lens has nonuniformity and the S/N ratio varies significantly over the measurement region.

The concentration ratio of Rhodamine B and Rhodamine 110 was set to CR = 20, which was known as the optimum concentration ratio of these dyes [6]. A color camera (Nikon D5100, 14bit, 4928 × 3264 pixels) was used to record the visualized image.

The temperature of the atmosphere was T_c = 25°C, and a buoyant plume was generated from the circular pipe

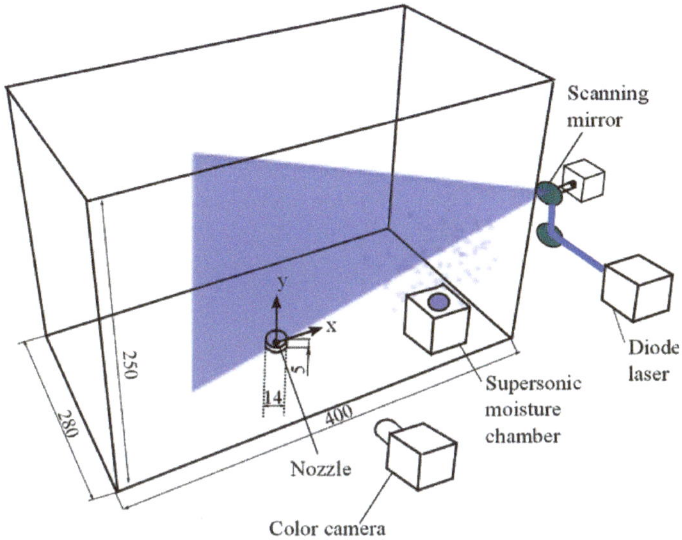

Figure 1. Experimental setup (unit in mm).

(inside diameter: D = 14 mm) at the bottom surface of the test section. The temperature of hot air was T_h = 35°C at y = −10 mm in the nozzle, and the average velocity was V_0 = 10 mm/s. Therefore, the Froude number of the vertical buoyant plume was Fr_0 = 4.23, and the Reynolds number was Re = 1670. The above conditions were selected because the temperature distribution of the thermal plume was obtained suitably [7]; the temperature distribution was predicted to be stable. These conditions were suitable for evaluating this measurement method.

RESULTS

Application to Temperature Distribution in Thermal Buoyant Plume

Figure 2(a) shows images of the sprayed fluorescent mist as visualized by a diode laser and color camera at various temperatures. The image at lower temperature (25°C) has an orange color, and the color changes slightly at a higher temperature (35°C). **Figure 2**(b) shows the variation of R/G with respect to temperature; the R/G value constantly decreased with an increase in temperature, which was the same result as that of the two-color LIF method using a single-color camera [4]. This result suggests that the R and G filters of a color camera can be used as filters for the two-color LIF technique even though the object to be visualized has changed from a liquid to a spray. In this study, the temperature measurement range was limited by the specifications of the experimental setup; however, this measurement range is expected to extend to be the same as that of the two-color LIF technique for water flow (0°C - 60°C) [6].

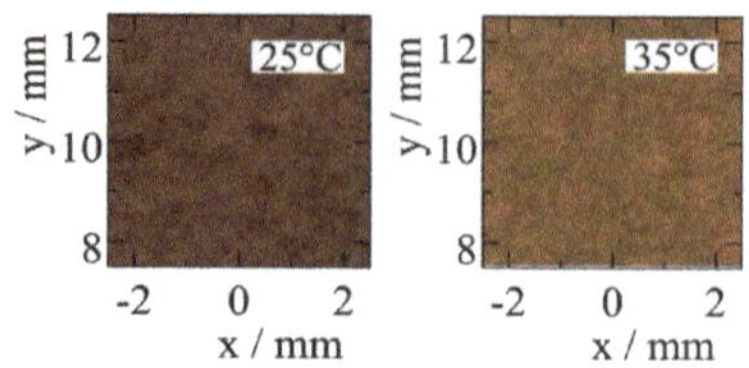

(a)

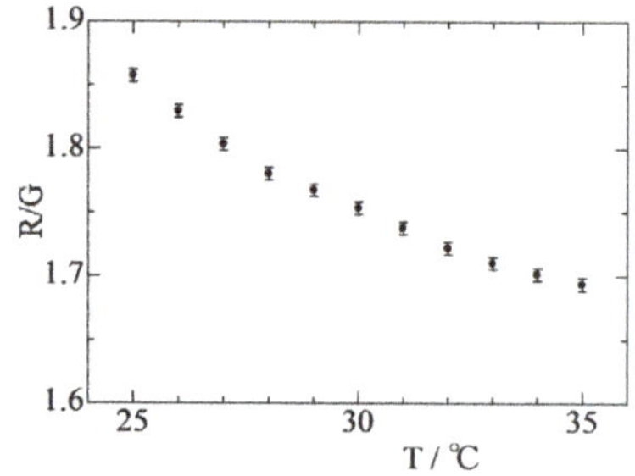

(b)

Figure 2. Visualized image and optical characteristics of two-color LIF image. (a) Visualized images; (b) Relationship between temperature and R/G.

Figure 3 shows the temperature distributions in the thermal buoyant plume. It shows the instantaneous structure of the thermal plume that was generated from a circular nozzle and that interacted with the stagnant surrounding cold air. In the previous experimental and numerical study of a vertical buoyant jet, it was shown that the jet became stable and the decrease in temperature by the inflow did not occur [7]. The temperature distribution in this study shows good agreement with that in the previous study. We found two-color LIF thermometry to be very effective for the study of the thermal structure of plumes. Furthermore, we found that the two-color LIF technique was well suited for measuring the temperature field of an airflow.

Uncertainty Analysis

As a part of our assessment of the proposed visualization technique, we investigated its temperature measurement uncertainty. The uncertainty originates from the error sources, such as the temperature calibration by a thermocouple (bias error: ±0.05 K), color-to-temperature transformation, image processing, fluctuation of the particle density of the fluorescent mist, and tracing accuracy of the fluorescent mist. The measurement uncertainty in color-to-temperature transformation and image processing can be evaluated from a comparison of the experimental data of the temperature measurement of two-color LIF with that of a thermocouple in a uniform temperature field.

A uniform temperature field was generated by heating and mixing the atmosphere in the test section by using a heater and an electric fan. The result indicates that the standard deviation of temperature was ±0.30 K (precision error).

The non-uniform tracer particle diameter of the fluorescent mist was considered to influence the evaluated temperature, but it was not kept steady because the mist was always dropping and mixing with surrounding flows. Therefore, the influence of the dropping and mixing of the fluorescent mist was also evaluated by spraying the

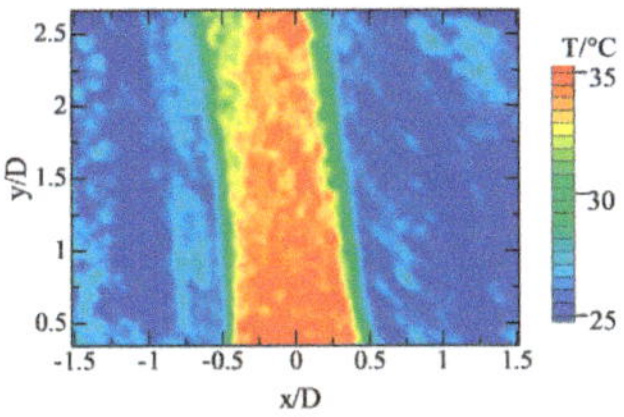

Figure 3. Temperature distribution of vertical thermal plume.

mist and recording the visualized images and evaluating the time variation of the evaluated temperature for 10 min (**Figure 4**). The result shows that the evaluated standard deviation of the temperature fluctuation due to nonuniform tracer particles was ±0.41 K (precision error).

It was also necessary to evaluate the tracing accuracy of the fluorescent mist as it is related to the temperature of the atmosphere. The temperature variance of a propylene glycol mist surrounded by air was evaluated by numerical simulation using an unsteady one-dimensional thermal diffusion equation of spherical coordinates which is expressed by the following Equation [8]:

$$\rho c\frac{\partial T}{\partial \tau}=\frac{1}{r^2}\frac{\partial}{\partial r}\left(kr^2\frac{\partial T}{\partial r}\right) \quad (1)$$

where, r is the radial coordinate, τ is the time, T is the temperature, k, r and c is the thermal conductivity, density and specific heat at constant pressure, respectively. These thermal properties are given as a function of temperature.

The tracking time of the propylene glycol mist to reach the temperature of air was evaluated at various diameters of the mist (D_p: 1 - 5 μm, 25°C) and the temperature difference between the mist and surrounding air (2.5 - 40 K). The mist temperature is defined as volume average temperature. **Table 1** lists these results. The tracking time is defined as the time until the temperature difference becomes smaller than 0.01 K. In these conditions, the tracking time is less than 0.1 ms.

The latent heat of vaporization of propylene glycol is

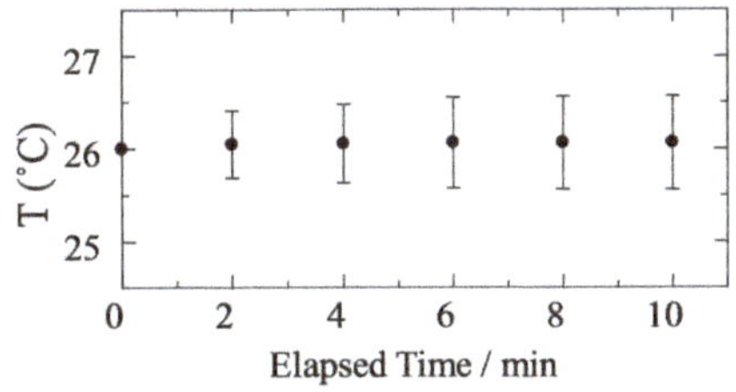

Figure 4. Time variation of evaluated temperature.

Table 1. Tracking time of propylene glycol mist without vaporization.

D_p/μm	ΔT/K	Tracking time
1	10	0.0028
2		0.0118
3	2.5	0.0209
	5	0.0240
	10	0.0272
	20	0.0309
	40	0.0355
4	10	0.0491
5		0.0773

Tracking time: unit of milliseconds.

300 times the heat capacity of the liquid state. The tracking time increases with the increasing heat quantity during the phase change. However, this increasing time decreases when the magnitude of the heat flow from the surrounding air to the mist increases. For example, if the temperature difference is 1 K and all the mist vaporizes, the tracking time is about 301 times longer than that in the non-vaporized case. This is because the required heat quantity is the sum of the increase in the amount of internal energy and the heat quantity during the phase change. However, under normal conditions, the temperature difference is larger than 1 K, and only a part of the mist is vaporized. Therefore, the tracking time did not become so large. We assume that the tracking time was less than 1ms.

This evaluation did not take into account the influence of convective heat transfer by the difference between the speed of the mist and the surrounding air. Therefore, the actual tracing accuracy will be higher than is indicated by this result, because the influence of the convective heat transfer leads to an improvement in the tracing accuracy. Therefore, the total uncertainty interval of temperature at 95% coverage was ±0.99 K and the tracking time was less than 1 ms.

CONCLUSIONS

The two-color LIF technique was used for measuring the temperature of an airflow by spraying a mist of propylene glycol and by using a 445 nm diode laser. The proposed measurement technique was successfully applied to a vertical buoyant plume. The total uncertainty interval of temperature at 95% coverage was ±0.99 K and the tracking time was less than 1 ms.

The expansion of the temperature measurement range and a detailed study of a vertical buoyant plume for various Froude numbers will be conducted in the next study.

ACKNOWLEDGEMENTS

This research was supported by the Grant-in-Aid for Young Scientists (B) No.24760131 of the Japan Society for the Promotion of Science (JSPS).

REFERENCES

1. W. G. Bessler, C. Schulz, T. Lee, D.-I. Shin, M. Hofmann, J. B. Jeffries, J. Wolfrum and R. K. Hanson, "Quantitative NO-LIF Imaging in High-Pressure Flames," Applied Physics B, Laser and Optics, Vol. 75, No. 1, 2002, pp. 97-102. doi:10.1007/s00340-002-0946-0
2. W. G. Bessler and C. Schulz, "Quantitative Multi-Line NO-LIF Temperature Imaging," Applied Physics B, Laser and Optics, Vol. 78, No. 5, 2004, pp. 519-533.doi:10.1007/s00340-004-1421-x
3. C. Hecht, A. Abdali, T. Dreier and C. Schulz, "GasTemperature Imaging in a Microwave-Plasma Nanoparticle-Synthesis Reactor Using Multi-Line NO-LIF Thermometry," Zeitschrift Fur Physikalsche Chemie—International, Vol. 225, 2011, pp. 1225-1235.doi:10.1524/zpch.2011.0162
4. M. C. Thunber, F. Grisch and R. K. Hanson, "Temperature Imaging with Singleand Dual-Wavelength Acetone Planar Laser-Induced Fluorescence," Optics Letters, Vol. 22, No. 4, 1997, pp. 251-253. doi:10.1364/OL.22.000251
5. T. Handa, M. Masuda, M. Kashitani and Y. Yamaguchi, "Measurement of Number Densities in Supersonic Flows Using a Method Based on Laser-Induced Acetone Fluorescence," Experiments in Fluids, Vol. 50, No. 6, 2011, pp. 1685-1694. doi:10.1007/s00348-010-1029-3
6. S. Funatani, N. Fujisawaand and H. Ikeda, "Simultaneous Measurement of Temperature and Velocity Using TwoColour LIF Combined with PIV with a Colour CCD Camera and Its Application to the Turbulent Buoyant Plume," Measurement Science and Technology, Vol. 15, No. 5, 2004, pp. 983-990. doi:10.1088/0957-0233/15/5/030
7. T. Maeda, N. Fujisawa, T. Syuto and T. Yamagata, "Experimental and Numerical Study on Onset of Inflow in Near Field of Buoyant Jet at Low Froude Number," Journal of Visualization, Vol. 15, No. 1, 2012, pp. 67-75. doi:10.1007/s12650-011-0104-z
8. J. P. Holman, "Heat Transfer," 10th Revised Edition, McGraw Hill Higher Education, New York, 2009.

3

CHAPTER

LIF MEASUREMENT OF THE DILUTING EFFECT OF SURFACE WAVES ON TURBULENT BUOYANT PLUMES

David B. Sharp, Alistair Shawcross, Clive A. Greated

Keywords: Buoyant Plumes, Surface Waves, Laser Induced Fluorescence, Multi-Port Diffuser, Outfall Pipe

INTRODUCTION

Waste fluid is commonly discharged into the marine environment by means of an outfall pipe. In order to limit the impact on the surrounding area, a high initial dilution must be obtained and this is usually achieved by terminating the outfall with a multi-port diffuser. As effluent is often less dense than the receiving water, it is released as a turbulent, buoyant plume [1] .

While the mechanics of such plumes have been well studied [2] , the effects of surface waves on buoyant plumes are less well understood. In recent years, experimental measurements [3] -[10] have tended to focus on the effect of waves on the flow characteristics of neutrally buoyant jets. Such work has included full field Laser Induced Fluorescence (LIF) measurements of a neutrally buoyant jet discharged vertically into a wave environment [11] . In contrast, however, the work that has been conducted on buoyant plumes emerging into a wave environment [12] [13] has tended to use point measuring techniques and has investigated only the surface dilution.

In this paper, a full field Laser Induced Fluorescence (LIF) technique is used to investigate the diluting effect of surface waves on buoyant plumes. The technique allows concentration measurements from diffuser to surface, not only permitting any increase in dilution due to the presence of surface waves to be measured, but also showing the region where this increased mixing takes place. The full field nature of the

Citation: Sharp, D. , Shawcross, A. and Greated, C. (2014) LIF Measurement of the Diluting Effect of Surface Waves on Turbulent Buoyant Plumes. Journal of Flow Control, Measurement & Visualization, 2, 77-93. doi:10.4236/jfcmv.2014.23010.

technique also allows measurement of other plume characteristics, such as centre line or profile concentrations. Similar LIF measurements have previously been carried out on a buoyant plume discharged into stagnant receiving water [14] -[18] .

In the next two sections, dimensionless parameters are introduced that describe the properties of a discharge emerging from a diffuser and the effect of surface waves acting on that discharge. Then, in Section 4, these dimensionless parameters are combined and used to inform the design of laboratory experiments concerned with investigating the diluting effect of surface waves on buoyant plumes. Finally, in Sections 5 and 6, details of the wave tank and the other apparatus used in the laboratory experiments are provided; the methodology is described; and results are presented and analysed.

DISCHARGE PROPERTIES

The behaviour of a discharge emerging from the port of a diffuser can be classified by two dimensionless quantities, the Reynolds and Froude numbers. The former is given by the ratio of inertial to viscous forces in the flow

$$R = \frac{u_o d}{v_o} \quad (1)$$

where u_0 is the discharge velocity at the port, d is the port diameter and v_0 the kinematic viscosity (given by μ_o / ρ_o where μ_0 is the initial viscosity of the plume and ρ_0 its initial density). In order for the flow to be fully turbulent, as it is from a real diffuser, the discharge must have a Reynolds number greater than 4000 [19] .

The Froude number, which determines how soon the discharge evolves from a jet to a plume, is defined as the ratio of inertial to buoyancy forces in the flow

$$F = \frac{u_o}{\sqrt{g_o d}} \quad (2)$$

where g_0, the reduced gravity, is given by

$$g_o = \frac{\Delta\rho}{\rho_o} g \quad (3)$$

g being the acceleration due to gravity and $\Delta\rho = \rho_a - \rho_0$, the difference between the density of the pure plume solution ρ_0 and the receiving water ρ_a. A typical outflow, in which the discharge becomes plume-like very quickly, has a Froude number of 16 [12] . The discharge used in the laboratory experiments in this study was designed to have a Froude number of a similar order.

If the receiving water is sufficiently deep, then the discharges from a multi-port diffuser will merge into a single plume. The dilution along the centre line of this plume is then the same as that from a discharge by a line source of buoyancy flux only and is given by [20] [21]

$$\frac{S_m}{F} = 0.5\left(\frac{2Z}{s}\right)^{-\frac{2}{3}}\left(\frac{Z}{dF}\right)^{\frac{5}{3}} \quad (4)$$

where S_m is the minimum centre line dilution, s is the port spacing along the diffuser, and Z is the vertical distance above the diffuser.

WAVE INTERACTION

When describing the effect of sinusoidal waves on a discharge, two further dimensionless quantities are required. The first,

$$h/L \tag{5}$$

where h is the total depth of the water and L is the wavelength of the waves, determines the effect of the waves on the movement of the column of water below. For values less than 0.05 the wave induces motion uniformly throughout the whole depth of the water column below it, while for values between 0.05 and 0.5 the motion at the bottom of the column is smaller than that near the surface. This is the situation normally encountered at the sites of outfalls. Finally, for values greater than 0.5, the induced movement does not reach right to the bottom of the column [12] . The experiments in this study were conducted with h/L = 0.16, producing motion through the full depth of the tank.

The second dimensionless quantity, a measure of the effect of the waves on the discharge, is numerically equal to the ratio of the maximum horizontal wave induced velocity at the port to the port discharge velocity and is formed from two characteristic length scales. Following Fischer et al.[1] the discharge volume flux Q, momentum flux M and the buoyancy flux B are first defined

$$Q = \frac{\pi}{4} d^2 u_o \tag{6}$$

$$M = Q u_o \tag{7}$$

$$B = Q g_o \tag{8}$$

The following length scales can then be defined: L_Q, the length over which port geometry has an influence over the flow, L_M the distance from the port at which buoyancy forces begin to dominate the flow and Z_M, the rise height for the plume momentum to be of the order of wave induced momentum.

$$L_Q = \frac{Q}{M^{1/2}} \tag{9}$$

$$L_M = \frac{M^{3/4}}{B^{1/2}} \tag{10}$$

$$Z_M = \frac{M^{1/2}}{u_{max}} \tag{11}$$

where u_{max} is the maximum horizontal wave induced velocity at the port, given by [22]

$$u_{max} = \frac{agk}{\sigma \cosh kH} \tag{12}$$

where H is the discharge depth, a is the wave amplitude, k the wave number (defined as k = 2π/L, where L is the wavelength) and σ the wave frequency (defined as σ = 2π/T, where T is the wave period).

The dimensionless parameter

$$\frac{L_Q}{Z_M} \tag{13}$$

then gives the wave effect at the port. This study focuses on values of L_Q/Z_M in the range of 5.1×10^{-2} to 9.5×10^{-2} constrained by the tank dimensions and wave paddle, but of the order of real outflows [12] .

EXPERIMENTAL DESIGN

In order to realistically model the plumes produced by outfall pipes, the discharged solution used in the laboratory experiments had to fulfill a number of criteria; it had to be less dense than the receiving fluid so as to be buoyant, it had to have a similar viscosity to the receiving fluid to allow the Reynolds number at the port to be such that the plume was turbulent, and it also had to be transparent and miscible in the receiving fluid.

Cost and ease of filling the wave tank dictated using tap water as the receiving fluid and therefore a fluid less dense than, miscible in and transparent in water needed to be found. Initially it was thought that an ethanol/water solution would fulfill these criteria. To investigate the necessary discharge parameters, the Froude number equation (Equation (2)) was substituted into the Reynolds number equation (Equation (1)) to yield an equation for the port diameter required for a discharge of given Froude and Reynolds numbers

$$d = \left(\frac{Rv_o}{Fg_o^{1/2}}\right)^{\frac{2}{3}} \tag{14}$$

This equation was then substituted into equation (2) to find the term for the required port discharge velocity

$$u_o = \left(Rv_0\right)^{\frac{1}{3}}\left(Fg_o^{1/2}\right)^{\frac{2}{3}} \tag{15}$$

From the above two equations it was then possible to find expressions for the total discharge, D_T, and discharge of ethanol, D_E, required, the latter being plotted so that the most economical as well as practical solution could be found

$$D_T = u_o\pi\left(\frac{d^2}{4}\right) \tag{16}$$

$$D_E = \left(\frac{P_E}{100}\right)D_T \tag{17}$$

where P_E is the percentage by weight of ethanol. These give the discharge rate per port and must be multiplied by the total number of ports in the multi-port diffuser to find the total discharge rate.

Despite pure ethanol having a similar viscosity to water, when the two are mixed the viscosity can substantially increase (see Figure 1); the density of the mixture will also not be constant (as shown in Figure 2, plotted from data given in [23]). Solving Equations (14) through to (17) for ethanol discharged with a Froude number of 14 (a value of the order of a realistic outfall), showed that the required properties for a turbulent discharge were both impractical and uneconomical, with both the total discharge and ethanol discharge rates being too high. A similar solution but with a lower viscosity needed to be found.

Methanol having very similar properties to ethanol, but a lower viscosity also looked a good candidate. Even though the viscosity and density of a methanol/water solution changes in a similar way to an ethanol/water mix (see Figure 3 and Figure 4), the combined viscosity is not as high [24][25] . Equations (14) to (17) were again solved but this time for a methanol-based discharge. Figures 5-8 are graphs showing respectively the required port diameter, port velocity, total discharge rate and methanol discharge rate for a discharge with Froude number of 14 and for a range of Reynolds numbers.

Ideally the best plume solution is that which gives a high Reynolds number and a low methanol discharge rate. From Figure 8, it can be seen that this condition is met at both very high methanol concentration and very low methanol concentration. However, Figure 7 indicates that for a low methanol concentration discharge the required total discharge rate is much higher than for a high methanol concentration discharge. As a discharge rate of around 3 l/min seemed more achievable than a rate of 20 l/min, it was decided to use a solution of pure methanol. Figures 9-11 show the port diameter, velocity and discharge rate required for a given Reynolds number for a discharge of pure methanol. Using Figure 9 a port diameter of 2.7 mm was chosen as this gives a discharge with a Reynolds number greater than 4000 and therefore a turbulent plume. The appropriate discharge rate was then found from Figure 11.

The extent of the effect of the wave at the port is governed by the maximum horizontal wave induced velocity at the port (Equation (12)) and the wave parameters that control this are the wave amplitude, period and length. To produce an effect on the water column under the wave similar to that experienced at the sites of real outflows, the wavelength of the waves and therefore the period had already been fixed. The remaining wave variable that could be altered to vary the effect of the waves on the plume was the amplitude. However, the wave maker available on the tank used to perform the experiments was only able to produce waves with a maximum amplitude of 3 cm. Controlling the wave effect purely by altering the wave amplitude therefore offered a very limited number of cases. However Equation (12) shows that the wave induced velocities increase as the surface is approached, therefore the discharge depth can also be used to vary the effect of the wave on the port. It was therefore decided to discharge the solution from three different depths, 30, 45 and 60 cm, and vary the wave amplitude at each of these depths. In this way plumes under a wider range of wave conditions could be studied.

To measure the concentration of the plume a Laser Induced Fluorescence (LIF) technique was used. By dissolving a fluorescent dye in the discharge and illuminating it with laser light of the correct frequency, the plume can be made to fluoresce. The plume fluoresces at a different frequency to the illuminating laser light allowing the background intensity to be filtered out, leaving only the fluorescent intensity. As this is proportional to the concentration of dye present and therefore the concentration of plume solution, measurements of intensity

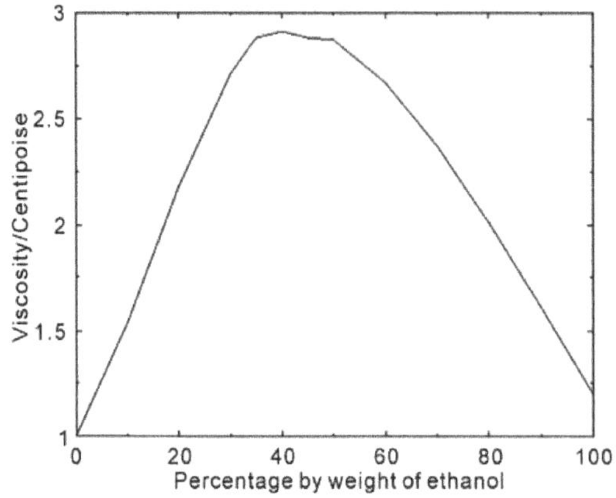

Figure 1. Variation of viscosity μ_o of ethanol/water solution with percentage weight of ethanol P_E.

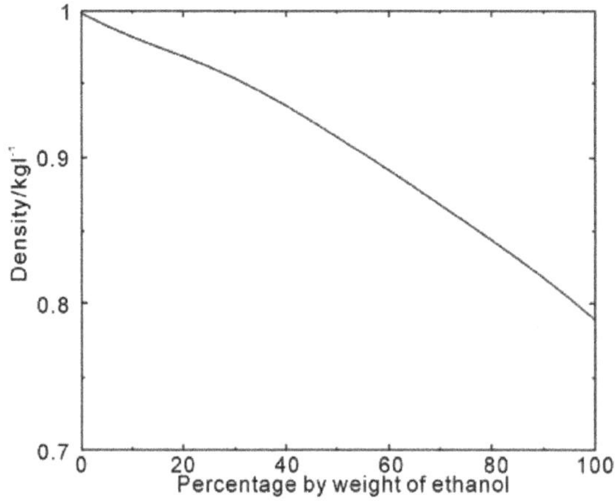

Figure 2. Variation of density ρ_o of ethanol/water solution with percentage weight of ethanol P_E.

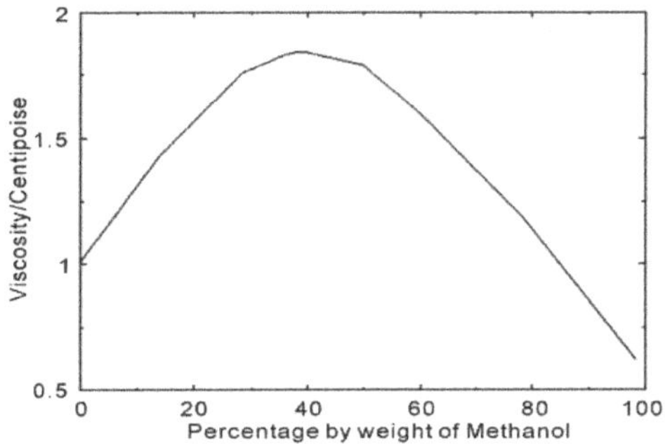

Figure 3. Variation of viscosity μ_o of methanol/water solution with percentage weight of methanol P_M.

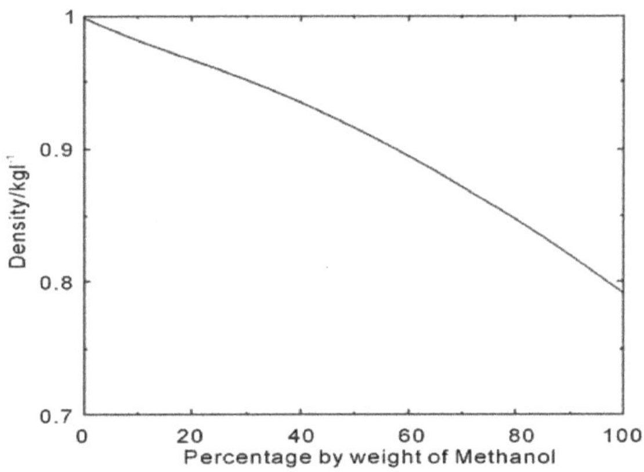

Figure 4. Variation of density ρ_o of methanol/water solution with percentage weight of methanol P_M.

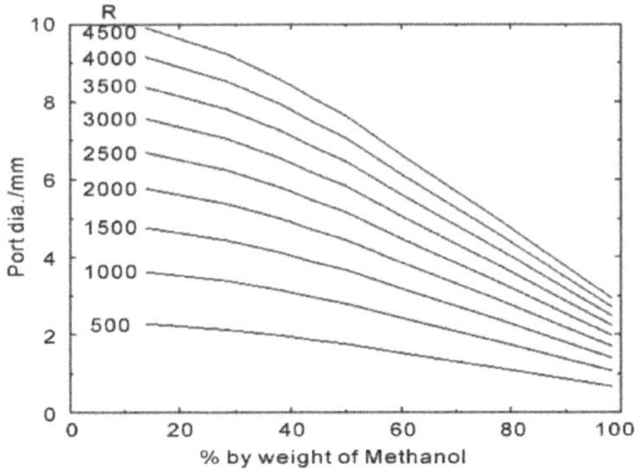

Figure 5. Port diameter d required for a discharge with Froude number F = 14, plotted as a function of percentage weight of methanol P_M(individual curves correspond to different Reynolds numbers).

can, by calibration with known concentrations, be converted to measurements of concentration [26] .

EXPERIMENTAL METHOD

Experiments were conducted in a glass sided and bottomed wave tank 7.5 m long, 0.4 m wide and 1.0 m deep. Sinusoidal waves were generated by a computer controlled paddle at one end of the tank, and absorbed by a foam "beach" at the other, allowing progressive waves to be produced. The tank was filled with tap water into which pure methanol (99.8 percent, ρ_o = 0.790 kg/l), dyed with Rhodamine B was discharged (approximately 4 mg of Rhodamine B powder was used per litre of methanol). The effluent was discharged through a diffuser with seven, evenly spaced, horizontally orientated ports along its length. The ports were positioned so that discharge from the end ports, reflected from the tank walls, could be regarded as discharge from virtual ports spaced with the same port spacing further along the diffuser. The seven individual plumes then merge to form a plume equivalent to that from a line source.

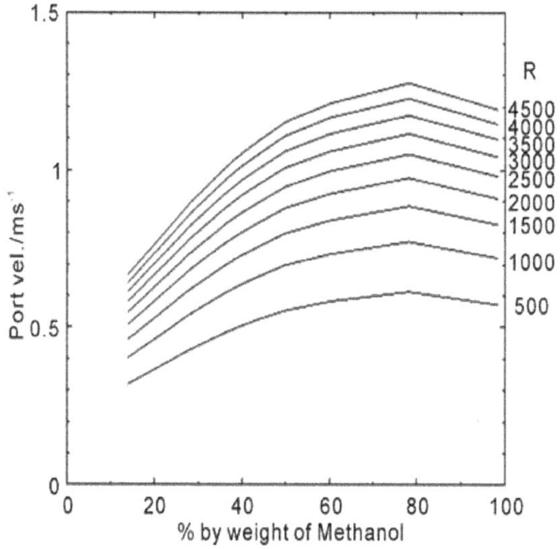

Figure 6. Port discharge velocity u_o required for a discharge with Froude number F = 14, plotted as a function of percentage weight of methanol P_M(individual curves correspond to different Reynolds numbers).

Illumination of the plume was achieved with a pseudo light sheet, created using a scanning beam box [27] . The beam from an Argon ion laser (λ = 514.5 nm, 15 W) was directed onto a spinning octagonal mirror, which in turn reflected the beam onto a parabolic mirror. This produced a light sheet, 0.7 m wide, which was projected vertically through the glass bottom of the tank (Figure 12).

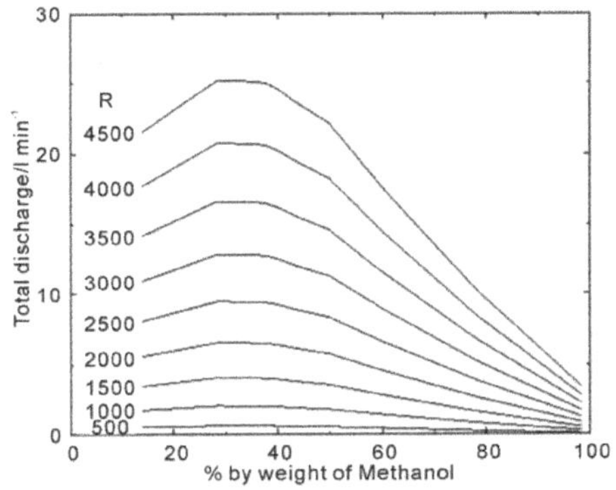

Figure 7. Total discharge rate D_T required for a discharge with Froude number F = 14, plotted as a function of percentage weight of methanol P_M(individual curves correspond to different Reynolds numbers).

Images of the fluorescing plume were captured using an 8 bit monochrome Cohu CCD camera (array size 576 × 768 pixels) connected to a frame grabber card in a PC. The camera was fitted with an orange filter to filter out the blue laser light so that only the orange fluorescent light was detected. The system allowed a maximum of ninety nine greyscale images to be captured at regular time intervals. All results shown in this paper were obtained from images taken at half second intervals. The target area for each of the images was approximately 60 cm wide by 80 cm high.

Before each experiment, a series of calibration measurements were taken to find the greyscale pixel values (on a scale from 0 to 255) of known concentrations of plume solution. The concentrations ranged from 0.25% up

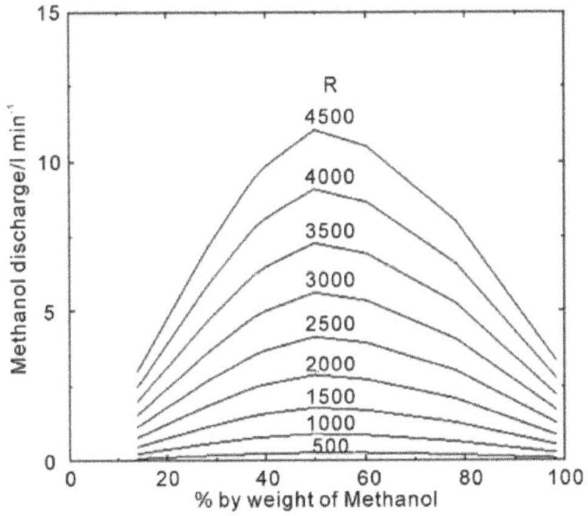

Figure 8. Methanol discharge rate D_M required for a discharge with Froude number F = 14, plotted as a function of percentage weight of methanol P_M (individual curves correspond to different Reynolds numbers).

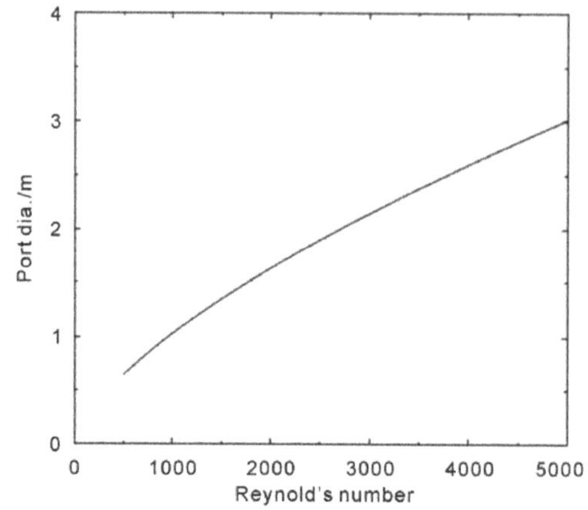

Figure 9. Port diameter d required for a pure methanol discharge ($P_M = 100$) with Froude number F = 14, plotted as a function of Reynolds number.

to 2.5% in 0.25% intervals, and from 2.5% up to 5.0% in 0.5% intervals. For each calibration measurement, an image was taken of a small transparent rectangular container filled with plume solution of a particular concentration, placed into the light sheet. The average greyscale pixel value over a selected area of the image was then determined.

A background image was also taken with the wave tank completely filled with dyed water. By removing this background image from images taken during the experiments, and also during the calibration measurements, imperfections in the light sheet could be compensated for.

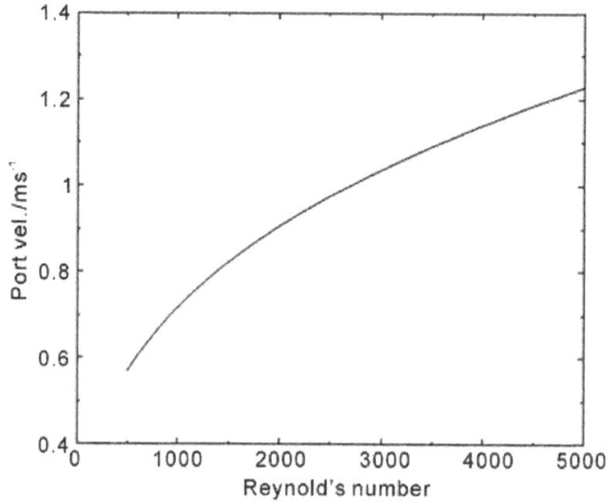

Figure 10. Port discharge velocity u_o required for a pure methanol discharge ($P_M = 100$) with Froude number F = 14, plotted as a function of Reynolds number.

Table 1 shows the parameters for each of the twelve experiments performed. Effluent was discharged from three different depths under a series of different wave amplitudes. By varying both the wave amplitude and discharge depth it was possible to produce a range of L_Q/Z_M values and investigate the effect of this on the dilution of the plume. For all experiments conducted with waves, the wavelength was 4.28 m.

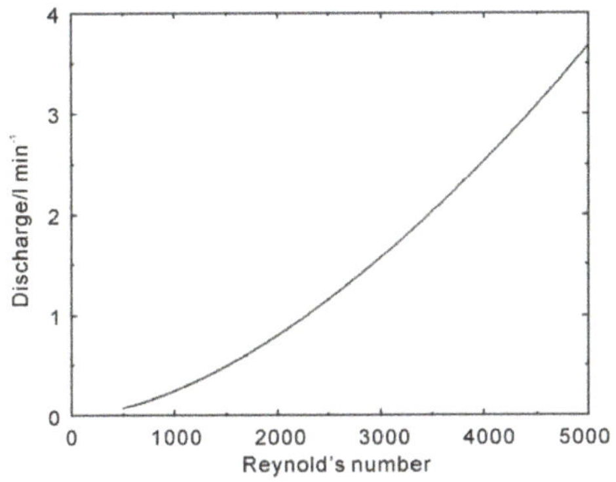

Figure 11. Total discharge rate D_T required for a pure methanol discharge ($P_M = 100$) with Froude number F = 14, plotted as a function of Reynolds number.

Table 1. Values of parameters for experiments performed. R = 4400, F = 14 and h/L = 0.16 for all experiments.

Experiment number	Discharge depth/m	Wave amplitude/m	L_Q/Z_M
1	0.30	0.0	0.0
2	0.30	0.015	0.051
3	0.30	0.019	0.068
4	0.30	0.020	0.095
5	0.45	0.0	0.0
6	0.45	0.020	0.061
7	0.45	0.022	0.067
8	0.45	0.030	0.091
9	0.60	0.0	0.0
10	0.60	0.020	0.053
11	0.60	0.025	0.066
12	0.60	0.030	0.079

RESULTS AND ANALYSES

In order to compare the experimental results with theoretical values, instantaneous and time-averaged concentration maps were produced. For each of the twelve experiments performed, 50 of the captured greyscale CCD images (Figure 13 shows an example reverse video image as an illustration) were converted to instantaneous concentration maps. This was done by comparing the individual greyscale pixel values in each image with the greyscale pixel values for known concentrations of plume solution (established during the calibration measurements). The error in the instantaneous concentration measurements was estimated to be ±5% based on an analysis of the calibration procedure and measurement repeatability (so that, for example, a measured instantaneous concentration value of 1% could actually lie anywhere in the range 0.95% to 1.05%).Figure 14 shows a typical instantaneous concentration map of a plume. Meanwhile, for each experiment a time-averaged concentration map was determined by first averaging the individual greyscale pixel values across the 50 CCD images, and then comparing these averaged values with the greyscale pixel values for known concentrations of plume solution.

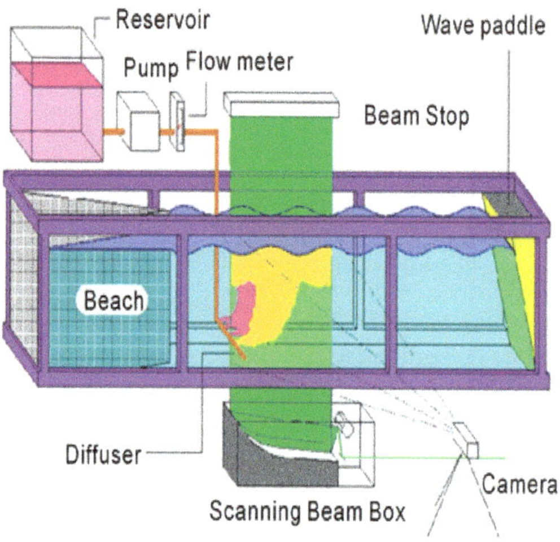

Figure 12. Wave tank equipped for LIF study of buoyant plumes.

Figure 13. Typical instantaneous reversed CCD image of the plume.

Figures 15-17 show time-averaged concentration maps for three of the twelve experiments carried out. From these images alone an increase in dilution due to the presence of waves can be seen.Figure 15 shows the situation with no waves present ($L_Q/Z_M = 0$), Figure 16 shows the discharge when small amplitude waves are present ($L_Q/Z_M = 0.053$) and Figure 17 shows the situation when the waves have a larger amplitude ($L_Q/Z_M = 0.066$). As L_Q/Z_M increases, the dilution of the plume has clearly also increased.

Analysis of each time-averaged concentration map with in-house software allows the centre line of the plume to be located and the change in concentration along that centre line to be found. Figure 18 shows the centre line concentrations (expressed as a percentage of the pure plume solution) plotted as a function of the normalized rise height (Z/H) for the time-averaged concentration maps of Figures 15-17. It can be seen that the "no waves" centre line concentration profile only follows the theoretical curve (derived from Equation (4)) in approximately the middle third of the rise height. The reason for this is that the theoretical curve is the solution for a line source and is therefore only valid once the individual plumes have merged, only then will the centre line concentration tend to the theoretical curve. Also, as can be seen from Figure 13, the final third of the rise height is through the effluent wastefield that forms at the surface. Within this region the analysis software was unable to accurately determine the position of the plume centre line and therefore the data will not be that for the true centre line.

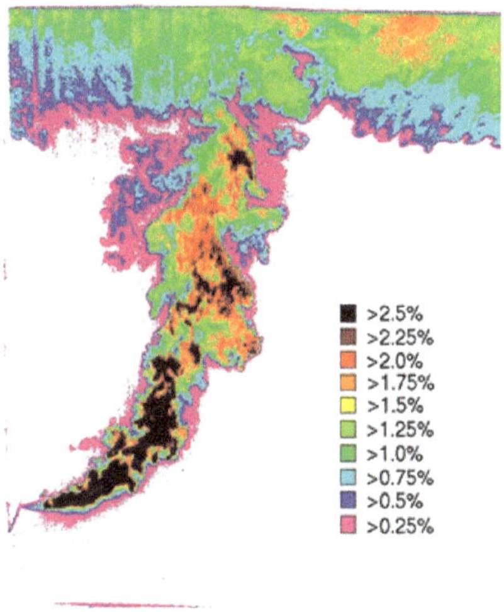

Figure 14. Typical instantaneous concentration map of the plume.

The increased dilution due to the effect of the waves is again visible in Figure 18, with lower centre line concentrations observed when waves are present and with the greatest dilution occurring during the first fifth of the rise height. After this initial increase, dilution increases at a similar rate for both wave and no wave cases. This was observed by Chin [12] who noted the discharged jets "exploding" close to the port when the wave induced velocity opposed that of the discharge, but the plume merely being translated from side to side further above the port.

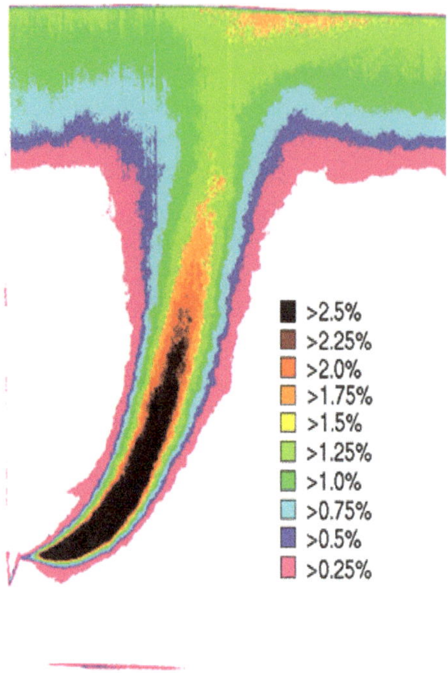

Figure 15. Concentration map for experiment 9 (see Table 1); no waves present.

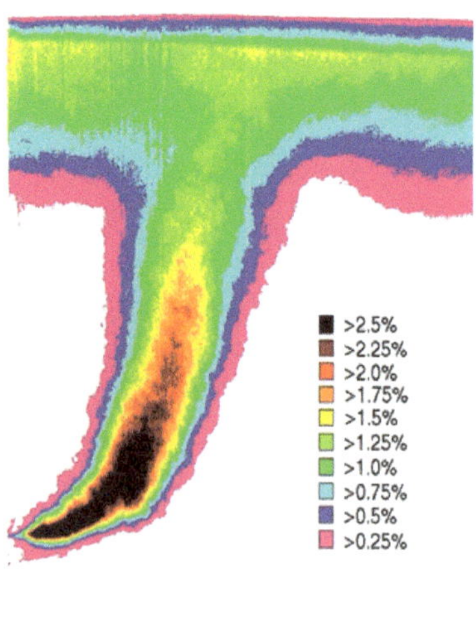

Figure 16. Concentration map for experiment 10 (see Table 1); small amplitude waves.

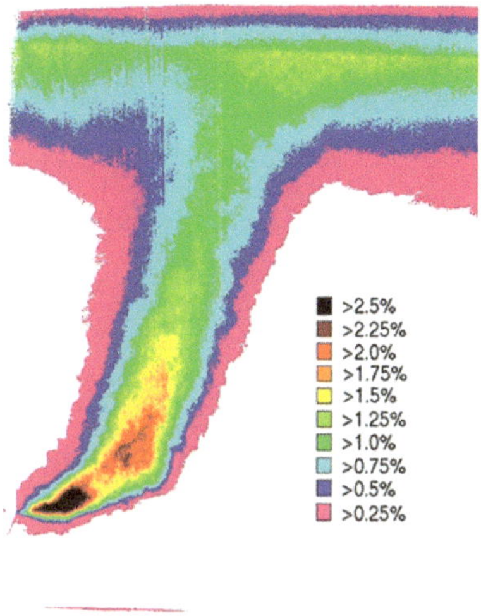

Figure 17. Concentration map for experiment 11 (see Table 1); large amplitude waves.

Chyan and Hwung [11] , although investigating a non-buoyant jet orientated vertically, also noted that the rise height could be divided into a number of regions. The first region (similar to Chin's "exploding" region) they named the "deflection region", as a vertical jet's trajectory is deflected by the wave motion in this region. They also then identified a region of lower or decreased dilution as observed in the present results and noted that the increase in concentration was higher in stronger wave conditions. They named this the 'transition region' and argued that here the initial jet momentum had decayed to the extent that the jet flow was no longer significantly deflected. Indeed in the present results this increase in concentration takes place at around $2Z_M$ in the rise height, the point at which the jet momentum becomes of the order of half the wave induced horizontal momentum. In this region the influence of the initial jet momentum has decreased to such an extent that it no longer influences the plume's trajectory and motion is solely due to the wave effect. Above this point, as with the present results and with Chin's results, they observed a "developed region" in which the plume is translated horizontally by the wave motion and identified by a decreased rate of dilution. The existence of these three different regions was confirmed by further experiments carried out by Mossa [5] on a vertical non-buoyant jet.

Koole and Swan [3] made related observations while investigating the effect of waves on a non-buoyant horizontal jet. In their work, they refer to a "zone of flow establishment" close to the jet source where the rate of dilution is high (analogous to Chin's "exploding region") and then a "zone of established flow" further away from the source where the dilution rate is substantially reduced. They suggest that wave motion promotes a transfer of momentum from the jet to the turbulent components of the flow field and that the effect is greatest within the "zone of flow establishment". This leads to a significant increase in the turbulent Reynolds normal stresses in this region resulting in the high rate of dilution. Further evidence for this hypothesis is provided by Mossa [6] who also notes the contribution of the wave Reynolds normal stresses.

In order to quantify the effect of the waves on increasing plume dilution in the present work, a program was used to find the average increase in dilution in the region $0.1 < Z/H < 0.6$ between wave and no wave cases (this region was chosen as this is the area in which the plume centre line detection program worked best). The increase in dilution can be expressed as S/S_o, where S is the centre line dilution with waves present and S_o is the centre line dilution under stagnant conditions. This increase S/S_o was then plotted against L_Q/Z_M (Figure 19). From the gradient of the line of best fit the increase in dilution due to the effect of the waves can be written as

$$\frac{S}{S_o} = 1 + 6\frac{L_Q}{Z_M} \qquad (18)$$

in good agreement with Chin's equation [12] derived from measurements of dilution taken as the surface:

$$\frac{S}{S_o} = 1 + 6.11\frac{L_Q}{Z_M} \qquad (19)$$

The exploding effect observed by Chin also leads to a broadening of the plume as can be clearly seen by comparing Figures 15-17. This broadening can be quantified and further demonstrated as will now be shown.

First of all, additional analysis of each time-averaged concentration map using the in-house software enables the variation in concentration over the cross-section of the plume (i.e. normal to the centre line of the plume) to be determined at different heights above the diffuser. These cross-sectional concentration profiles are Gaussian in nature, as predicted by [1] . Therefore, by fitting a theoretical Gaussian curve to each cross-sectional concentration profile, it is possible to take 2 × the standard deviation corresponding to the fitted curve as being a measure of the plume width w at that particular height Z. For any given height, by subtracting the plume width when no waves are present from the plume width under the influence of surface waves (w_{n-w}), the broadening of the plume due to the waves can be quantified. A non-dimensional parameter representing the broadening of the plume can be defined as $(w_{n-w}/a_Z)(L/H)$, where a_Z is the horizontal wave induced amplitude at height Z. Figure 20 and Figure 21, nondimensional plots of the broadening of the plume against rise height, show that the rate of broadening is increased for rise heights less than $2Z_M$, Chyan and Hwung's deflection region. In this region the rate of broadening is 1.6 times higher than that for rise heights greater than $2Z_M$, with a line of best fit gradient of 1.1×10^6 compared with a value of 0.69×10^6 in the latter developed region.

CONCLUSIONS

Using an LIF technique it has been shown that, under wave conditions of the type present at the sites of outfalls,

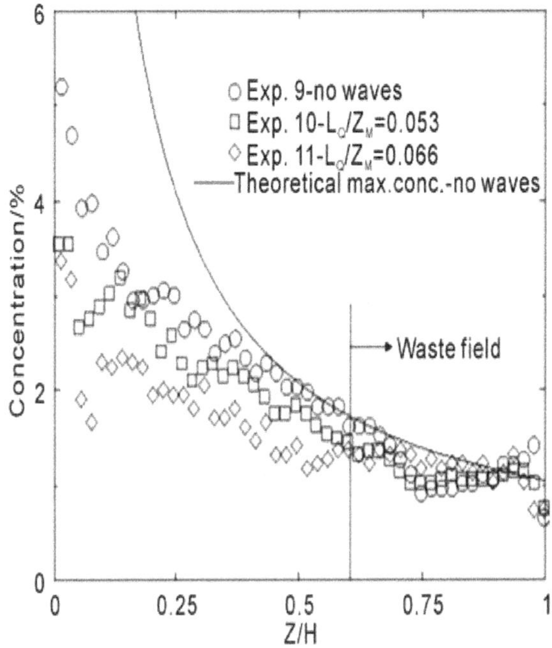

Figure 18. Plume centre line concentrations plotted as a function of normalized rise height.

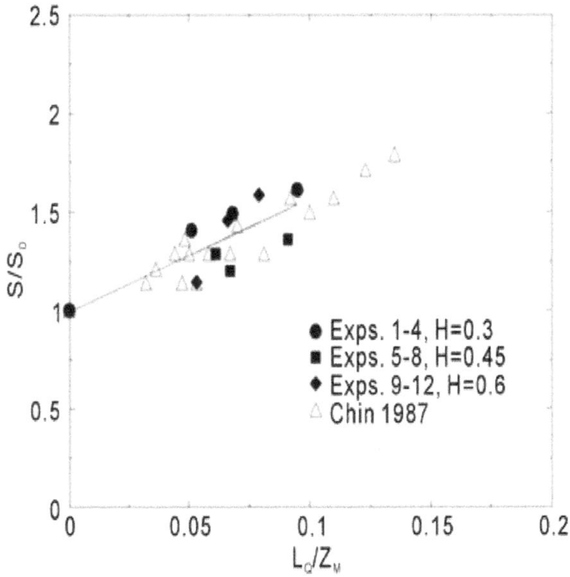

Figure 19. Increase in centre line dilution of plume due to effect of waves plotted against L_Q/Z_M.

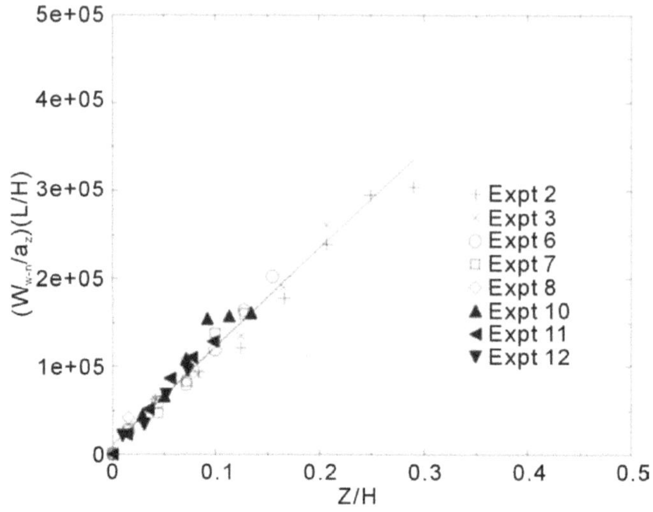

Figure 20. Non-dimensional plot of the broadening of the plume with increasing rise height (for $Z < 2Z_M$).

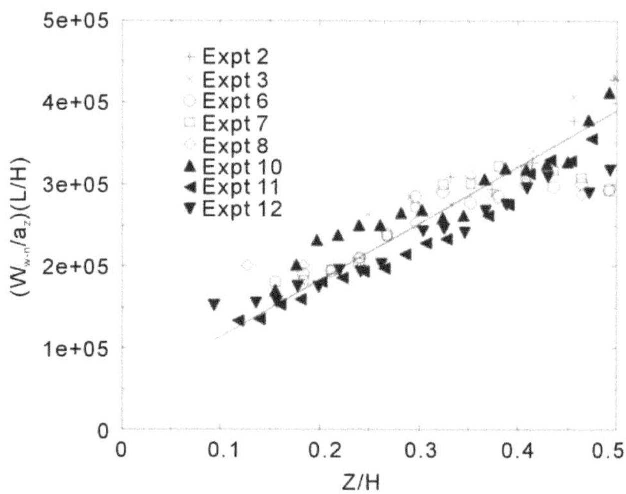

Figure 21. Non-dimensional plot of the broadening of the plume with increasing rise height (for $Z > 2Z_M$).

the dilution of a buoyant plume discharged from a multiport diffuser can be increased by wave effects. The full field nature of the technique has allowed not only the increase in dilution to be measured but also the region in which the enhanced mixing takes place to be clearly identified. The average increase in dilution of the plume due to wave effects was found to be similar to the increase in surface dilution reported previously by other researchers. This is to be expected as the greatest increase in dilution was found to be in the region near to the diffuser and before individual plumes had merged. In this region, where the plume's horizontal momentum is still greater than that of the wave induced motion, an "exploding" of the plume is seen. This results in an enhanced rate of broadening of the plume in this region. After this point, as the plume's momentum decreases to half that of wave motion, the plume is merely oscillated by the wave motion and plume dilution is comparable to that of a plume released into quiescent conditions.

In future, a coupling of LIF and Particle Image Velocimetry techniques (such as demonstrated in[16] -[18]) could be used to simultaneously produce instantaneous and time-averaged, full field concentration and velocity maps. This would allow a more complete study of the mechanics of a buoyant plume under the influence of surface waves.

ACKNOWLEDGEMENTS

This study was funded by the UK research councils EPSRC and NERC, whose support is gratefully acknowledged.

REFERENCES

1. Fischer, H.B., List, E.J., Koh, R.C.Y., Imberger, J. and Brooks, N.H. (1979) Mixing in Inland and Coastal Waters. Academic Press, Waltham.
2. List, E.J. (1982) Turbulent Jets and Plumes. Annual Review of Fluid Mechanics, 14, 189-212. http://dx.doi.org/10.1146/annurev.fl.14.010182.001201
3. Koole, R. and Swan, C. (1994) Measurements of a 2-D Non-Buoyant Jet in a Wave Environment. Coastal Engineering, 24, 151-169. http://dx.doi.org/10.1016/0378-3839(94)90031-0
4. Mori, N. and Chang, K.-A. (2003) Experimental Study of a Horizontal Jet in a Wavy Environment. Journal of Engineering Mechanics, 129, 1149-1155.http://dx.doi.org/10.1061/(ASCE)0733-9399(2003)129:10(1149)
5. Mossa, M. (2004) Experimental Study on the Interaction of Non-Buoyant Jets and Waves. Journal of Hydraulic Research, 42, 13-28.http://dx.doi.org/10.1080/00221686.2004.9641179
6. Mossa, M. (2004) Behaviour of Nonbuoyant Jets in a Wave Environment. Journal of Hydraulic Engineering, 130, 704-717. http://dx.doi.org/10.1061/(ASCE)0733-9429(2004)130:7(704)
7. Ryu, Y., Chang, K.-A. and Mori, N. (2005) Dispersion of Neutrally Buoyant Horizontal Round Jet in Wave Environment. Journal of Hydraulic Engineering, 131, 1088-1097.http://dx.doi.org/10.1061/(ASCE)0733-9429(2005)131:12(1088)
8. Tam, B.Y.-F. and Li, C.-W. (2008) Flow Induced by a Turbulent Jet under Random Waves. Journal of Hydraulic Research, 46, 820-829.http://dx.doi.org/10.1080/00221686.2008.9521926
9. Chang, K.-A., Ryu, Y. and Mori, N. (2009) Parameterization of Neutrally Buoyant Horizontal Round Jet in Wave Environment. Journal of Waterway, Port, Coastal,

and Ocean Engineering, 135, 100-107. http://dx.doi.org/10.1061/(ASCE)0733-950X(2009)135:3(100)

10. Hsiao, S.-C., Hsu, T.-W., Lin, J.-F. and Chang, K.-A. (2011) Mean and Turbulence Properties of a Neutrally Buoyant Round Jet in a Wave Environment. Journal of Waterway, Port, Coastal, and Ocean Engineering, 137, 109-122.http://dx.doi.org/10.1061/(ASCE)WW.1943-5460.0000073
11. Chyan, J.-M. and Hwung, H.-H. (1993) On the Interaction of a Turbulent Jet with Waves. Journal of Hydraulic Research, 31, 791-810.http://dx.doi.org/10.1080/00221689309498819
12. Chin, D.A. (1987) Influence of Surface Waves on Outfall Dilution. Journal of Hydraulic Engineering, 113, 1006-1018. http://dx.doi.org/10.1061/(ASCE)0733-9429(1987)113:8(1006)
13. Shuto, H. and Ti, L.H. (1974) Wave Effects on Buoyant Plumes. Proceedings of the 14th International Conference on Coastal Engineering, Copenhagen, June 1974, 2199-2208.
14. Tian, X. and Roberts, P.J.W. (2003) A 3D LIF System for Turbulent Buoyant Jet Flows. Experiments in Fluids, 35, 636-647. http://dx.doi.org/10.1007/s00348-003-0714-x
15. Webster, D.R., Rahman, S. and Dasi, L.P. (2003) Laser Induced Fluorescence Measurements of a Turbulent Plume. Journal of Engineering Mechanics, 129, 1130-1137.http://dx.doi.org/10.1061/(ASCE)0733-9399(2003)129:10(1130)
16. Funatani, S., Fujisawa, N. and Ikeda, H. (2004) Simultaneous Measurement of Temperature and Velocity Using Two-Colour LIF Combined with PIV with a Colour CCD Camera and Its Application to the Turbulent Buoyant Plume. Measurement Science and Technology, 15, 983-990. http://dx.doi.org/10.1088/0957-0233/15/5/030
17. Watanabe, Y., Hashizume, Y. and Fujisawa, N. (2006) Simultaneous Measurement of Temperature and Velocity in Turbulent Buoyant Plume by Combined LIF and PIV Technique. Proceedings of the 14th International Conference on Nuclear Engineering, Miami, 17-20 July 2006.
18. GrafsrØnningen, S. and Jensen, A. (2012) Simultaneous PIV/LIF Measurements of a Transitional Buoyant Plume above a Horizontal Cylinder. International Journal of Heat and Mass Transfer, 55, 4195-4206.http://dx.doi.org/10.1016/j.ijheatmasstransfer.2012.03.060
19. Isaacson, M.S., Koh, R.C.Y. and Brooks, N.H. (1983) Plume Dilution for Diffusers with Multi-Port Risers. Journal of Hydraulic Engineering, 109, 199-219.http://dx.doi.org/10.1061/(ASCE)0733-9429(1983)109:2(199)
20. Roberts, P.J.W. and Toms, G. (1988) Ocean Outfall System for Dense and Buoyant Effluents. Journal of Environmental Engineering, 114, 1175-1191.http://dx.doi.org/10.1061/(ASCE)0733-9372(1988)114:5(1175)
21. Roberts, P.J.W. (1994) Jets and Plumes in Ocean Outfall Design. In: Davies, P.A. and Valente Neves, M.J., Eds., Recent Research Advances in the Fluid Mechanics

of Turbulent Jets and Plumes, Kluwer Academic Publishers, Dordrecht, 441-464. http://dx.doi.org/10.1007/978-94-011-0918-5_28

22. Ippen, A.T. (1966) Estuary and Coastline Hydrodynamics. McGraw-Hill, New York.
23. Lide, D.R., Ed. (1993) CRC Handbook of Chemistry and Physics. 74th Edition, CRC Press, Boca Raton.
24. Irving, J.B. (1977) Viscosities of Binary Liquid Mixtures: The Effectiveness of Mixture Equations. Report No. 631, National Engineering Laboratory, Department of Industry, London. [Paper reference:1]
25. Perry, R.H. and Green, D.W. (1997) Perry's Chemical Engineers' Handbook. 7th Edition, McGraw-Hill, New York.
26. Hann, D.B., Davies, P.A. and Greated, C.A. (1998) Laser-Induced Fluorescence Measurements of the Effect of Breaking Waves on a Density Interface. Proceedings of the International Conference on Optical Methods and Data Processing in Heat and Fluid Flow, London, 16-17 April 1998, 397-406.
27. Gray, C., Greated, C.A., McCluskey, D.R. and Easson, W.J. (1991) An Analysis of the Scanning Beam PIV Illumination System. Measurement Science and Technology, 2, 717-724. http://dx.doi.org/10.1088/0957-0233/2/8/003

4
CHAPTER

QUANTITATIVE MEASUREMENT OF FLY ASH CONTENTS USING LASER-INDUCED BREAKDOWN SPECTROSCOPY

Zhenzhen Wang, Yoshihiro Deguchi, Hiroaki Watanabe, Ryoichi Kurose, Junjie Yan, Jiping Liu

Keywords: Laser-Induced Breakdown Spectroscopy, Fly Ash, Temperature Correction, CO_2 Effect

INTRODUCTION

The coal provides the largest share of world electricity generation and the thermal power plant share is dominant in total electric power supply in the world. Operating characteristics of coal-fired boilers are heavily influenced by factors such as differences in fuel properties and the combustion conditions. Accordingly, in order to achieve optimal operation of multiple coal-fired boilers, it is necessary to accurately understand the state of combustion and to adjust the control parameters. The research presented here focused on the quantity of unburned carbon in fly ash as an indicator of the combustion state, considering the measured value as a basis for optimal and stable boiler combustion control, as well as fly ash quality control.

Recently, Laser-Induced Breakdown Spectroscopy (LIBS) technique has been applied to various industrial fields [1] - [7] because of the fast response, high sensitivity, real-time and non-contact features. For the purpose of improving the economy and security of power plant, the content measurement of unburned carbon and other materials is essential for adjusting the boiler combustion. LIBS is suitable for composition measurement in gas, liquid, and solid materials [8] - [16]. This method is also suited to the analysis of heterogeneous materials such as fly ash and coal [17] - [20] and has been widely applied to analyze the compositional characterization of coal.

Citation: Wang, Z. , Deguchi, Y. , Watanabe, H. , Kurose, R. , Yan, J. and Liu, J. (2015) Quantitative Measurement of Fly Ash Contents Using Laser-Induced Breakdown Spectroscopy. Journal of Flow Control, Measurement & Visualization, 3, 10-21. doi: 10.4236/jfcmv.2015.31002.

For example, inorganic species in coal were analyzed quantitatively under different conditions [21] [22] . The solid sample features higher homogeneity than powdered counterparts. LIBS has also been applied for detection of unburned carbon in fly ash, char and pulverized coal under high-pressure and high-temperature conditions. The calibration difficulty of aerosol sample was surpassed by the correction factors. This automated LIBS apparatus was also applied in a boiler-control system of a power plant with the objective of achieving optimal and stable combustion [23] [24] , which enabled real-time measurement of unburned carbon in fly ash. This apparatus featured a detection time capability of less than 1 minute, which showed better real-time capability compared to most of the studies with the preparation of samples and analysis procedures [18] [25] . The boiler control in the real power plant was also demonstrated to achieve an optimized operation without time consumption. The contents of size-segregated particles depending on particle diameter have been measured using LIBS. The plasma temperature correction method was introduced to the size-segregated fly ash and pulverized coal to detect their quantitative contents [26] .

In this study, improvement on quantification of LIBS has been demonstrated in terms of plasma temperature and surrounding gas effects to further applications. These factors are important for practical applications; however, they have not been fully discussed in previous studies [23] [24] . Dependence of elemental signals of LIBS on the plasma temperature attributes to a very complex process in plasma. Several studies have reported the LTE (Local Thermodynamic Equilibrium) condition of plasma in several types of plasmas [27] . The plasma temperature is a very important factor for the quantification of the LIBS measurement [1] [24] . The correction factor depends on the experimental parameters such as the optical setups, pressure, buffer gas composition, and so on. Because the plasma induced by the LIBS process is not uniform and complex, LTE can be assumed in the limited range of these conditions. The effects of non-equilibrium were discussed using the intensity ratio of the emission pair of magnesium atom as a plasma temperature indicator and the appropriate range of the plasma temperature indicator was shown for quantification. The CO_2 effects were also shown and the requirement for quantification was discussed for the real-time measurement of unburned carbon in fly ash for the application of coal-fired power plants.

THEORY

A calibration of the LIBS signal is necessary for quantitative analysis. Despite the fact that the LIBS processes involved are complex, the emission intensity from the atomized species can be described by the following equation with the assumption of uniform plasma temperature [1] [23] [24] [28] :

$$I_i = n_i K_{i,j} g_{i,j} \exp\left(-\frac{E_{i,j}}{kT}\right). \qquad (1)$$

In the above expression, I_i is the emission intensity of species i, n_i is the concentration of species i, $K_{i,j}$ is a variable that includes the Einstein A coefficient from the upper energy level j, $g_{i,j}$ is the statistical weight of species i at the upper energy level j, $E_{i,j}$ is the upper level energy of species i, k is the Boltzmann constant and T is the plasma temperature. In Equation (1), there are several factors that affect the emission intensity I_i, including plasma temperature, plasma non-uniformity, and matrix effects, etc. The appropriate correction factors must be contained in $K_{i,j}$ to obtain quantitative results.

The main components of fly ash are Si, Al, Fe, Ca and C (carbon). The carbon content in fly ash can be calculated using the emission intensity ratios of Al, Fe, Ca and C to Si that is the main element in fly ash, from the following relation:

$$\text{Carbon content} = \frac{\alpha_C I_C / I_{Si}}{1 + \alpha_{Al} I_{Al}/I_{Si} + \alpha_{Fe} I_{Fe}/I_{Si} + \alpha_{Ca} I_{Ca}/I_{Si} + \alpha_C I_C/I_{Si}}. \quad (2)$$

α_i is a variable factor related to species i, which contains the plasma temperature correction factor. These parameters have to be determined in the experimental operating conditions. Other species contents besides carbon, such as Al, Fe and Ca, can also be calculated using the similar relation.

In Equation (1), a uniform plasma temperature is assumed. Actually, emission intensity I_i is a function of concentration of species i and plasma temperature. The emission intensity for each element also fluctuates with plasma temperature, which introduces error into quantitative measurements. There are several methods of plasma temperature correction. The following equation is used in this study. α_i is defined by use of the intensity ratio of $I_{m,j1}/I_{m,j2}$:

$$\alpha_i = D_{i,j}\left(I_{m,j1}/I_{m,j2}\right)^{b_{i,m}}. \quad (3)$$

$D_{i,j}$ is the emission intensity correction factor from the upper energy level j of species i. $I_{m,j1}$ is the emission intensity from the upper energy level j1 of species m, $I_{m,j2}$ is the emission intensity from the upper energy level j2 of species m, and $b_{i,m}$ is the temperature correction factor of the emission pair of $I_{m,j1}$ and $I_{m,j2}$ for species i.

$\left(I_{m,j1}/I_{m,j2}\right)^{b_{i,m}}$ is dependent on plasma temperature and is used for plasma temperature correction. Different spectral lines from the same species are selected to correct the plasma temperature dependence of the emission intensity, and the intensity ratio of the emission pair is determined from Equation (1):

$$\frac{I_{m,j1}}{I_{m,j2}} = \frac{K_{m,j1} g_{m,j1}}{K_{m,j2} g_{m,j2}} \exp\left(-\frac{E_{m,j1} - E_{m,j2}}{kT}\right). \quad (4)$$

Applying the intensity ratio to Equation (3), α_i becomes:

$$\alpha_i = D_{i,j}\left(\frac{K_{m,j1} g_{m,j1}}{K_{m,j2} g_{m,j2}}\right)^{b_{i,m}} \exp\left[b_{i,m}\left(-\frac{E_{m,j1} - E_{m,j2}}{kT}\right)\right]. \quad (5)$$

Taking C (carbon) from the upper energy level jC for example, the concentration ratio is as follows:

$$\alpha_C \frac{I_C}{I_{Si}} = \alpha_C \frac{n_C}{n_{Si}}\left(\frac{K_{C,jC}g_{C,jC}}{K_{Si,jSi}g_{Si,jSi}}\right)\exp\left[-\frac{E_{C,jC}-E_{Si,jSi}}{kT}\right]$$
$$= D_{C,jC}\frac{n_C}{n_{Si}}\left(\frac{K_{C,jC}g_{C,jC}}{K_{Si,jSi}g_{Si,jSi}}\right)\exp\left[-\frac{E_{C,jC}-E_{Si,jSi}}{kT}\right]\left(I_{m,j1}/I_{m,j2}\right)^{b_{C,m}}$$
$$= D_{C,jC}\frac{n_C}{n_{Si}}\left(\frac{K_{C,jC}g_{C,jC}}{K_{Si,jSi}g_{Si,jSi}}\right)\left(\frac{K_{m,j1}g_{m,j1}}{K_{m,j2}g_{m,j2}}\right)^{b_{C,m}}\cdot\exp\left[-\left(b_{C,m}\frac{E_{m,j1}-E_{m,j2}}{kT}+\frac{E_{C,jC}-E_{Si,jSi}}{kT}\right)\right]. \quad (6)$$

This ratio becomes a function of species concentration and does not depend on the plasma temperature on condition that $b_{C,m}$ is given as follows:

$$b_{C,m} = -\frac{E_{C,jC}-E_{Si,jSi}}{E_{m,j1}-E_{m,j2}}. \quad (7)$$

From this analytical result, the correction factor can be calculated using the above equation. However, the correction factor depends on the experimental parameters such as the optical setups, pressure, buffer gas composition, and so on. The plasma induced by the LIBS process is not uniform and complex and LTE can be assumed in the limited range of these experimental conditions. In the case of ion emission intensities it is also necessary to consider the degree of ionization which can be calculated by the Saha equation as a function of plasma temperature and electron number density in a LTE condition. Therefore the correction factor cannot be determined by the theory and should be evaluated under actual experimental conditions. In this study the ranges of LTE conditions were categorized using the intensity ratio of the emission pair of magnesium atom as a plasma temperature indicator and Equation (3) was used for the plasma temperature correction.

EXPERIMENTAL APPARATUS

Figure 1 shows the experimental apparatus used in this study. A Q-switched Nd:YAG laser (LOTIS TII, LS-2137U, 6 - 8 ns, energy stability: 2.5%, beam diameter: 8 mm, 10 Hz) was operated at 1064 nm with pulse energy of 30 - 90 mJ/p and laser beam was focused into the measurement area to make a plasma using a f = 200 mm lens. The samples were introduced to the measurement area by the 1.8 mm (inner diameter) stainless steel pipe and the breakdown of the particles was occurred along the laser path around the focusing point. The emission signals were separated from the laser path using a splitter and focused into abundled optical fiber with 24 fibers. The spectra were dispersed using a spectrometer (JASCOCT-10S) and detected by an ICCD camera (iStar 334T Series, Andor) with the wavelength resolution of 0.072 nm/pixel, FWHM of 0.5 nm and spectral bandwidth of 65 nm. The ICCD camera consisted of 1024 × 1024 active pixels with an effective active area of 13.3 mm × 13.3 mm and was cooled to −30°C. The effective CCD pixel size was 13 μm × 13 μm. The frame rate was 4 image/sec. The row of the ICCD camera was used as wavelength axis and the column as the position axis of 24 fibers. Each spectrum was evaluated by summing the signal intensity of each column. The temperature correction factors were detected using laser power of 30 mJ/p and different delay time from 800 ns to 3000 ns by the measurement of a fly ash sample. Temperature correction factors and LTE conditions were determined using a fly ash sample by the intensity ratio of the emission pair of magnesium atom as a plasma temperature indicator. The data was acquired by setting the accumulation of 10 s each time. Pulverized coal (Newlands Coal) was added to combustion-supporting gas of methane and air using a feeder (Nisshin Engineering Inc. Feedcon-μ Mtype) and then introduced to the Bunsen-type burner. The flow rates of methane, air and coal were 3.9 l/min, 15.6 l/min and 0.41 g/min, respectively. A sampling pipe above the burner was employed to connect with the vacuum chamber of LIBS analyzer. The compositions of coal and fly ash used in this study are summarized in Table 1.

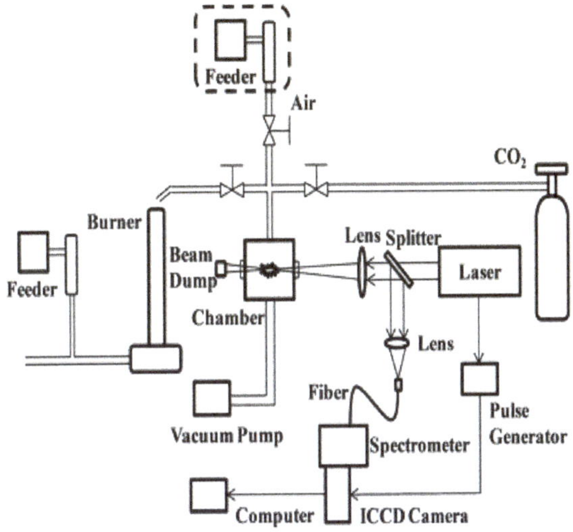

Figure 1. Schematic diagram of the experimental apparatus.

Table 1. Compositions of coal and fly ash samples.

Species	Coal (%)	Fly ash (%)
C	56.2	17.25
SiO_2	7.31	45.01
Al_2O_3	5.19	25.13
Fe_2O_3	1.05	6.17
CaO	0.54	0.75

RESULTS AND DISCUSSION

Temperature Correction Factors

In LIBS applications the plasma generation process contains several physical phenomena such as electron impact ionization and ion recombination, and its plasma temperature intrinsically fluctuates causing the fluctuation of evaluated concentrations from the LIBS signal intensities. Therefore the plasma temperature correction becomes important to assure the quantitative measurement, especially for the particle measurement in gases. The fly ash sample was used to determine the plasma temperature correction factors under different experimental conditions. Figure 2 shows LIBS spectra of fly ash in delay time of 800 ns, 1500 ns and 3000 ns with the laser power of 30 mJ/p. The intensity concerning species concentration varied with different delay time, which means the alteration of plasma temperature. The emission lines listed in Table 2 were used for the analysis of species concentration [29] . The ratio of I_{Mg1}/I_{Mg2} was used for the plasma temperature correction factor as shown in Equation (3). I_{Mg1}/I_{Mg2} was altered between 4 - 24 by changing the delay time between 800 ns - 3000 ns. I_{Mg1}/I_{Mg2}

Table 2. Upper level energy of detection species in fly ash.

Species	Upper energy (cm^{-1})	Wavelength (nm)
Si	39955.05	251.6
Si	40991.88	288.2
Fe	45206.45 - 36766.96 (ion and atom)	274.9 - 275
Al	32436.8	309.27
Ca ion	56839.25	315.9
Ca ion	56839.25	318.1
C	61982.2	247.9
Mg1 ion	35760.88	279.6
Mg1 ion	35669.31	280.3
Mg2	35051.27	285.2

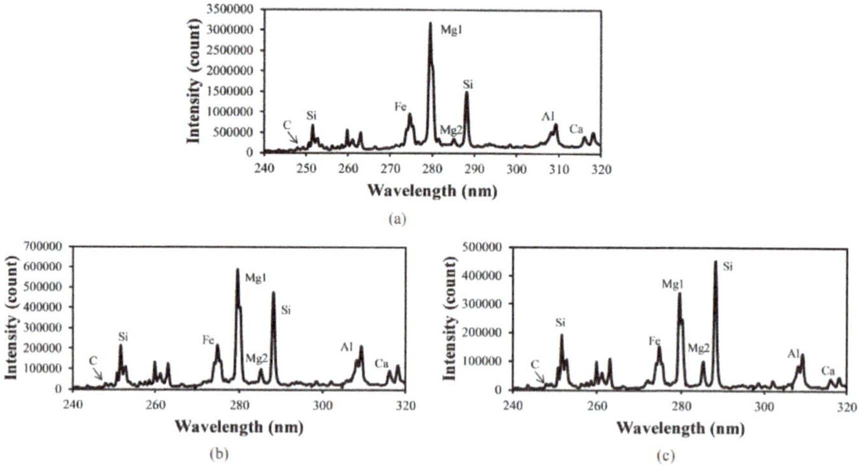

Figure 2. LIBS spectra of fly ash in different delay time (laser power: 30 mJ/p, gate width: 300 ns). (a) Delay time: 800 ns; (b) Delay time: 1500 ns; (c) Delay time: 3000 ns.

became bigger according to the decrease of the delay time. The typical temperature correction curves between I_{Mg1}/I_{Mg2}: 4 - 24 and temperature corrected results of I_C/I_{Si} and I_{Fe}/I_{Si} are shown in Figure 3. It is clear from Figure 3(a) and Figure 3(b) that the ratios are influenced by the plasma temperature, which is directly related to the ratio of I_{Mg1}/I_{Mg2}. By applying the temperature correction scheme described in Equations (2)-(7), the fluctuation of I_C/I_{Si} and I_{Fe}/I_{Si} became smaller at different plasma temperatures, i.e. I_{Mg1}/I_{Mg2}. However, the plasma temperature dependencies of I_{Mg1}/I_{Mg2} are apparently different between I_{Mg1}/I_{Mg2}: 4 - 8, which means the low plasma temperature condition, and I_{Mg1}/I_{Mg2}: 12 - 24, i.e. high plasma temperature condition. The high temperature condition corresponds to the initial process of the plasma generation from the fly ash particles and its expansion process to the surrounding air. While the low temperature condition is formed after the plasma expansion to the surrounding air. This result means that the plasma induced by the LIBS process is not uniform and complex and LTE can be assumed in the limited range of plasma temperature, i.e. I_{Mg1}/I_{Mg2}. The limitation of LTE assumption has been also discussed in the glow discharge plasma [27] , and the understanding of the phenomena is important for the quantitative measurement, which requires the accurate plasma correction.

The temperature correction curves between I_{Mg1}/I_{Mg2}: 4 - 8 and temperature corrected results of I_C/I_{Si} and I_{Fe}/I_{Si} are shown in Figure 4. Figure 5 shows those between I_{Mg1}/I_{Mg2}: 12 - 24. It is clear from Figure 4(a) and Figure 4(b) that the correction curves are much more accurate than those inFigure 3(a) and Figure 3(b) and Figure 5(a) and Figure 5(b), and $b_{i,m}$ in Equation (3) takes the different values depending on the range of I_{Mg1}/I_{Mg2}, which means the limitation of LTE assumption. In the high temperature plasma, i.e. I_{Mg1}/I_{Mg2}: 12 - 24, the plasma is not uniform due to the initial process of the plasma generation from the fly ash particles and its expansion process to the surrounding air. With the increase of delay time, the plasma becomes uniform compared to the initial process of the plasma generation and tends to follow the LTE condition. The fluctuations of the concentration ratios of C/SiO_2 and Fe_2O_3/SiO_2 corrected by the correction factor $b_{i,m}$ in the range of I_{Mg1}/I_{Mg2}: 4 - 24 are 17% and 8.9%, respectively. The fluctuations of the concentration ratios do not improve by limiting the range of I_{Mg1}/I_{Mg2}: 12 - 24. On the other hand the fluctuations of these ratios corrected by the correction factor $b_{i,m}$ in the range of I_{Mg1}/I_{Mg2}: 4 - 8 reduce to be 11% and 3.7%. This phenomenon also appears in other species

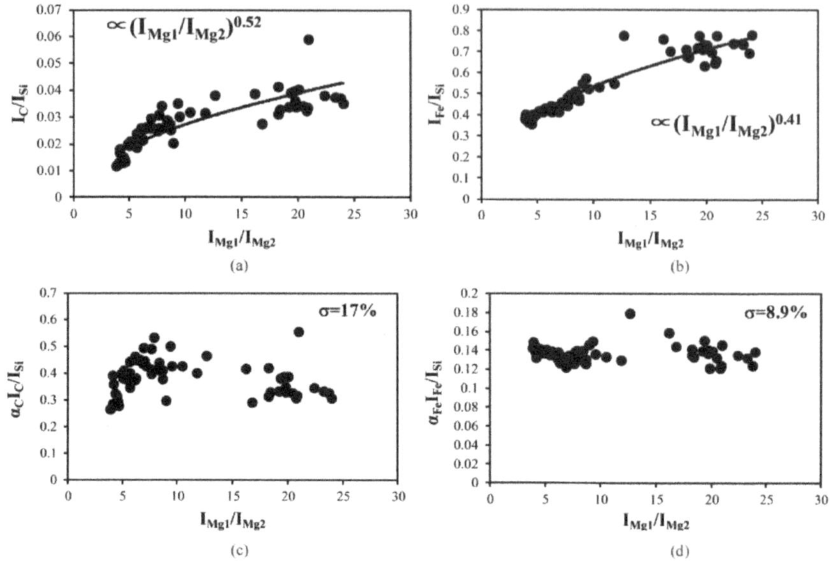

Figure 3. Correction curves and results of plasma temperature between I_{Mg1}/I_{Mg2}: 4 - 24. (a) Correction factor for the I_C/I_{Si} ratio; (b) Correction factor for the I_{Fe}/I_{Si} ratio; (c) Correction result for the I_C/I_{Si} ratio; (d) Correction result for the I_{Fe}/I_{Si} ratio.

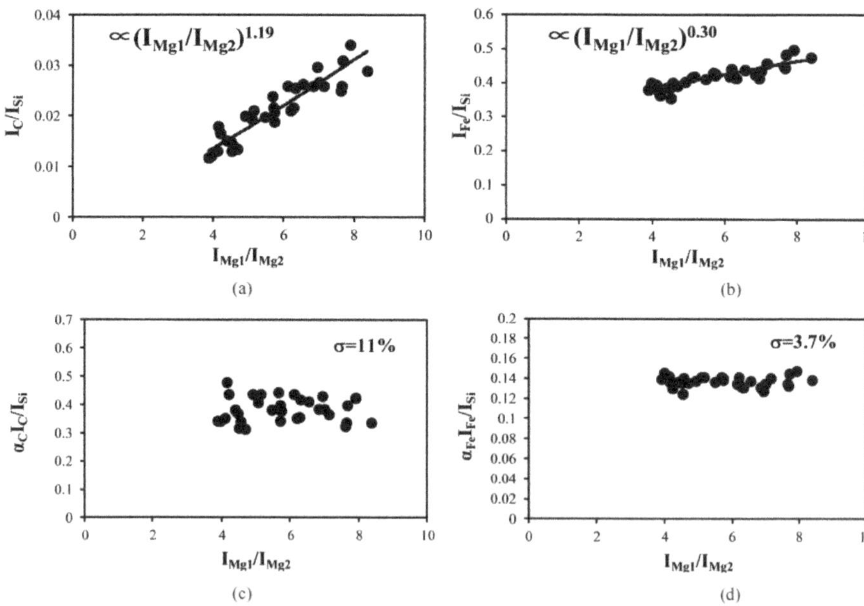

Figure 4. Correction curves and results of plasma temperature between I_{Mg1}/I_{Mg2}: 4 - 8. (a) Correction factor for the I_C/I_{Si} ratio; (b) Correction factor for the I_{Fe}/I_{Si} ratio; (c) Correction result for the I_C/I_{Si} ratio; (d) Correction result for the I_{Fe}/I_{Si} ratio.

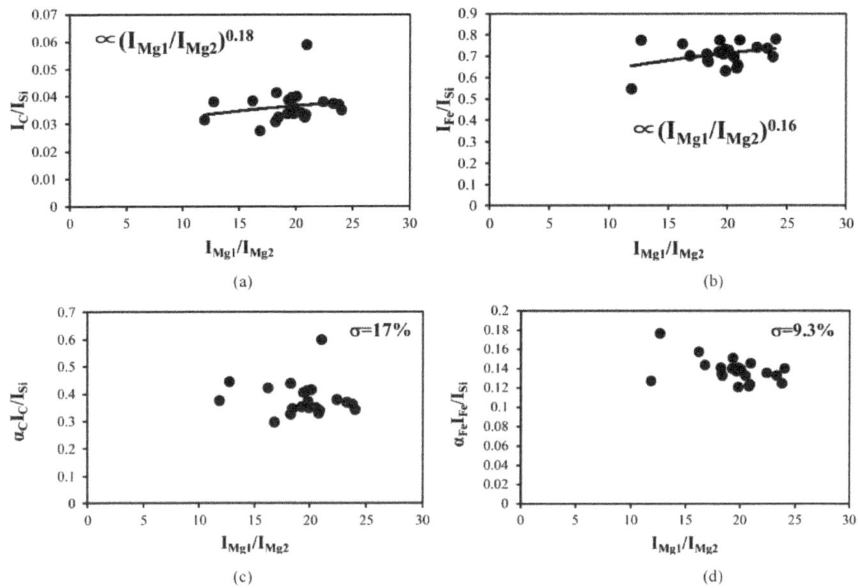

Figure 5. Correction curves and results of plasma temperature between I_{Mg1}/I_{Mg2}: 12 - 24. (a) Correction factor for the I_C/I_{Si} ratio; (b) Correction factor for the I_{Fe}/I_{Si} ratio; (c) Correction result for the I_C/I_{Si} ratio; (d) Correction result for the I_{Fe}/I_{Si} ratio.

such as I_{Al}/I_{Si} and I_{Ca}/I_{Si}. The results for C/SiO_2, Al_2O_3/SiO_2, Fe_2O_3/SiO_2 and CaO/SiO_2 are summarized in Table 3. The correction characteristics for these concentration ratios show the same tendency as described above and it is important to set measurement condition within an appropriate I_{Mg1}/I_{Mg2} range. The cause of fluctuation in the temperature corrected results is also attributed to the sample uniformity. Because the fly ash contents are not uniform in the microscopic sense. Using this set of experimentally determined correction factors by choosing the appropriate I_{Mg1}/I_{Mg2} range, the fluctuations of the concentration ratios such as Fe_2O_3/SiO_2, Al_2O_3/SiO_2, CaO/SiO_2, C/SiO_2, were greatly reduced.

There are several plasma temperature correction methods such as employing the Boltzmann plot method which uses several emission lines from the same atom from the wide-wavelength LIBS spectra. The calibration method using the Boltzmann, which often called "CF (calibration free) LIBS", is a systematic approach for the temperature correction. However it requires wide-wavelength LIBS spectra and causes the loss of the signal intensity and the complication of the systems [30] [31] . The method employed in this paper can be applied using the LIBS spectra with the wavelength range of 240 - 320 nm and keeps the sufficient signal intensity with sufficient plasma temperature correction capability for fly ash measurements. Considering the system simplicity and real-time measurement capability, the temperature correction method employing emission pair of I_{Mg1} and I_{Mg2} is preferable compared to others, which demonstrated that the method is workable and satisfactory for practical applications by choosing the appropriate plasma temperature correction range.

Influence of Coexisting Gases

In the case of fly ash measurement at actual conditions such as the real-time measurement of fly ash contents in exhausts of coal-fired boilers, influence of coexisting gases such as CO_2 must be considered to maintain the quantitative measurement of this method. In the case of particle contents measurement in gases using LIBS, the effects of coexisting gases become important because the gas phase materials also produce plasma caused by the particle breakdowns. Figure 6shows the LIBS signals from the burned gas of the burner without coal addition using different laser power. It is clear from Figure 6(a), the burned gas was broken by the laser input of 90 mJ/p and the strong C signal caused by the CO_2 breakdown was detected at 247.9 nm. When reducing the laser power, the C signal became small and the C signal disappeared at 37 mJ/p as shown in Figure 6(b). On the other hand, when the pulverized coal was added in the fuel and the fly ash in the burned gas was measured. The signals of

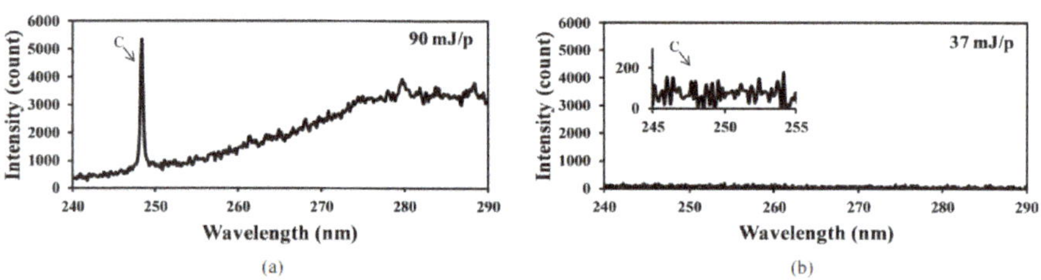

Figure 6. LIBS spectra of burned gas without pulverized coal addition (delay time: 1000 ns, gate width: 300 ns). (a) Laser power: 90 mJ/p; (b) Laser power: 37 mJ/p.

Table 3. Standard deviation of LIBS measurement depending on I_{Mg1}/I_{Mg2} correction range.

Composition ratio	Chemical analysis	LIBS standard deviation (%)			
		No correction (I_{Mg1}/I_{Mg2}: 4 - 24)	Corrected (I_{Mg1}/I_{Mg2}: 4 - 24)	Corrected (I_{Mg1}/I_{Mg2}: 4 - 8)	Corrected (I_{Mg1}/I_{Mg2}: 12 - 24)
C/SiO_2	0.383	34	17	11	17
Al_2O_3/SiO_2	0.558	48	20	4.9	14
Fe_2O_3/SiO_2	0.137	27	8.9	3.7	9.3
CaO/SiO_2	0.017	44	27	10	22

fly ash contents such as Si and Fe were measured along with the strong C signal as Figure 7(a) shows. This C signal is mainly produced by CO_2 in burned gas and it interferes with the measurement of carbon content in fly ash. Figure 7(b) shows that the C signal was also detected using the laser power of 37 mJ/p in the case of the fuel with pulverized coal, though it was not detected without pulverized coal as shown in Figure 6(b). This is because the particles such as fly ash became the core of the plasma and the plasma expanded to the surrounding gases. The influence of coexisting gases in LIBS depends on the measurement conditions, especially the particle conditions in gases, and its effects must be evaluated in the actual measurement applications.

Figure 8 shows the LIBS spectra of fly ash with different CO_2 concentration. In this experiment the fly ash was introduced into the measurement area with the addition of CO_2 in air. As CO_2concentration increases, the C signal increases compared to other signals while other signal ratios such as Fe and Si remain the same. This is because the breakdown of CO_2 added in air was induced by the fly ash plasma. As shown in Figure 6(b) the C signal does not appear without pulverized coal addition. Figure 9 shows the dependence of I_C/I_{Si} on CO_2 concentration that was measured two times at each CO_2 concentration. The CO_2 effect is rather sensitive to the concentration in surrounding gas and less than 1% CO_2 causes the spurious C signal in LIBS spectra of fly ash. The measured results also fluctuated under the conditions with the same CO_2 concentration because of the unstable breakdown phenomena of CO_2 induced by fly ash. Therefore it is important to reduce the CO_2 content in burned gas when applying LIBS for unburned carbon content in fly ash under burned gas conditions. The threshold level of CO_2 depends on the measurement conditions such as the range of carbon contents and particle density. CO_2 effect is highly laser power-dependent. An appropriate laser power setting is essential to minimize this spurious carbon signal. Increased pressure aggravates the problem due to the breakdown of CO_2 molecules. Careful attention is needed under high-pressure conditions [23] . The experimental parameters must be determined for the actual operation conditions. The sampling of fly ash using a cyclone employed in the coal-fired power plant is one of the effective methods to reduce this CO_2 effect [24] .

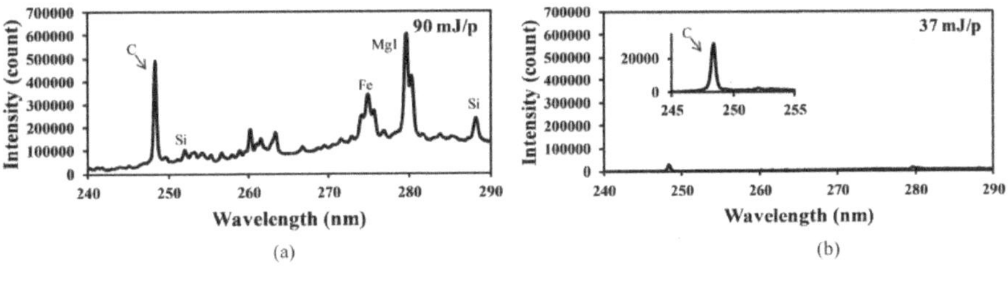

Figure 7. LIBS spectra of burned gas with pulverized coal addition (delay time: 1000 ns, gate width: 300 ns). (a) Laser power: 90 mJ/p; (b) Laser power: 37 mJ/p.

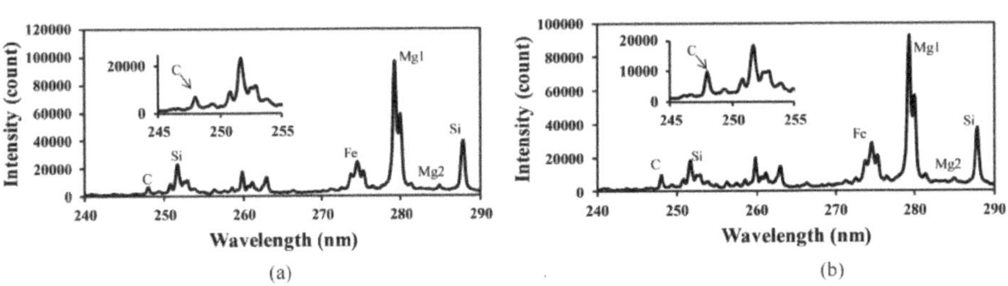

Figure 8. LIBS spectra of fly ash with different CO_2 concentration(laser power: 37 mJ/p, delay time: 1000 ns, gate width: 300 ns). (a) CO_2: 0.84%; (b) CO_2: 2.5%.

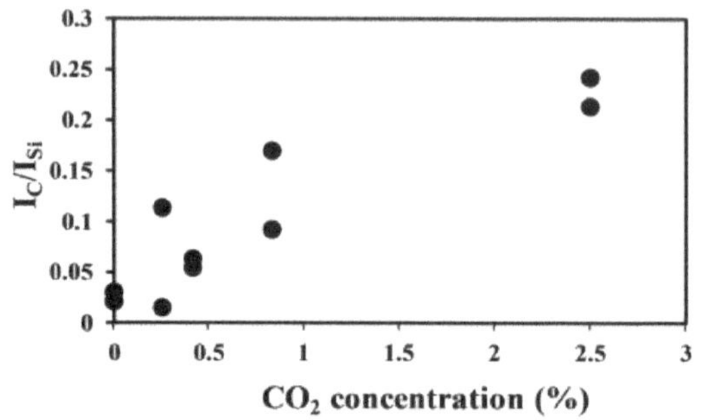

Figure 9. CO_2 concentration dependence of I_C/I_{Si} (laser power: 37 mJ/p, delay time: 1000 ns, gate width: 300 ns).

CONCLUSIONS

Contents of fly ash are important factors for the operation of coal-fired plants. This research has specified the effects of plasma temperature and coexisting materials to the quantitative measurement of contents of fly ash. Quantitative capability of LIBS was improved using the plasma temperature correction method by choosing the appropriate temperature correction range. The CO_2effect was also clarified to accurately evaluate unburned carbon in fly ash in exhausts.

1) LTE was assumed in a limited range of plasma temperature in laser-induced plasma and the quantitative measurement was improved by choosing the appropriate plasma temperature range. In the case of fly ash measurement, the emission intensity ratio of Mg: I_{Mg1}/I_{Mg2} became a plasma temperature indicator and the range of I_{Mg1}/I_{Mg2}: 4 - 8 showed better LTE condition resulting in the accurate plasma correction compared to the range of I_{Mg1}/I_{Mg2}: 12 - 24.

Zhenzhen Wang, Yoshihiro Deguchi, Hiroaki Watanabe, Ryoichi Kurose, Junjie Yan, Jiping Liu 2) In the measurement of fly ash contents, CO_2 caused the spurious C signal in LIBS spectra of fly ash. The CO_2 effect was rather sensitive to the concentration in surrounding gas and less than 1% CO_2caused the spurious C signal. It became important to reduce the CO_2 content in burned gas when applying LIBS for the measurement of unburned carbon content in fly ash under the burned gas conditions.

REFERENCES

1. Deguchi, Y. (2011) Industrial Applications of Laser Diagnostics. CRC Press, New York. http://dx.doi.org/10.1201/b11497
2. Noll, R., Mönch, I., Klein, O. and Lamott, A. (2005) Concept and Operating Performance of Inspection Machines for Industrial Use Based on Laser Induced Breakdown Spectroscopy. Spectrochimica Acta Part B, 60, 1070-1075. http://dx.doi.org/10.1016/j.sab.2005.05.025
3. Gaft, M., Sapir-Sofer, I., Modiano, H. and Stana, R. (2007) Laser Induced

Breakdown Spectroscopy for Bulk Minerals Online Analyses. Spectrochimica Acta Part B, 62, 1496-1503. http://dx.doi.org/10.1016/j.sab.2007.10.041

4. Boué-Bigne, F. (2008) Laser-Induced Breakdown Spectroscopy Applications in the Steel Industry: Rapid Analysis of Segregation and Decarburization. Spectrochimica Acta Part B, 63, 1122-1129. http://dx.doi.org/10.1016/j.sab.2008.08.014
5. St-Onge, L., Kwong, E., Sabsabi, M. and Vadas, E.B. (2002) Quantitative Analysis of Pharmaceutical Products by Laser-Induced Breakdown Spectroscopy. Spectrochimica Acta Part B, 57, 1131-1140. http://dx.doi.org/10.1016/S0584-8547(02)00062-9
6. Juvé, V., Portelli, R., Boueri, M., Baudelet, M. and Yu, J. (2008) Space-Resolved Analysis of Trace Elements in Fresh Vegetables Using Ultraviolet Nanosecond Laser-Induced Breakdown Spectroscopy. Spectrochimica Acta Part B, 63, 1047-1053. http://dx.doi.org/10.1016/j.sab.2008.08.009
7. Ma, Q.L., Motto-Ros, V., Lei, W.Q., Boueri, M., Zheng, L.J., Zeng, H.P., Bar-Matthews, M., Ayalon, A., Panczer, G. and Yu, J. (2010) Multi-Elemental Mapping of a Speleothem Using Laser-Induced Breakdown Spectroscopy. Spectrochimica Acta Part B, 65, 707-714. http://dx.doi.org/10.1016/j.sab.2010.03.004
8. Peter, L., Sturm, V. and Noll, R. (2003) Liquid Steel Analysis with Laser-Induced Breakdown Spectrometry in the Vacuum Ultraviolet. Applied Optics, 42, 6199-6204. http://dx.doi.org/10.1364/AO.42.006199
9. Wang, Z.Z., Deguchi, Y., Kuwahara, M., Yan, J.J. and Liu, J.P. (2013) Enhancement of Laser-Induced Breakdown Spectroscopy (LIBS) Detection Limit Using a Low-Pressure and Short-Pulse Laser-Induced Plasma Process. Applied Spectroscopy, 67, 1242-1251. http://dx.doi.org/10.1366/13-07131
10. Wang, Z.Z., Deguchi, Y., Kuwahara, M., Zhang, X.B., Yan, J.J. and Liu, J.P. (2013) Sensitive Measurement of Trace Mercury Using Low Pressure Laser-Induced Plasma. Japanese Journal of Applied Physics Part 2, 52, Article ID: 11NC05.
11. Yaroshchyk, P., Morrison, R.J.S., Body, D. and Chadwick, B.L. (2005) Quantitative Determination of Wear Metals in Engine Oils Using LIBS: The Use of Paper Substrates and a Comparison between Single- and Double-Pulse LIBS.
12. Yueh, F.Y., Sharma, R.C., Singh, J.P., Zhang, H.S. and Spencer, W.A. (2002) Evaluation of the Potential of LaserInduced Breakdown Spectroscopy for Detection of Trace Element in Liquid. Journal of the Air & Waste Management Association, 52, 1307-1315. http://dx.doi.org/10.1080/10473289.2002.10470860
13. Hubmer, G., Kitzberger, R. and Mörwald, K. (2006) Application of LIBS to the In-Line Process Control of Liquid High-Alloy Steel under Pressure. Analytical and Bioanalytical Chemistry, 385, 219-224. http://dx.doi.org/10.1007/s00216-006-0321-9
14. Laville, S., Sabsabi, M. and Doucet, F.R. (2007) Multi-Elemental Analysis of Solidified Mineral Melt Samples by Laser-Induced Breakdown Spectroscopy Coupled with a Linear Multivariate Calibration. Spectrochimica Acta Part B, 62, 1557-1566.

15. Goueguel, C., Laville, S., Vidal, F., Sabsabi, M. and Chaker, M. (2010) Investigation of Resonance-Enhanced LaserInduced Breakdown Spectroscopy for Analysis of Aluminium Alloys. Journal of Analytical Atomic Spectrometry, 25, 635-644. http://dx.doi.org/10.1039/b927013b
16. Diwakar, P.K., Loper, K.H., Matiaske, A.M. and Hahn, D.W. (2012) Laser-Induced Breakdown Spectroscopy for Analysis of Micro and Nanoparticles. Journal of Analytical Atomic Spectrometry, 27, 1110-1119. http://dx.doi.org/10.1039/c2ja30012e
17. Panne, U., Haisch, C., Clara, M. and Niessner, R. (1998) Analysis of Glass and Glass Melts during the Vitrification Process of Fly and Bottom Ashes by Laser-Induced Plasma Spectroscopy. Part I: Normalization and Plasma Diagnostics. Spectrochimica Acta Part B, 53, 1957-1968. http://dx.doi.org/10.1016/S0584-8547(98)00238-9
18. Ctvrtnickova, T., Mateo, M.P., Yañez, A. and Nicolas, G. (2009) Characterization of Coal Fly Ash Components by Laser-Induced Breakdown Spectroscopy. Spectrochimica Acta Part B, 64, 1093-1097. http://dx.doi.org/10.1016/j.sab.2009.07.032
19. Gaft, M., Dvir, E., Modiano, H. and Schone, U. (2008) Laser Induced Breakdown Spectroscopy Machine for Online Ash Analyses in Coal. Spectrochimica Acta Part B, 63, 1177-1182. http://dx.doi.org/10.1016/j.sab.2008.06.007
20. Yuan, T.B., Wang, Z., Lui, S.L., Fu, Y.T., Li, Z., Liu, J.M. and Ni, W.D. (2013) Coal Property Analysis Using LaserInduced Breakdown Spectroscopy. Journal of Analytical Atomic Spectrometry, 28, 1045-1053.
21. Mateo, M.P., Nicolas, G. and Yañez, A. (2007) Characterization of Inorganic Species in Coal by Laser-Induced Breakdown Spectroscopy Using UV and IR Radiations. Applied Surface Science, 254, 868-872. http://dx.doi.org/10.1016/j.apsusc.2007.08.043
22. Yuan, T.B., Wang, Z., Li, L.Z., Hou, Z.Y., Li, Z. and Ni, W.D. (2012) Quantitative Carbon Measurement in Anthracite Using Laser-Induced Breakdown Spectroscopy with Binder. Applied Optics, 51, B22-B29.
23. Noda, M., Deguchi, Y., Iwasaki, S. and Yoshikawa, N. (2002) Detection of Carbon Content in a High-Temperature and High Pressure Environment Using Laser-Induced Breakdown Spectroscopy. Spectrochimica Acta Part B, 57, 701-709. http://dx.doi.org/10.1016/S0584-8547(01)00403-7
24. Kurihara, M., Ikeda, K., Izawa, Y., Deguchi, Y. and Tarui, H. (2003) Optimal Boiler Control through Real-Time Monitoring of Unburned Carbon in Fly Ash by Laser-Induced Breakdown Spectroscopy. Applied Optics, 42, 6159-6165. http://dx.doi.org/10.1364/AO.42.006159
25. Ctvrtnickova, T., Mateo, M.P., Yañez, A. and Nicolas, G. (2010) Laser Induced Breakdown Spectroscopy Application for Ash Characterisation for a Coal Fired Power Plant. Spectrochimica Acta Part B, 65, 734-737. http://dx.doi.org/10.1016/j.sab.2010.04.020
26. Wang, Z.Z., Deguchi, Y., Kuwahara, M., Taira, T., Zhang, X.B., Yan, J.J., Liu,

J.P., Watanabe, H. and Kurose, R. (2013) Quantitative Elemental Detection of Size-Segregated Particles Using Laser-Induced Breakdown Spectroscopy. Spectrochimica Acta Part B, 87, 130-138. http://dx.doi.org/10.1016/j.sab.2013.05.034

27. Zhao, Y.H. and Horlick, G. (2006) A Spectral Study of Charge Transfer and Penning Processes for Cu, Zn, Ag, and Cd in a Glow Discharge. Spectrochimica Acta Part B, 61, 660-673. http://dx.doi.org/10.1016/j.sab.2006.05.010

28. Goddard, B.J. (1991) Materials Analysis Using Laser-Based Spectroscopic Techniques. Transactions of the Institute of Measurement and Control, 13, 128-139. http://dx.doi.org/10.1177/014233129101300304

29. Payling, R. and Larkins, P. (2000) Optical Emission Lines of the Elements. John Wiley & Sons, New York.

30. Unnikrishnan, V.K., Mridul, K., Nayak, R., Alti, K., Kartha, V.B., Santhosh, C., Gupta, G.P. and Suri, B.M. (2012) Calibration-Free Laser-Induced Breakdown Spectroscopy for Quantitative Elemental Analysis of Materials. PramanaJournal of Physics, 79, 299-310. http://dx.doi.org/10.1007/s12043-012-0298-1

31. Cavalcanti, G.H., Teixeira, D.V., Legnaioli, S., Lorenzetti, G., Pardini, L. and Palleschi, V. (2013) One-Point Calibration for Calibration-Free Laser-Induced Breakdown Spectroscopy Quantitative Analysis. Spectrochimica Acta Part B, 87, 51-56. http://dx.doi.org/10.1016/j.sab.2013.05.016

5
CHAPTER

THE FAILURE OF THE CROSS CORRELATION MEASUREMENT TECHNIQUE

Ken McGill, Kathryn Ham, Kris Schock

Keywords: Acoustic Array, Cross Correlation, Flow Measurement

INTRODUCTION

Since the establishment of the first oil transporting pipeline system in 1879, half a million miles of pipe line networks have been built across the United States [1] . With the rising nationwide demand for various commodities such as water, petroleum, and natural gas, the expansion of the pipeline system was a necessity as a means of transportation for millions of dollars of valuable merchandise. With the growing number of pipelines nationwide and the increasing volume of liquids transported daily via pipeline, a newfound need has arisen for technology that accurately calculates the rate of flow of these liquids. In order to control the amount of a substance sent through a pipe, the measure of its flow velocity must be determined to a high level of accuracy. Various techniques and technologies have been created in an attempt to provide an accurate measurement for flow in a closed conduit; however, they offer results with varying degrees of inaccuracies. The Cross Correlation Method has become a popular method in determining rate of flow through a pipeline [2] . A proposed acoustic method employs a set of transducers along the measured pipe which detect time delay for an acoustic wave propagating through the pipe.

Citation: McGill, K. , Ham, K. and Schock, K. (2015) The Failure of the Cross Correlation Measurement Technique. Open Journal of Acoustics, 5, 25-28. doi: 10.4236/oja.2015.51003.

THEORY

The Wave Equation

A solution to the wave equation for a sound wave propagating through a pipe is:

$$\Psi = A\sin\left(k\left(x+ct\right)\right) = A\sin\left(\frac{2\pi}{\lambda}\left(x+\lambda ft\right)\right) = A\sin\left(2\pi\left(\frac{x}{\lambda}+ft\right)\right) = A\sin\left(\theta+2\pi ft\right) \tag{1}$$

where A is the normalization constant, k is $(2\pi)/\lambda$, f is the frequency of the wave, t is the time, x is the position, λ is the wavelength of the wave, and θ is the phase angle of the wave propagating through space [3] . By comparing the arguments of the functions sin ( ) in equality (1), one can see that the phase plotted versus the position will include the speed of sound [3] .

$$\theta = 2\pi\frac{x}{\lambda} = 2\pi\frac{fx}{c} \tag{2}$$

$$m = \frac{2\pi f}{c} \tag{3}$$

where m is the slope of the line. Thus, the speed of sound, c, can be found as follows:

$$c = \frac{2\pi f}{m} \tag{4}$$

PROCEDURE

In an experiment to test this proposed method, 450 distinct frequencies of sound waves from 500 to 5000 Hz were sent down a 7.50 m long metal pipe with a 0.1524 m diameter Figure 1. The range of 500 to 5000 Hz was chosen since any frequency below 500 Hz has interference due to a large amount of ambient noise, which could interfere with the results and 5000 Hz was the maximum frequency possible with the sample rate of 10,000

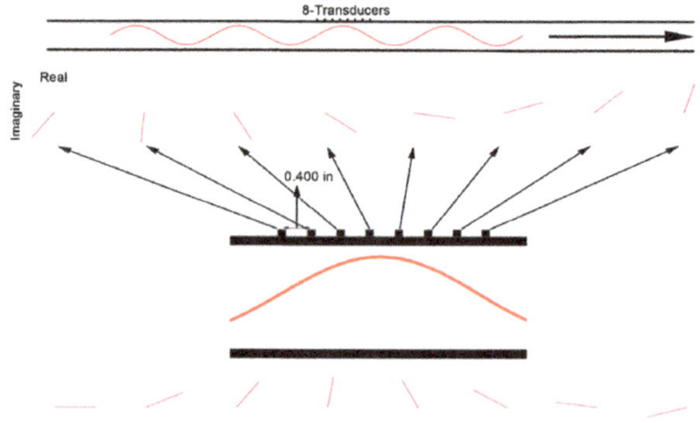

Figure 1. The acoustic array with 8 microphones.

samples/s due to the Nyquist limit. The sound wave was measured using 8 transducers 0.01016 m apart. Since the minimum frequency used was 500 Hz, the distance between the microphones needed to be larger than the maximum wavelength; thus, 0.01016 m (equivalent to 0.4 in) was chosen due to the precision of the machinery available. The 450 data points taken from the instrument were then converted using an analog-digital converter (ADC) as seen in the block diagram in Figure 2. In order to determine the real and imaginary components of the phase, a Fourier Analysis was conducted on the data obtained from experimentation. In order to obtain a value for the speed of sound, the phase was plotted versus the position and the slope was calculated (Figure 2). The speed of sound along with the correlation coefficients for all frequencies from 500 to 5000 Hz was then plotted.

Data and Results

While the results obtained from experimentation yield a high correlation coefficient between the phases of the sound wave and the position along the pipe, the value obtained for the speed of sound is incorrect even at correlation coefficients of 0.999. The proposed cross-correlation method for determining the speed of a sound wave propagating through a pipe has been proven invalid for its intended purposes.

CONCLUSIONS

Based on the data obtained from experimentation, the Cross-Correlation Phase Measurement Technique has proven itself an inaccurate method in determining the speed of a sound wave propagating through a pipe at various frequencies. Even with a consistently high correlation coefficient between the phases of the sound wave and the position along the pipe, across the range of tested frequencies, the calculated speed of sound was inaccurate at a majority of the points (Figure 3). As seen in Figure 3, the range of frequencies: 1797 Hz - 2197 Hz, has a correlation coefficient close to 1; however, the speed of sound measured in m/s has a range of values from 207.1 m/s to 634.6 m/s. The speed of sound in air at room temperature should be 343 m/s [4] . The correlation coefficient and the incorrect data suggest a flaw in the Cross-Correlation Measurement Technique. The flawed methodology behind this Cross-Correlation Phase Measurement Technique has neglected the presence of the reflected wave moving through the pipe in its calculations, thus resulting in inaccurate results. The inaccurate results are due to the superposition of the waves.

This technique does not take the superposition theory into account, but assumes that the input is coming from only one direction. Superposition of a sound wave describes a state of interference between multiple sound waves. This interference can be either constructive, where the waves are additive, or destructive, where the waves are reductive. This effect can be seen in the reflection of a sound wave in the pipe when two waves combine to form one superimposed sound wave. To illustrate the magnitude of effect superposition has on the results, consider the wave shown below:

$$\Psi = \cos\left(2\pi f\left(t-\frac{x}{c+v}\right)\right)+0.01\sin\left(2\pi f\left(t-\frac{x}{c+v}\right)\right) \qquad (5)$$

where f is the frequency; c is the speed of sound; and v is the flow velocity, which adds to c when the flow velocity is in the same direction as the propagating wave. The equation assumes the presence of the same wave from another source, 90° out of phase with 1% amplitude of the above wave. The resulting wave would have

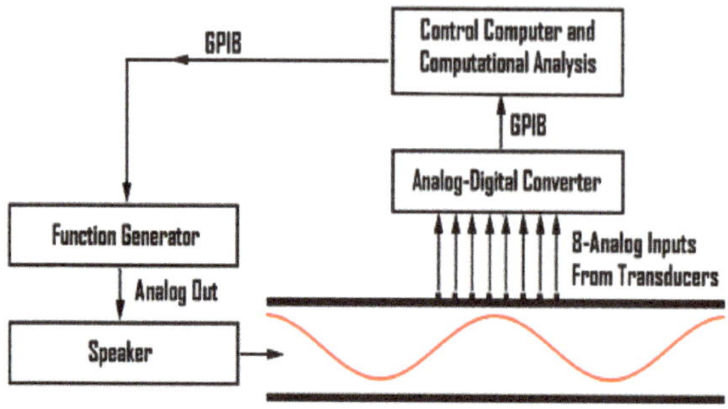

Figure 2. Block diagram of the experiment.

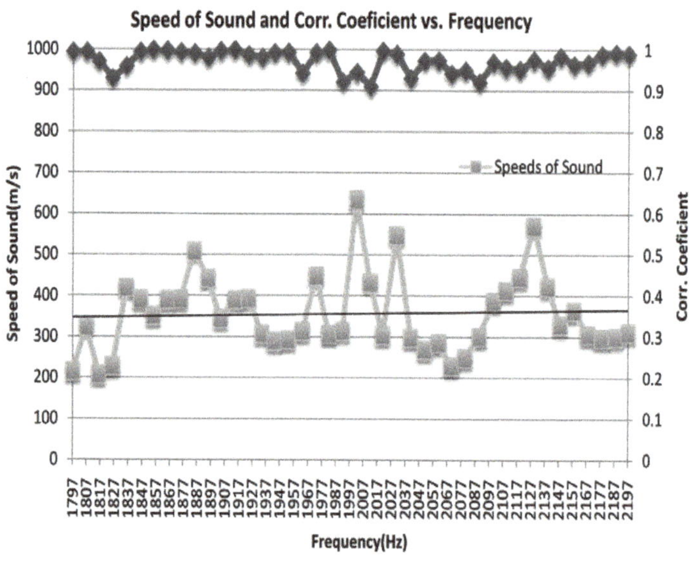

Figure 3. Speed of sound and correlation coefficient versus frequency.

a phase difference of 0.57° = $\tan^{-1}(0.01)$ or 0.16% full scale. While 0.16% may seem small, this slight discrepancy translates into a notable error of 0.56 m/s for a speed of sound of 350 m/s, and given a full range of flow velocity from 0 m/s to 10 m/s, it translates to 5.6% error in flow velocity.

To give an idea of magnitude by converting the interfering wave amplitude to the input wave amplitude, a 40 dB difference is revealed in sound. This is equivalent to the sound level in a library. It is very likely that noise of this level, or more, would be present from the ambient noise in the pipe or from reflected waves.

$$\frac{I}{I_o}=\left(\frac{\Delta A}{A}\right)^2=0.0001,\ \text{or}\ \beta=10\log\left(\frac{I}{I_o}\right)=-40\ \text{dB} \tag{6}$$

To employ acoustics to measure this phase phenomenon, it will be necessary to detect and separate out any miscellaneous background sound waves in order to more accurately analyze the sound wave present in the pipe. If the sound can be separated, it may be possible to use the ambient noise for the phase measure, and thereby remove the need for a sound source.

Acknowledgements

Ken McGill, Kathryn Ham, Kris Schock We would like to thank Georgia College and State University for providing the materials and space needed for this research.

REFERENCES

1. Kennedy, J.L. (1993) Oil and Gas Pipeline Fundamentals. 2nd Edition, PennWell Books, Oklahoma, 3-5.
2. Beck, M.S. and Plaskowski, A. (1987) Cross Correlation Flowmeters, Their Design and Application. Taylor & Francis, UK.
3. Taylor, J.R. (2005) Classical Mechanics. University Science Books, Sausalito, 688.
4. Kirkpatrick, L. and Francis, G. (2010) Physics: A Conceptual World View. 7th Edition, Brooks/Cole, Cengage Learning, California, 331.

6

CHAPTER

COMPARISON OF TWO ELECTRONIC ROOT CANAL LENGTH MEASUREMENT DEVICES

Endang Suprastiwi, Ratna Meidyawati

Keywords: Electronic Root Canal Measurement Devices

INTRODUCTION

It is generally accepted that root canal treatment procedures should be confined within the root canal system. The success rate could reach up to 90% - 94% [1] [2] . In order to achieve the successful root canal treatment, it is important to know the canal terminus which is the limit of root canal and periodontal tissues. There are various methods in determining the canal terminus. Radiographic and electronic root canal length measuring device are often used to measure the root canal length [3] . Radiography is an image of the results of a photographic film irradiated x-ray as it passes through an object on the surface of the radiosensitive. The result of measuring the length of dental radiograph has its limitations because of distortion so as to provide a two-dimensional picture of a threedimensional structure of the tooth with the result that less representative. Due to limitation of the results, the radiographic method was invented an electronic device for the measurement of root canal length. Electronic root canal length measurement device has a more accurate measurement results when compared with the radiographic measurement results [3] [4] .

Electronic root canal length measurement device is used to determine the position of the apical foramen which is the meeting point between root canal and periodontal tissues. Root canal dentin and cement area insulator to electric currents. However, on

Citation: Suprastiwi, E. and Meidyawati, R. (2014) Comparison of Two Electronic Root Canal Length Measurement Devices. Open Journal of Stomatology,4, 263-267. doi: 10.4236/ojst.2014.45037.

the minor apical foramen, there is a small hole which is the conductive material is electrically connected to the periodontal ligament it is a conductor for electricity. The basic assumption with all electronic length measuring devices is that human tissues have certain characteristics that can be modeled by a combination of electrical components. Since 1962, electronic root canal length measurement device began to be used, but there was no explanation how these devices work. Classifying and describing electronic root canal length measuring device could be based on the generation time of the issuance of such a device [5] . Nekoofar et al. (2006) states grouping based on the operating of the instrument. While the simple device measure resistance, other devices measure impedance using either high frequency, two frequencies or multiple frequencies. In addition, some systems use low frequency oscillation and/or a voltage gradient method to detect the canal terminus [2] . The impedance ratio-based device using two frequencies is independent of the electrolyte liquid in root canals, and according to Jenkins et al. (2001) this device could detect the apical foramen in the range of 0.4 mm without disturbing existing irrigation fluid [6] . While according Plotino et al. (2006), the accuracy of this device in measuring the length of work is still in the range of ±0.5 mm from the apical constriction [3] .

There have been efforts to further increase the accuracy of electronic root canal length measurement device. One concept was to measure the impedance characteristics using multiple frequencies. This device uses five different frequencies to measure both components (phase and amplitude) impedance at each frequency so then can know the location of the minor apical foramen (apical constriction). This new generation device has some advantages of more accurate measurement because it measures the characteristic impedance using more than two frequencies. Plotino et al. (2006) stated that the precision instrument with five frequencies is 100% while the device impedance ratio of the two frequencies with accuracy just reached 97.37% [3] .

Based on the background above, this study will compare and test the accuracy between two electronic root canal length measurement devices: two frequencies impedance ratio and multi frequencies with actual tooth length as a reference. The research results can be used as a reference for endodontist to perform a root canal treatment, especially at the stage of working length measurement.

MATERIALS AND METHODS

Forty anterior teeth, closed apices with no defect on root surface were used as a sample study. All samples are cleaned up from any calculus or stain then soaked in saline solution until the study perform. All samples included in these criteria were sectioned at cervical area with carbarundum disc and low speed hand piece. Flattened surface at cervical area was used as reference point. Samples were numbered and ready to be measured.

A sample was fixed on imaging plate Digora (Juusula, Finland). Then, sample was exposed with x-ray from bucco-lingual direction. The measurement was obtained from the calculation of the tooth length from reference point to radiographic apex minus 1.0 mm. The result image radiographic was measured with Windows version 2.5 Digora. After that, each sample was embedded in container filled with sponge and NaCl 0.9%

solution. The root canal was flooded with NaCl 0.9% solution and then dried with a small cotton pellet. Measuring root canal length with two frequencies impedance ratio devices (Root ZX mini, J. Morita Mfg Corp Kyoto, Japan) was performed using a #15 K file and entered the root canal until sound beep heard for at least five seconds. Electronic measurement with two frequencies impedance ratio device was recorded. The same way was done by using multi frequencies device (Woodpex I, Guilin WoodPecker Medical Instrument Co. Ltd., Guilin, China). After measuring all samples with electronic device, all samples were vertically split and measured the actual length from reference point to apical constriction with K file #15 under 3× magnifying loupe.

The measurement results of radiographs, two frequencies impedance ratio and multi frequencies device were subtracted with the measurement results of actual length in order to obtain the value of difference. Positive value showed the measurement was more than the actual length. As negative value show the measurement was less than the actual length. Descriptively refer to the Real et al. (2011) that makes three categorical made to determine the accuracy of the measurement results are: categorical (−): <0.5 mm (shorter than the actual length of the teeth), (0) ± 0.5 mm (length approaching actual length) and (+) > 0.5 mm (longer than the actual length of the teeth) [7] . The data obtained was analyzed with Chi-square test ($p < 0.05$) and the comparative test Kolmogorov-Smirnov.

RESULTS

To assess the accuracy of tooth length measurement, the actual length used as the reference measurement. The result of each method of measurement was calculated the difference from the actual length. The results of the three groups were performed with comparative analysis Chi-square test, and for significance for each group were performed with comparative analysis Kolmogorov-Smirnov. Distribution of number and percentage of samples measured with two frequencies impedance ratio electronic device, multi frequencies electronic device and radiograph can be seen in Table 1 and Figure 1.

The determining of the location of the apical constriction on category (0); two frequencies impedance ratio electronic device present in as many as 20 samples of 40 samples 50% and electronic device based on multi frequencies were 19 samples from 40 samples 47.5%, while the radiograph measurements were only 5%.

In Table 2, statistic analysis using the Kolmogorov-Smirnov test for the three groups, the value of significance-(p) between the two frequencies-based electronic devices and multi frequencies-based is 1.000 ($p > 0.05$). These results illustrate that the two groups did not have a significant difference. As the two groups compared with radiographs there were a significant difference value.

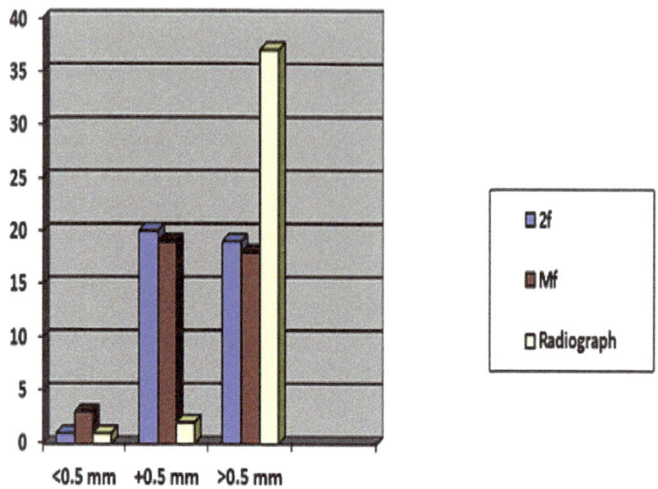

Figure 1. Diagram of the distribution method of measuring the root canal length.

Table 1. Distribution and percentage of electronic and radiographic measurements.

Methods of measurement	(−)		(0)		(+)		Total	
	n	%	n	%	n	%	n	%
2f	1	2.5	20	50	19	47.5	40	100
Mf	3	7.5	19	47.5	18	45	40	100
Radiograph	11	2.5	2	5	37	92.5	40	100

(−) = <0.5 mm from actual length; (0) = ±0.5 mm from actual length; (+) = >0.5 mm from actual length; 2f = Two ratio impedance frequencies; Mf = Multi frequencies.

Table 2. Test of significance between groups.

Method of Measurement	*p* Values
2f vs. Mf	1.00
2f vs. Radiographs	0.00*
Mf vs. Radiographs	0.01*

*Significantly different, significance analysis by Kolmogorov-Smirnov test.

DISCUSSION

To simplify and uniformity of the sample in this study was used anterior teeth as a sample because they have single root. In order condition of the sample in accordance with the biological situation in the oral cavity, the sample was immersed in saline solution prior to the study. Measurement of the teeth length usually starts from a reference point that is located on the highest tip the crown. For simplify determining location of the reference point, all samples cut perpendicular to the axis of the teeth in cervical area, that it is located a stable reference point [8] . The number of samples used in this study, refer to the Rules of Thumb that suggest the nonparametric-test or multivariate tests on samples that are not given treatment, the number of samples used at least 40 samples [9] .

Radiographic measurements performed with parallel technique, because it can get images with the same angle so it can download the picture better visualization [10] . Radiograph measurements done by measuring the distance between the reference point to the apex end. Apical constriction location is determined by subtracting 1 mm of tooth length measured value. It refers Vieyra et al. (2010) which stated that location of the apical constriction is in position 1 mm from the radiographic apex end [11] .

Tooth length measurement by electronic means using the file K number 10 or 15 because the maximum tactile sensations tool was available [12] . A length measurement with electronic device is programmed only to the extent of the apical constriction [7] -[13] . Thus electronic root canal length device can only detect the presence of apical constriction. According to Huang (1987) principle of electronic working length device is on electric properties of the tooth itself [14] . To achieve clinical situations, samples were placed in a medium that has an electrical resistance similar to the periodontal tissue. The samples were embedded in saline-soaked sponge that has properties similar to the alveolar bone and fluid [7] .

Actual tooth length measurements made from the reference point to the apical constriction. Location of apical constriction was used as the zero point for calculating the difference in measurements with radiographic and electronic methods. The results were then grouped into 3 categorical outcomes: (0) = ± 0.5 mm (near the apical constriction), (−) = <0.5 mm (shorter than apical constriction) or (+) => 0.5 mm (longer than apical constriction).

The results in Figure 1 and Table 1 indicate that two frequencies electronic device at category (0) has a value of 50% and (+) has a value of 47.5%, the multi frequencies device at category (0) has a value of 47%, 5%, and (+) has a value of 45%, but the radiographic method at category (+) has a value of 92.5%. It is proved that all three methods of measuring the length of the teeth, radiographic measurement method is a method with the results of the lowest accuracy as it is always longer than the actual length. This result is opposite to the results of Real et al. (2011) who observed the digital radiographic measurements with accuracy of 64.9%. The accuracy of electronic root canal length measurement device in this study is only about 50% of the value. It can be concluded that the ability of the device to measure the length of the teeth have only limited accuracy rate of 50% [7] .

In Table 2 the three groups were compared, there is significant difference between the two frequency impedance ratio electronic device with radiographic and multi frequencies electronic device with radiographic, whereas if two electronic devices compared there is no significant difference between them. It can be concluded that system difference in the two electronic devices do not affect the accuracy of length measurement. This finding is in agreement with the statement of the manufacturer of electronic measuring devices which stated that the ability of at two frequencies electronic device is 53% - 74% accuracy, and the results of research supported by the statement by Javidi et al. (2009) that both two frequencies and multi frequencies electronic measuring device are not affected by the presence of pulp, pus or irrigation fluid. It can be concluded that the electronic root canal length measurement device measuring tooth length cannot be relied a hundred percent accuracy [15] .

CONCLUSION

The ratio two frequencies impedance root canal length measurement device has same accuracy level with multi frequencies but no significant difference.

REFERENCES

1. Ricucci, D. and Langeland, K. (1998) Apical Limit of Root Canal Instrumentation and Obturation Part 2. A Histological Study. International Endodontic Journal, 31, 399-409.
2. Nekoofar, M., Ghandi, M., Hayes, S. and Dummer, P. (2006) The Fundamental Operating Principles of Electronic Root Canal Length Measurement Devices. International Endodontic Journal, 39, 595-609. http://dx.doi.org/10.1111/j.1365-2591.2006.01131.x
3. Plotino, G., Grande, N., Brigante, L., Lesti, B. and Somma, F. (2006) Ex Vivo Accuracy of Three Electronic Apex Locators: Root ZX, Elements Diagnostic Unit and Apex Locator and ProPex. International Endodontic Journal, 39, 408- 414.
4. Dimitrov, S. and Dimitur, R. (2009) Sixth Generation Adaptive Apex Locator. Journal of IMAB-Annual Proceeding (Scientific Papers).
5. Gordon, M. and Chandler, N. (2004) Electronic Apex Locators. International Endodontic Journal, 37, 425-437. http://dx.doi.org/10.1111/j.1365-2591.2004.00835.x
6. Jenkins, J.A., Walker Iii, W.A., Schindler, W.G. and Flores, C.M. (2001) An in Vitro Evaluation of the Accuracy of the Root ZX in the Presence of Various Irrigants. Journal of Endodontics, 27, 209-211. http://dx.doi.org/10.1097/00004770-200103000-00018
7. Real, D.G., Davidowicz, H., Moura-Netto, C., Zenkner Cd, L.L., Pagliarin, C.M.L., Barletta, F.B., et al. (2011) Accuracy of Working Length Determination Using 3 Electronic Apex Locators and Direct Digital Radiography. Oral Surgery, Oral Medicine, Oral Pathology, Oral Radiology, and Endodontology, 111, e44-e49.http://dx.doi.org/10.1016/j.tripleo.2010.10.033
8. Assuncao, F. (2010) Ex Vivo Evaluation of the Accuracy and Coefficient of Repeatability of Three Electronic Apex Locator Using a Simple Mounting Model: A Preliminary Report. International Endodontic Journal, 43, 269-274. http://dx.doi.org/10.1111/j.1365-2591.2009.01674.x
9. Madiyono, B., Moeslichan, S., Sastroasmoro, S., Budiman, I. and Purwanto, S. (2002) Perkiraan Besar Sampel. In: Sastroasmoro, S. and Ismael, S., Eds., Dasar-Dasar Metodologi Penelitian Klinis, Sagung Seto, Jakarta, 285.
10. Walton, R. (2008) Diagnostic Imaging A. Endodontic Radiography. In: Ingle, J., Bakland, L. and Baumgartner, J.C., Eds., Endodontics, 6th Edition, People's

Medical Publishing House, 554-557.

11. Vieyra, J., Acosta, J. and Mondaca, J. (2010) Comparison of Working Length Determination with Radiographs and Two Electronic Apex Locators. International Endodontic Journal, 43, 16-20. http://dx.doi.org/10.1111/j.1365-2591.2009.01620.x
12. Meredith, N. and Gulabivala, K. (1997) Electrical Impedance Measurement of Root Canal Length. Endodontics and Dental Traumatology, 13, 126-131.http://dx.doi.org/10.1111/j.1600-9657.1997.tb00025.x
13. Welk, A.R., Baumgartner, J.C., Marshall, J.G. (2003) An in Vivo Comparison of Two Frequency-Based Electronic Apex Locators. Journal of Endodontics, 29, 497-500.http://dx.doi.org/10.1097/00004770-200308000-00002
14. Huang, L. (1987) An Experimental Study of the Principle of Electronic Root Canal Measurement. Journal of Endodontics, 13, 60-64. http://dx.doi.org/10.1016/S0099-2399(87)80156-5

7

CHAPTER

PRECISE MEASUREMENT OF MOVING OBJECT BY MOIRÉ-BASED IMAGE PROCESSING TECHNIQUE

Md. Tawhidul Islam Khan, Maeda Kazuhiko, Kenbu Teramoto, Md. Mahbub Hasan

Keywords: Moiré Imaging, Single Grating Method, Precise Measurement

INTRODUCTION

The word moiré refers to the meaning as watered or wavy appearance which evolved from the ancient French word mouaire. The scientific earlier application of moiré phenomena was for testing the quality of replicated diffraction gratings [1]. However, modern scientific research into the moiré phenomenon and its application has been started at the last of the 19th century. Since then the theoretical analysis of moiré phenomena has been based on purely geometric or algebraic approaches. Based on these approaches many special purpose mathematical developments have been devised for the needs of specific applications such as strain analysis, surface profilometry in metrology, etc. More recently several new approaches have been proposed for studying moiré phenomena, based, respectively, on no-standard analysis, on elementary geometry and potential theory, or on algebraic geometry. It is, however, undoubtedly the Fourier-based approach that most significantly contributed to the theoretic investigation of the moiré phenomenon.

Moiré fringes are the results of the interference produced by the superimposing two sets of repetitive gratings. These patterns are used in vibration analysis as well as in 3D surface reconstruction as well. Moiré images are normally obtained using a camera to capture the patterns generated by superimposing of two alternating opaquetransparent Ronchi gratings or two projected light patterns [2].

Citation: M. Khan, M. Kazuhiko, K. Teramoto and M. Hasan, "Precise Measurement of Moving Object by Moiré-Based Image Processing Technique," Open Journal of Fluid Dynamics, Vol. 2 No. 4A, 2012, pp. 202-207. doi:10.4236/ojfd.2012.24A022.

The passive-pixel sensors (without their own amplifiers) were being investigated as a solid-state alternative to vacuum-tube imaging devices, when the MOS passivepixel sensor used just a simple switch in the pixel to read out the photodiode integrated charge [3], which suffered from many limitations, such as high noise, slow readout, lack of scalability etc. The CMOS sensor (an active pixel sensor or APS sensor which was first devised by Tsutomu Nakamura, when he was in charge at Olympus and more broadly defined by Eric Fossum for developing the charge modulation device active pixel sensor [4]) is well established as a well controlled stable baseline for almost all logic and microprocessors. There was a resurgence in the use of passive-pixel sensors for low-end imaging applications, and active-pixel sensors for low-resolution high-function applications such as retina simulation and high energy particle detector [5].

In the present work a single repetitive grating is used instead of two gratings. The imaging device itself plays the rote of one of the grating with its regular 2D repetitive arrangements of the CMOS sensor (camera). This CMOS camera is then used to observe another grating. The interactions between these two gratings result in the formation of moiré patterns, which can be simply captured by the camera itself.

The advantage of the proposed research is its simplicity in a precise measurement of different mechanical as well as bio-medical engineering systems such as deformation measurement of aerofoil surface in a simple snapshot technique. As only one camera and one grating film is necessary in the present system, the overall measurement system becomes easier and simpler compared to the conventional projection technique of moiré imaging [6]. Moreover, the complexity of parameters evaluation in moiré-images has also been reduced by applying callback algorithm of image processing toolbox. Although, the present research is a basic work for developing the simple measurement technique based on moiré topography, its practical application in the different angular motion sensing system has been verified by using series of rotating model experiments in an optical bench test system.

MOIRÉ IMAGE WITH CMOS SENSORS

Measurements based on moiré topography are conducted by calculating the fundamental moiré parameters as its interval of two successive patterns and the orientation with vertical axis. The classical approach of calculating these parameters are documented by the following basic interferometric expressions [7].

When these two optical gratings are superimposed, the resulting intensity transmission function for moiré pattern is related as follows:

$$\begin{aligned}\phi(x,y) &= \phi_1(x,y)\phi_2(x,y)\\ &\approx \sum_{m=1}^{\infty}\sum_{n=1}^{\infty} A_{1m}A_{2n}\cos\left[m\psi_1(x,y)\right]\cos\left[n\psi_2(x,y)\right]\end{aligned} \tag{1}$$

where, f_i indicates the grating functions and $\psi_i(x,y)$ indicated the phase values, which show the basic shape of the grating lines; i = 1, 2. A is the coefficients determines the amplitude profile of the grating lines. After rewriting above Equation (1), moiré patterns for different conditions can be evaluated when the grating line functions (phases) are expanded with its sinusoidal functions as follows:

$$\psi_1(x,y)=\frac{2\pi}{d_1}(x\cos\alpha+y\sin\alpha) \tag{2}$$

$$\psi_2(x,y)=\frac{2\pi}{d_2}(x\cos\alpha-y\sin\alpha) \tag{3}$$

where, d_i indicates the line intervals of two gratings and a is the angle of each grating line created with y-axis of the coordinate system. As the sum and difference patterns are available in the above grating line functions, it represents the moiré patterns when it superimposed. Furthermore, the beat wavelength equation can be evaluated when the difference of these two grating functions (Equations (2)-(3)) is calculated. Thus, from the calculated beat wavelength functions, the general equations for calculating interval of moiré fringes as well as its angle with the vertical axis (y-axis) can be explained by the following expressions:

$$C=\frac{d_1d_2}{\sqrt{\sin^2\alpha(d_2+d_1)^2+\cos^2\alpha(d_2-d_1)^2}} \tag{4}$$

$$\tan\theta=\tan\alpha\left(\frac{d_2+d_1}{d_2-d_1}\right) \tag{5}$$

where, d_1 and d_2 are the intervals of two stripes. The classical examples of moiré patterns generated by the superposing of two gratings as explained above are shown in **Figure 1**. **Figure 1**(a) indicates the moiré patterns for the interference of two gratings with same grating-spaces where as, **Figure 1**(b) indicates the moiré fringe pattern for the interference of two gratings with different grating-spaces. The spacing of moiré fringes is indicated by C when, the angle that the moiré fringe generates with the vertical axis is indicated by q.

The rotational angle of the gratings is indicated by 2a. B and D indicate the bright and dark fringes respectively. From Equation (5), it is seen that if the intervals of two gratings are same, the angle that the moiré fringe generates with the vertical axis is 90 degrees, however, when they are not similar they generates acute angle with the vertical axis. These results can be seen in Figures 1(a) and (b) as well.

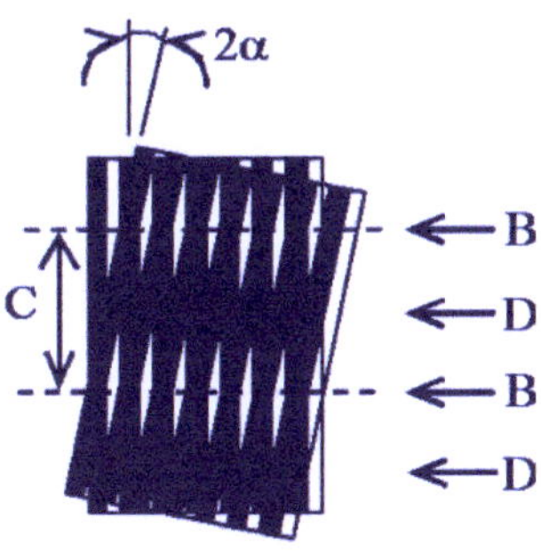

(a)

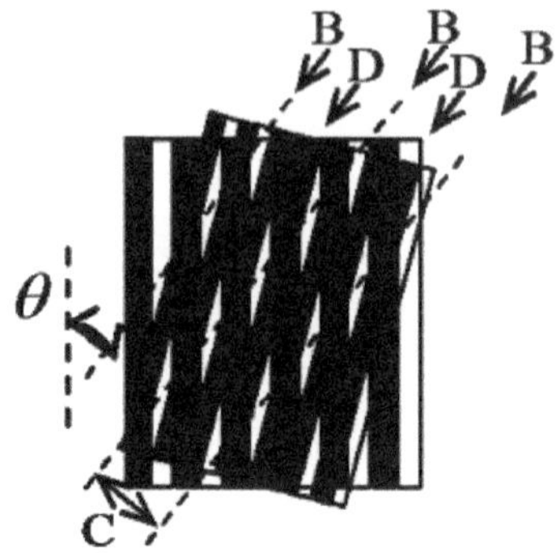

(b)

Figure 1. Moiré images with two gratings (2 wave image).

EXPERIMENTAL PROCEDURE

Experimental Set-Up

An experimental set-up has been developed capable of measuring the 6D motion of an object by using of a single gratings moiré fringe technique. Moiré fringes have been generated by applying a simple vertical straight line pattern with a grating interval of 0.71 mm. The optical gratings are designed by AR-CAD soft-wear and are printed by an inkjet canon printer (PIXUS iX7000) on to the adhesive printing paper which can be attached to an acryl board without making any air-gap. The acryl board was attached to a gonio which can move with 6 degrees of freedom (TSF60-EC, SIGMA KOKI Cor. Ltd., Japan). A small CMOS camera (MCM 4601; small CMOS sensor [5] of MICRO VISION Co. Ltd.) with 6.35 mm (1/4 inch) optical format and 2.8 μm × 2.8 μm pixel size has been used and mounted on a height adjustable stand and it is connected to a personal computer (PC) by USB cable for storage of images for further analysis. The camera-stand and the gonio-stand are mounted on an optical rail upon which all the base positions can be precisely measured. The center heights of both the camera and grating-object are kept same from the top of the rail at the beginning of the experiment. The panoramic view of the experimental setup is shown in **Figure 2**. The properties of CMOS camera, consisting of an integrated circuit and array of pixel sensors (photodetector and active amplifier), are shown in **Figure 3** and **Table 1** respectively.

The data acquisition and modulation of moiré images through the CMOS camera are conducted by OpenCV (Intel Open Source Computer Vision Library) algorithm inside the PC.

Experimental Procedure

According to the objective of the experimental set-up for motion sensing by moiré fringe with a CCD camera and a single strip, the experimental approach has been designed and executed. In the model experiment of the present approach, moiré fringes have been constructed by moving the striped-body through the angular range of 0 degree to 12 degrees with an interval of 1 degree. In all

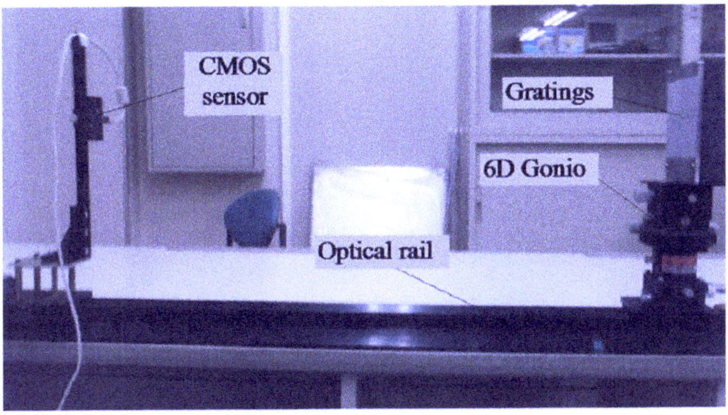

(a)

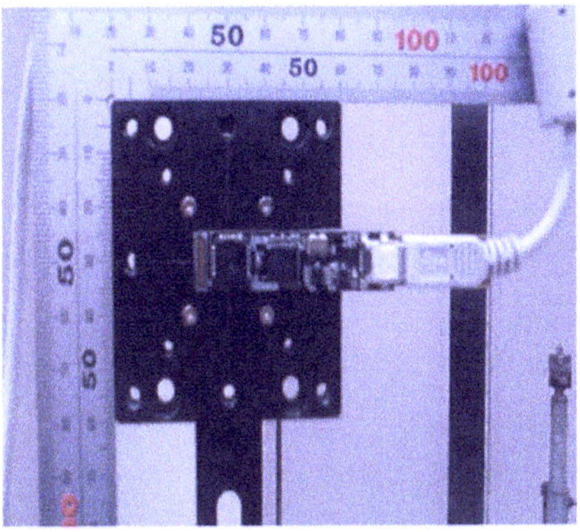

(b)

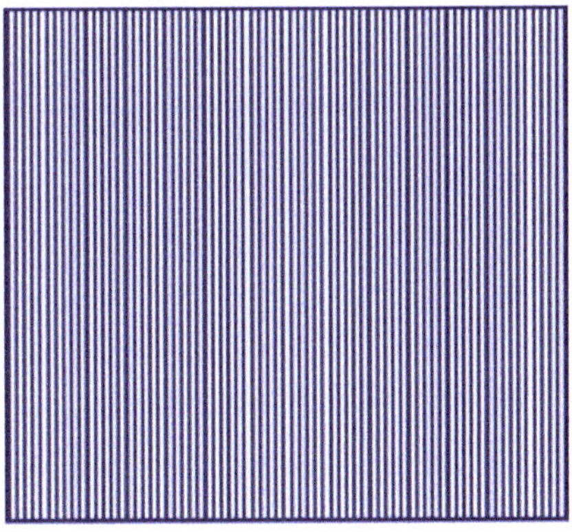

(c)

(d)

Figure 2. Experimental apparatus: (a) Aerial view of model experiment; (b) CMOS camera sensors; (c) Line gratings; (d) 6D gonio with gratings.

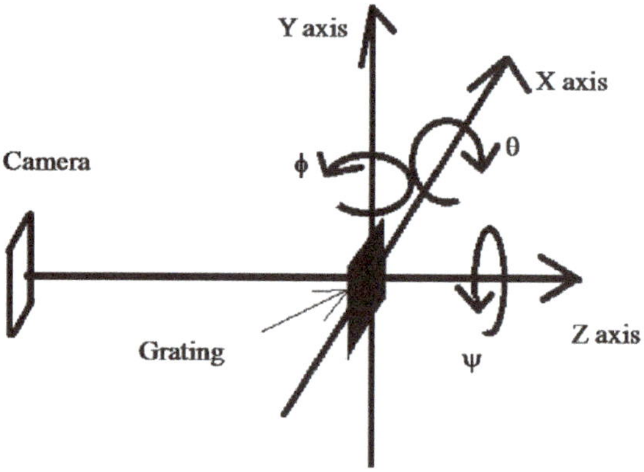

Figure 3. Coordinates specification for the 6D gonio's motions.

experiments, the position of the camera has been kept constant. The distance between the camera and the object has not been analyzed for its detail effects in the experimental results; however, the position of the gonio (the position of the object) has been verified for getting clear fringe images and overcoming the effects of camera artifacts on images.

Thus, experimental images for four positional distances as 750 mm, 900 mm, 950 mm and 1000 mm have been taken in the present measurement technique of angular motion.

Table 1. Specification of CMOS camera

Optical format	6.35 mm (1/4 inch)
Image size (pixel)	1280 (H) × 1024 (V)
Pixel size	2.8 μm × 2.8 μm
Filter	RGB Bayer pattern
Output	USB 2.0, 8 bit
Focal length	3.4 mm

Furthermore, the calculation and the measurement procedures of moiré fringe in the present short-cut method are summarized in following flowchart (**Figure 4**). Although, in conventional method, calculations of fringe parameters (C and q) are sometimes complicated and time consuming, in the present system it has become easier and simple. Only by selecting three points on two neighboring fringes (interval of which be calculated) of the experimental image in computer screen by mouse clicking options, the value of fringe interval as well as fringe inclination are saved in a separate result sheet automatically. The algorithm of this calculation is explained as follows.

First of all, the region of interest (ROI) has been cut off from the total experimental image. After that it has been magnified enough for getting accuracy in point selections for next calculations. The basic dimension of ROI in pixel is x = 80 and y = 95 and after magnification it has become 240 × 285.

From the ROI image, three points are selected randomly on two neighboring fringes; two points in on fringe and one on the other fringe. The pixel size of each point has been taken as 27 × 27 for getting average value while click point became the center. The coordinate of each point has been determined by calculating the center of gravity of each point based on moment method for pixel values as follows [8]:

$$M_{a,b} = \sum_{i=1}^{p} I(x,y)x^a y^b \qquad (6)$$

where, a, b indicate the orders of x and y respectively, I indicates the two dimensional image and p indicates the pixel values. If the values of a and b become zero, the

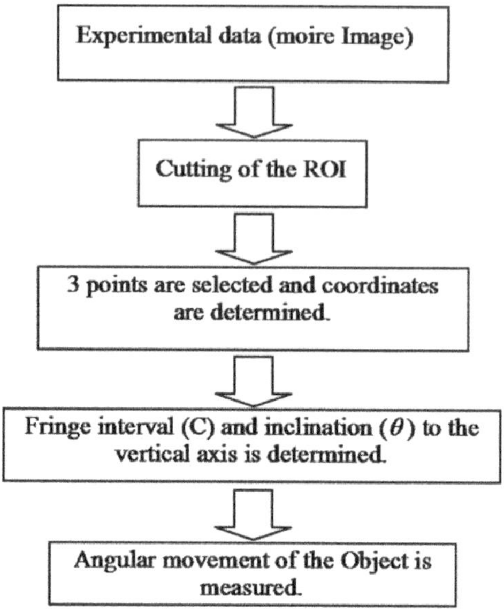

Figure 4. Algorithm of easy measurements systems of moiré parameters (C and q).

value of moment (M) becomes the length in pixels of the contour.

After determining the coordinates, two points on the same fringe are used to generate straight line by simple geometrical equation, and the third point of the neighboring fringe is used to get the perpendicular distance upon the generated straight line, and it has been used to express the interval (C) of two fringes. Again the inclination (q) of fringe has been calculated from the tangent of the straight line. After plotting the curves with these two fringe parameters, the rotational movement of the rotated object has been understood.

EXPERIMENTAL RESULTS AND DISCUSSION

Experiments have been conducted focusing on the objective of developing a single wave moiré imaging technique for multidimensional motion sensing. In that respect the images of moiré fringe have been taken for different angular positions of the gratings acryl board. However, the camera has been kept constant in its original position. Therefore, according to the theory as explained before, the moiré fringes have been generated for different angular positions of acryl body. The angular positions have been varied from 0 degree to 12 degrees with an angular interval of 1 degree. Several images of moiré fringe for the angular positions of 2 degrees, 5 degrees, 8 degrees for the camera-body distance of 750 mm and 950 mm are shown in **Figure 5**.

The changes of moiré fringe patterns for the changes in angular positions are remarkably identified in **Figure 4**. The major changes are noticed in the interval (C) of moiré fringe and the angle (q) that the moiré fringe generates with its vertical axis. It is found, as the angular movement of the body with gratings increases, the interval

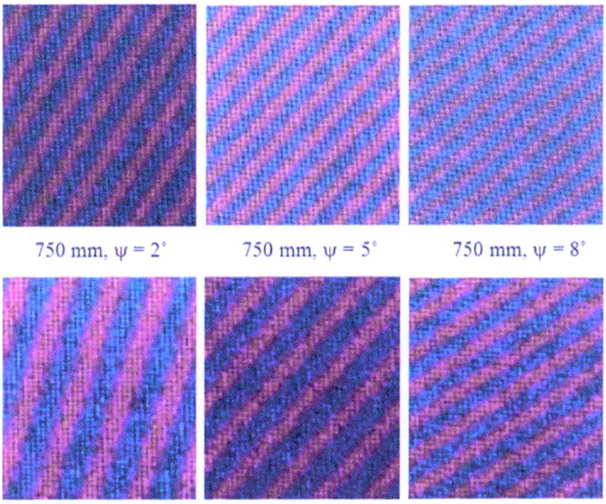

Figure 5. Moiré fringes for rotational to z-axis (y) at 2°, 5° and 8° at a distance of 750 mm and 950 mm between camera and the rotating body.

of the moiré fringe decreases. However, concerning to the angle of moiré fringe (q), the result is opposite, which means that the fringe angle also increases when the gratings angular movement increases. Thus, these experimental results also satisfy the theoretical explanations. Therefore, by calculating fringe intervals and its angular positions simultaneously, the rotational movements of an object containing grating patterns can be identified by the proposed technique.

A lot of procedures are tried to be employed now a days to calculate these two parameters, the fringe interval as well as its angle for different places of application. For the present research, these parameters are calculated by applying the image processing techniques employing the Intell OpenCV libraries as well as some other statistical modeling. By applying the OpenCV algorithms, the accurate extraction of interested area and appropriate filtration of the moiré fringe images are conducted. The main advantage of the present method is its simplicity and quickness in moiré parameters evaluation as well as calculation. This semi-automatic (as three points are necessary to be selected manually, it is called as semi-automatic, although, all other operations for calculating the required parameters are done automatically by the proposed algorithm) algorithm based on OpenCV programming, only three points are needed to be selected on the interested moiré image. As mentioned earlier, after this selection, the result-sheet of moiré parameters will be appeared in the screen as well as will be saved to the desired location. In parameters calculation as mentioned before, cautions have been taken to its robustness in application. For example, to avoid the errors in point selection by call-back function key, averaging data of 27 × 27 pixel values have been taken. Moreover, for accuracy in selection, the ROI is magnified three times for getting accurate location. Therefore, small effects of spatial variations due to temporal variations in pointing have been truncated in the proposed algorithm. Furthermore, the errors caused by person to person in pointing has also been considered and truncated by averaging for different persons (four persons for different categories have joined randomly in getting sample data) and different timings. The uncertainly of the present auto-calculation system has also been calculated by the following methods and the result has shown that the uncertainty of calculating fringe intervals falls below 5% when the uncertainty of calculating fringe inclination to vertical axis falls below 1%. Thus it has found that the present quick and easy technique of fringe parameters calculation makes easier to measure a rotating object correctly and precisely.

The results of calculations for the intervals and angles of moiré fringes for the angular movements of the rotating body (acryl board at gonio in the experiment) are shown in Figures 6 and 7, from where the angular position of the object can be calculated as well. Moreover, it is found in the result that the calculated value differs with respect to the distance of the object from camera. In the result this effect is largely influenced for the distance of 900 mm. Although, the detail effect of this property has not been analyzed in the present research, however, it is understood that the effect of camera lens (ray theory) has influenced the result.

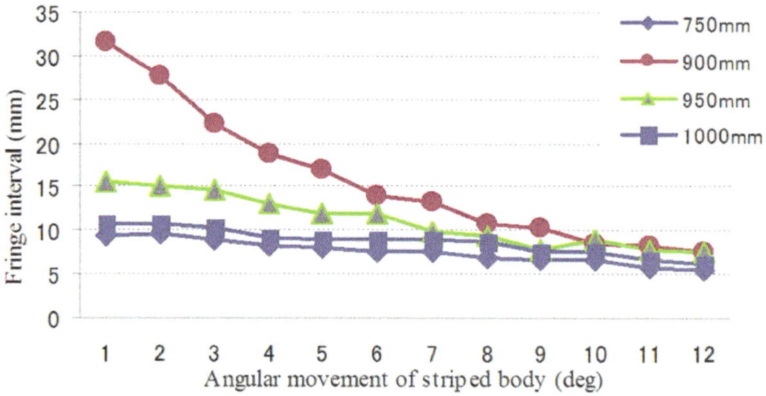

Figure 6. Moiré fringe intervals with respect to the angular movement of the measured striped body.

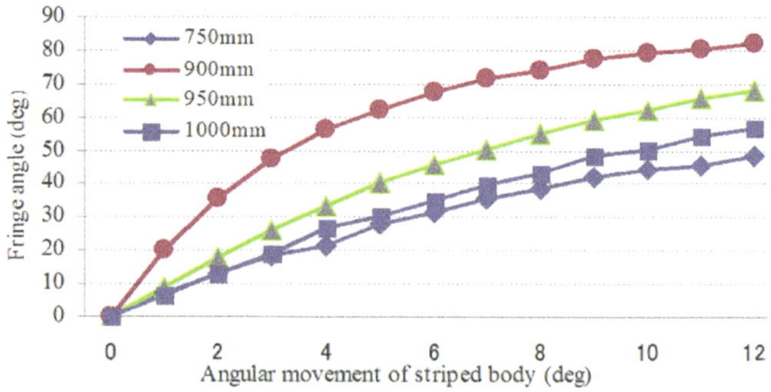

Figure 7. Moiré fringe intervals with respect to the angular movement of the measured striped body.

CONCLUSIONS

Precise measurement of angular moving body by a single grating based moiré technique has been performed. Although, some artifacts due to the differences in intervals of the camera and the specimen gratings are available, however, that effects are successfully truncated in calculations of the interval and the angle of moiré fringe in the present moiré technique.

As the proposed algorithm is successfully employed in calculating the moiré fringe parameters, it has made the moiré measurement technique simple, easy and semiautomatic as well.

Although, the proposed algorithm has been verified by the straight symmetric moiré fringes, however, the parameters of other curvilinear non-symmetric fringes can also be calculated by changing the required simple geometrical equation in step 4 of the proposed algorithm.

REFERENCES

1. S. A. Kamal, "Pattern Recognition Using Moire Fringe Topgraphy and Rasrerstereography," International Symposium on Biometrics and Security Technologies, Islamabad, 23-24 April 2008, pp. 1-7.
2. I. Amidror, "The Theory of the Moiré Phenomenon," 2nd Edition, Springer, 2009. doi:10.1007/978-1-84882-181-1
3. R. H. Dyck and G. P. Weckler, "Integrated Arrays of Silicon Photodetectors for Image Sensing," IEEE Transactions on Electron Devices, Vol. 15, No. 4, 1968, pp. 196- 201.
4. R. F. Eric, "Active Pixel Sensors: Are CCD's Dinosaurs?" Proceedings of SPIE Symposium on Electronic Imaging, San Jose, 12 July 1993, pp. 2-14.
5. D. Passeri, et al., "Characterization of CMOS Active Pixel Sensors for Particle Detection: Beam Test of the FourSensors RAPS03 Stacked System," Nuclear Instruments and Methods in Physical Research Section A: Accelerators, Spectrometers, Detectors and Associated Equipment, Vol. 617, No. 1-3, 2010, pp. 573-375.
6. S. S. Grothi and P. Rastogi, "Fringe Projection Techniques: Whither We Are?" Optics and Lasers in Engineering, Vol. 48, No. 2, 2010, pp. 133-140.doi:10.1016/j.optlaseng.2009.09.001
7. D. Malacara, "Optical Shop Testing," 3rd Edition, John Wiley & Sons, New Jersey, 2007.doi:10.1002/9780470135976
8. J. Flusser, "Refined Moment Calculation Using Image Block Representation," IEEE Transactions on Image Processing, Vol. 9, No. 11, 2000, pp. 1976-1978.

8

CHAPTER

DYNAMIC VISCOSITY MEASUREMENT IN NON-NEWTONIAN GRAPHITE NANOFLUIDS

Fei Duan, Ting Foong Wong and Alexandru Crivoi

Keywords: Dynamic viscosity, Graphite nanofluids, Nanoparticle aggregation, Non-Newtonian flow

BACKGROUND

Nanofluids, consisting of suspended nanoscale solid particles, can improve thermal conductivity and heat transfer coefficient from the base fluids [1–10]. The effectiveness of thermal property enhancement of nanofluids depends on nanoparticle amount, particle size, particle materials, particle shape, base fluids, etc. However, since nanofluids are suspensions with nanoparticles in their base fluids, achieving a stable dispersion in nanofluids would benefit industrial applications. Nanoparticles are expected to stabilize the fluids more effectively than the microparticles, the fluid properties including the dynamic viscosity would change accordingly. The viscosity of nanofluids is important for nanofluid transport related to flow dynamics and heat transfer. The spherical shaped Al_2O_3, TiO_2, and the other nanoparticles have been widely studied in the nanofluids [2–5, 8]. The results showed an enhancement on the effective dynamic viscosity as an increase of concentrations. A strong correlation was indicated between the rheological behavior and the structure of nanoparticles in the nanofluids. The Al_2O_3-water nanofluid exhibited as a Newtonian flow after freshly prepared, but a shear thinning non-Newtonian flow after the aggregation was formed in the nanofluids. The dynamic viscosity had a significant increase as a result. However, the properties can be resumed after the re-ultrasonication process, in which the aggregates were dispersed again [11]. The main

Citation: Fei Duan, Ting Foong Wong, Alexandru Crivoi, Dynamic viscosity measurement in non-Newtonian graphite nanofluids, DOI: 10.1186/1556-276X-7-360.

nonspherical nanoparticles in suspensions under the study are carbon nanotubes (CNTs) and graphite on thermal conductivity [1, 9, 10]. However, there are limited reports on the nanofluid dynamic viscosity with the nanoparticles at different heights to width aspect ratios, especially for the graphite nanoparticles. Ding et al. measured the dynamic viscosity of CNTs-water nanofluids as a function of shear rate, showing the fluids with a non-Newtonian property [10]. The viscosity of the nanofluids was found to increase with the increasing concentrations of CNTs and the decreasing temperature. Yang et al. investigated the rheological behavior of poly(α-olefin) solutions dispersed by the rodlike CNTs with an aspect ratio of about 30, or the disklike graphite nanoparticles an aspect ratio of about 0.025 [9]. The nanofluids acted as a shear thinning fluids. The above studies suggest that the nanoparticles may aggregate in the base fluids, and the aggregation would affect the rheological properties. To explicate the phenomena, the investigation of the dynamic viscosity is carried out in the graphite-water nanofluids for the potential application. The morphology of the nanoparticle aggregates and the structure in molecular vibration are demonstrated in this paper by using a transmission electron microscope and a Raman spectroscope, respectively.

METHODS

In the experiments, a two-step method was used to prepare the graphite water-based nanofluids. The graphite nanoparticles were supplied by SkySpring Nanomaterials, Inc. (Houston, TX, USA) with a reported average size of 3 to 4 nm. We dispersed the nanoparticles in the 40 mL deionized water to prepare the nanofluids with the volume concentrations at 1 %, 2 %, 3 %, and 4 % without adding any surfactant in order to study the effect of nanoparticle aggregation. The next step for the nanofluids was to undergo a mechanical stirring process with a magnetic stirrer at a rotation speed 540 rpm for about 7 h. Then, the nanofluids were performed under ultrasonication by using the ultrasonic bath (Elmasonic E 15H, Singen, Germany) continuously for about 1.5 h to prevent the well-dispersed fluids from aggregation initially.

The effective dynamic viscosity of the graphite-water nanofluids was measured directly with a standard controlled shear rate rheometer (Contraves LS 40, Mettler-Toledo, Greifensee, Switzerland) which has a cup and bob geometry. This instrument requires only a volume of liquid of approximately 5 mL. The instrument was calibrated by measuring the dynamic viscosity of the deionized water. The calibration results showed the measurement error within ±1% from the viscosity value of 0.000891 Pa·s. All measurements in this study were performed at 1 atm and 298.15 K. The effective dynamic viscosity of the nanofluids was measured instantly after the ultrasonication agitation. Thereafter, the same nanofluids were measured again after 3 days, which is determined by the experimental observation with the stratified fluids. Before the measurement, the fluids were shaken to prevent the possible particle sediment in the measurement. The relative effective dynamic viscosity is calculated with a reference value of the base fluid (pure water).

A transmission electron microscope (TEM, JEOL, JEM-2010, Tokyo, Japan) is applied to reveal the microstructural morphology of the graphite particles in the dried samples from the nanofluids. In preparation, the nanofluids in the volume fraction of 1%

were diluted so as to reduce the possibility of the particle agglomeration in preparing the TEM samples in the drying process [12]. Then, a little drop of the nanofluid samples was dried naturally by placing on the copper grid coated with carbon film. The TEM instrument was used at an operating voltage of 200 kV in the graphite-water nanofluids instantly after preparation (fresh) and 3 days after, respectively. The Raman spectra were to disclose the molecular structure of materials. A Renishaw inVia Raman spectroscope (Wotton-under-Edge, UK) was applied by using the 514 nm He-Ne laser source with a laser power setting at 10 mW to determine the structure of nanoparticle aggregation on the basis of the molecular vibration. The fresh nanofluids and the fluids held for 3 days were sampled and then measured at room temperature for the Raman spectra.

RESULTS AND DISCUSSION

The steady shear measurement was conducted at room temperature (298.15 *K*) for the series of nanofluids with the volume concentrations at 1%, 2%, 3%, and 4%. The effective dynamic viscosity of the fresh nanofluids, which were just prepared, is shown in Figure 1. It can be seen that the effective dynamic viscosity decreases with an increase of the shear rate in the nanofluids for a given concentration. The viscosity increases with the increasing loading of nanoparticles at the same shear rate, the effective dynamic viscosity has a higher value at 4 vol% than that at 1%. The dispersions with the graphite particles are shear thinning at low shear rates and approach a constant dynamic viscosity at high shear rates. The nanofluids act as the non-Newtonian flows.

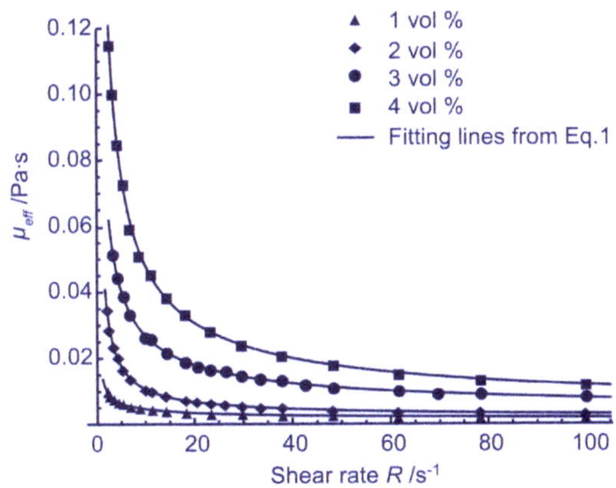

Figure 1: The effective dynamic viscosity as a function of steady shear rate in the fresh nanofluids.

With the assumption of a pseudoplastic flow, the modified Cross model [13], expressed in Equation 1, is applied to fit the experimental data,

$$\frac{\mu_{eff} - \mu_{\infty}}{\mu_0 - \mu_{\infty}} = \frac{1}{1 + \alpha R^n} \quad (1)$$

where R is the shear rate, μ_0 is the dynamic viscosity at zero rate of shear, μ_∞ is the dynamic viscosity at infinite rate of shear, α and n are constant. The fitting parameters are listed in Table1. The effective viscosities, μ_∞ and μ_0, increase with an increase of the volume concentration from 1% to 4%. The dynamic viscosity at infinite shear rate of the 4 vol% nanofluids is over 2.68 times that of the 1 vol% nanofluids.

Table 1: Fitting parameters of the steady shear measurement for the nanofluids held 3 days (*nff*)

Experiment:	μ_0/Pa· s	μ_∞/Pa· s	a	n
nff 1	5.02911×10^{10}	0.000968736	3.64600×10^{12}	0.577278
nff 2	1.07711×10^{11}	0.00165829	1.81188×10^{12}	0.814235
nff 3	4.85370×10^{11}	0.00218935	4.57560×10^{12}	0.632872
nff 4	1.11869×10^{13}	0.00259891	5.03442×10^{13}	0.689077

Similarly, the effective dynamic viscosity of the nanofluids held for 3 days is shown with a non-Newtonian behavior as well, illustrated in Figure 2. The effective dynamic viscosity decreases dramatically under low shear rates and approaches a constant dynamic viscosity at high shear rates. At a given shear rate, the dynamic viscosity increases with the increasing loading of particles, similar to that in the fresh nanofluids. However, the enhancement of the effective dynamic viscosity is much higher than the fresh nanofluids under the same volume concentration. The experimental data were also fitted to obtain the parameters of μ_0, μ_∞, α, and n, listed in Table 2. The dynamic viscosity at infinite shear rate, μ_∞ increases to 1.34 times for the 2 vol% of 3-day fluids, 3.39 times at 3 vol%, 11.21 times at 4 vol% in comparison with the 1 vol% nanofluids three days after the preparation.

Table 2: Fitting parameters of the steady shear measurement for the nanofluids held 3 days (*nfo*)

Experiment:	μ_0/Pa· s	μ_∞/Pa· s	a	n
nfo 1	6.78665×10^{11}	0.00197601	1.28804×10^{13}	0.786853
nfo 2	3.76320×10^{12}	0.00265571	4.77138×10^{13}	0.664450
nfo 3	1.17756×10^{13}	0.00670576	4.46724×10^{13}	0.679144
nfo 4	5.93012×10^{13}	0.02214670	1.01070×10^{14}	0.826530

The fitting errors were analyzed for each of the effective dynamic viscosity measurements. The average fitting errors were calculated from Equations 2 and 3. The mean of absolute fitting errors and the absolute average fitting error are listed in Table 3.

$$\overline{\left(\frac{\Delta\mu}{\mu}\right)}_{MA,j} = \frac{\sum|(\mu_{fit} - \mu_{eff})/\mu_{eff}|}{k} \tag{2}$$

where $\overline{\left(\frac{\Delta\mu}{\mu}\right)}_{MA}$ is the average of absolute fitting errors, *MA*, of the effective dynamic viscosity of the nanofluids, j is for the fresh nanofluids (*nff*) and or the 3-day fluids (*nfo*), μ_{fit} is the fitted dynamic viscosity at the measured rate of shear, μ_{eff} is the measured dynamic viscosity, and k is the number of readings at a run of viscosity measurement.

$$\overline{\left(\frac{\Delta\mu}{\mu}\right)}_{AM,j} = \left|\frac{\sum(\mu_{fit} - \mu_{eff})/\mu_{eff}}{k}\right| \quad (3)$$

where $\overline{\left(\frac{\Delta\mu}{\mu}\right)}_{AM}$ is the absolute average fitting error, *AM*.

Table 3: The fitting errors of the effective dynamic viscosity of fresh nanofluids and 3-day fluids

	1%	**2%**	**3%**	**4%**
(Δμμ)—MA,nff $\overline{\left(\frac{\Delta\mu}{\mu}\right)}_{MA,nff}$	0.49%	0.83%	1.23%	1.28%
(Δμμ)—AM,nff $\overline{\left(\frac{\Delta\mu}{\mu}\right)}_{AM,nff}$	0.00%	0.02%	0.05%	0.11%
(Δμμ)—MA,nfo $\overline{\left(\frac{\Delta\mu}{\mu}\right)}_{MA,nfo}$	0.98%	1.27%	1.33%	1.87%
(Δμμ)—AM,nfo $\overline{\left(\frac{\Delta\mu}{\mu}\right)}_{AM,nfo}$	0.04%	0.22%	0.24%	0.08%

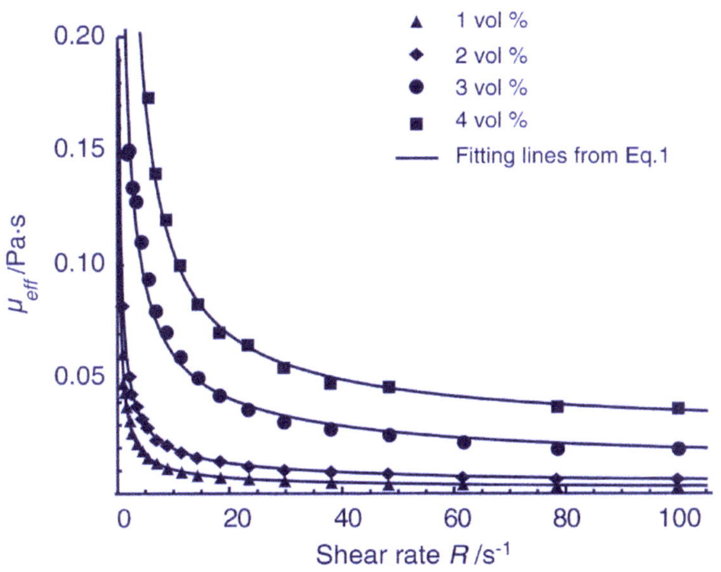

Figure 2: The effective dynamic viscosity as a function of steady shear rate in the 3-day fluids.

As listed in Table 3, the average of absolute fitting errors is 0.49% for the 1 vol% fresh graphite-water nanofluids compared with the value at 0.98% in the 1% nanofluids held for 3 days. For a higher volume concentration at 4%, the mean of absolute fitting errors is 1.28% for the fresh nanofluids, but 1.87% for the 3-day fluids. The absolute average fitting error has a smaller value compared with the mean of the absolute fitting errors in the same nanofluids. The maximum absolute average fitting error is 0.11% for the fresh nanofluids while it is 0.24% for the 3-day fluids. It can be seen that the fitting curve is very close to the experimental data.

Figure 3 shows the relative effective dynamic viscosity, μ_∞/μ_f, as a function of the nanoparticle volume concentration, in which μ_f is the dynamic viscosity of the base fluid. It is found that the viscosity ratio increases monotonically as a function of the volume concentration for both the fresh nanofluids and the 3-day fluids. The relative effective dynamic viscosity at infinite shear rate increases gradually to 2.92 in the fresh nanofluids, but up to 24.86 in the nanofluids held for 3 days from 1% to 4% in volume fraction. Yang et al. showed a similar trend in the nanofluids of graphite and poly(α-olefin) solutions, the relative effective viscosity increased from 1.15 at the volume fraction at 0.39% to 1.33 at 0.78% [9]. As seen from Figure 3, the relative effective dynamic viscosity shows a significant higher value in the 3-day fluids than that in the fresh nanofluids under the same concentration at the volume fraction of 1%, 2%, 3%, or 4%, respectively. The relative effective dynamic viscosity at infinite shear rate is 1.09 for the fresh nanofluids at the volume fraction at 1%, but 2.22 for the 1% 3-day fluids. As shown in the insert of 3, the increasing dynamic viscosity gradient reported by Yang et al. [9] is shown in between the fresh nanofluids and the fluids held for three days in this study. The enhancement can be mainly explained that a higher aggregation of nanoparticles, induced by a higher concentration, results in a higher effective dynamic viscosity [11]. Even though the nanofluids held for 3 days were remixed before the measurement, the shaking process cannot totally break down the aggregation formed with the time.

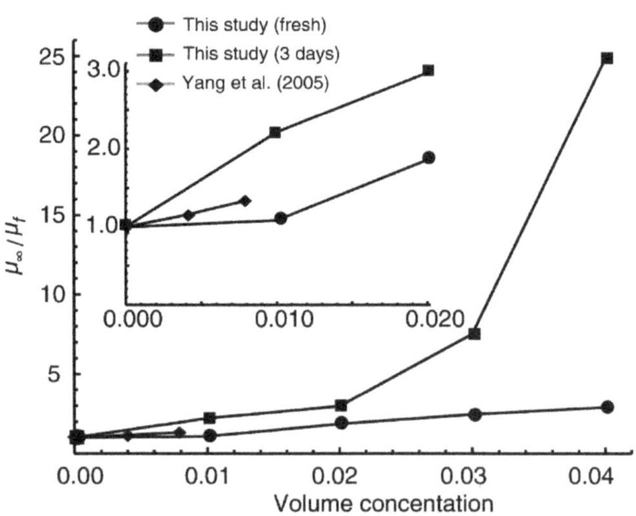

Figure 3: The relative dynamic viscosity, μ_∞/μ_f, as a function of particle volume concentration.

Similarly, Kim et al. reported that the CNT-based nanofluids had such the phenomenon by pointing out that the high surface effect and the strong van der Waals force drive the nanoparticles to form the aggregation in the suspensions [14]. Since most of aggregates might be destroyed under high shear rates, the nanofluids are shown in shear thinning non-Newtonian behaviors [9, 11, 15]. We have to mention that the graphite-water nanofluids have a different flow property from the Al_2O_3-water nanofluids, which are Newtonian flows if the nanofluids are freshly prepared, but non-Newtonian flows if the fluids have large aggregates [11]. The different results might be from the various nanoparticle sizes, species, and configuration. The graphite-water nanofluids might have the particle aggregation just after the preparation, and show the non-Newtonian flow properties. The size of aggregation in nanofluids would increase with the increase of the particle volume concentration and the holding time. Thus, it would take a higher force to break the ligand structure among particles in the aggregated fluids [9, 11], as a result, a high effective dynamic viscosity ratio can be observed in Figure 3.

The effective dynamic viscosity enhancement can also be qualitatively explained with the viscosity ratio between a nanofluid (*nf*) and its base fluid (*f*) in the model [16], $\frac{\mu_{nf}}{\mu_f} = (1 - \frac{\phi_a}{\phi_m})^{-[\mu]\phi_m}$, in which μ is the intrinsic viscosity for spherical particles with a value of 2.5, ϕ_a is the volume fraction of aggregates, and ϕ_m is the volume fraction of densely packed spheres. The volume fraction of the aggregates can be expressed as $\phi_a = \phi(\frac{d_a}{d})^{3-d_f}$, in which d_f is the fractal dimension of the aggregates. When the nanoparticle aggregation size, d_a, increases, the magnitude of $\frac{d_a}{d}$ increases. Thus, the volume fraction of the aggregates increases, and the viscosity ratio increases based on the model of Krieger and Dougherty [16]. As shown in the microstructure in Figure 4, the shape of graphite nanoparticles is not spherical, and the aggregates of particles are more complex in the 3-day fluids. The intrinsic viscosity, μ, consequently changes larger with the complicated shape [17]. This could also account for the rise in the effective dynamic viscosity as the volume concentration or the holding time increases. The relative effective viscosity of the 3-day fluids at the volume concentration of 4% is as 22.86 times as the value of the fresh nanofluids at 1 vol%, as shown in Figure 3.

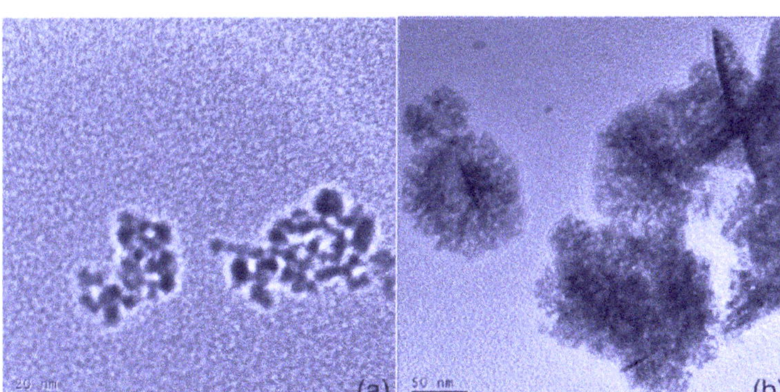

Figure 4: TEM images of graphite nanoparticle aggregation from the fresh nanofluids (a) and 3-day fluids (b).

The TEM images of the graphite particles dried from the nanofluids are shown in Figure 4. It is found that the average diameter of graphite particles dispersed in the nanofluids just after the ultrasonic agitation is up to 50 nm shown in Figure 4a. The nanoparticles are still larger than those specified by the supplier in the powder form. It suggests that the graphite nanoparticles have aggregated into a certain size even in the fresh nanofluids, resulted from the high surface effect of nanoparticles and the inter-particle attraction [11, 18]. However, the graphite particles were significantly aggregated if the nanofluids were held for 3 days, as shown in Figure 4b. The size of aggregated graphite particles is larger than 150 nm at least. The highly fragmented aggregation clusters were found in the microstructure analysis. The aggregation of nanoparticles in the graphite-water nanofluids increased with a longer holding time. It supports the aforementioned discussion that the effective dynamic viscosity increases at the 3-day fluids. Note that the nanoparticle boundary is detectable in the aggregates shown in Figure 4a. We can estimate that the largest dimension of the particle is at about 18 nm. Thus, if the 3 to 4 nm is treated as the thickness of the graphite nanoparticle, the height to width aspect ratio could be up to 0.17, which is much larger than those used by Yang et al. [9].

Figure 5 illustrates the Raman spectra at various volume concentrations from 1% to 4% for the fresh nanofluids and the 3-day fluids. Two characteristic peaks are observed in the graphite-based nanofluids, locating in the range of 1,570 to 1,594 cm^{-1} and 1,330 to 1,360 cm^{-1}. These features could be characterized as the *G* peak and the *D* peak [19–21], marked in Figure 5. In the crystallization analysis, the *G* peak of graphite at 1,575 cm^{-1}indicates the sp^2vibration of the carbon atoms in the structure, and the *D* peak of graphite at 1,355 cm^{-1}, suggesting the sp^3hybridization of carbon atoms, is resulted from defects, disorder, and impurities in the materials. As illustrated in Figure 5, the intensities of the *D* and *G* peaks increase as an increase of the volume concentration in the fresh nanofluids, the strength of the *D* peaks increases to about 10 times as the volume concentration is 2%, and 20 times as the concentration is 4%. Then, the intensities acutely increase with an increase of the particle volume concentration in the 3-day fluids. A dramatic increase of the intensity at the *D* peak is found between the fresh nanofluids and those held for 3 days for a given concentration, i.e., the intensity increases over 70 times in the nanofluids with 1 vol%. The other three increase the intensity of the *D* peak above 100 times compared with the 1% fresh nanofluids. It is indicated that the effective intensity of the *D* peak in the nanofluids qualitatively reflects the size of the aggregation clusters of graphite nanoparticles in the nanofluids. The size of clusters became larger at a higher concentration and a longer holding time. The results from the Raman spectra are consistent with the dynamic viscosity measurements and the TEM microstructures. In addition, the intensity ratio at about a unit between the *D* peak and the *G* peak also suggests the cluster and chain groups formation [22].

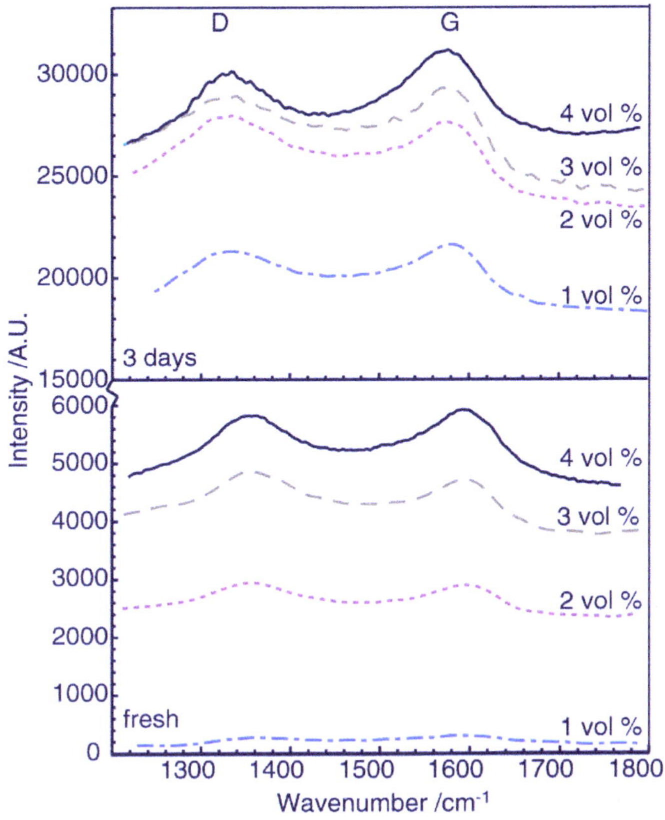

Figure 5: Raman spectra of the fresh nanofluids and 3-day fluids.

CONCLUSIONS

The effective dynamic viscosity of the graphite-water nanofluids is experimentally found to decrease with an increase of shear rate in a given particle volume fraction (Figures 1 and 2). The nanofluids act as the shear thinning non-Newtonian flows. The data of the effective dynamic viscosity in nanofluids are fitted numerically, the relative effective dynamic viscosity at infinite rate of shear increases to 2.92 in the fresh nanofluids at 4 vol% in comparison of the base fluid, but 24.86 for the nanofluids held for 3 days (Figure 3). The microstructure of the diluted nanofluids indicates that the aggregation of nanoparticles is significantly higher in the 3-day fluids than that in the fresh nanofluids, as shown in Figure 4. The Raman spectra are used for showing the formation of larger graphite nanoparticle aggregation with an increase of the volume concentration or the holding time of the nanofluids in Figure 5. This study suggests that the aggregation would happen in the nanofluids which have not been treated specially by adding the surfactant, controlling the pH value, etc. The aggregation would dramatically change the nanofluid properties including viscosity consequently.

ACKNOWLEDGEMENT

The authors acknowledge the support of MOE AcRT Tier 1 and the help from the staff of Materials Lab in Nanyang Technological University.

Competing interests

The authors declare that they have no competing interests.

REFERENCES

1. Choi SUS, Zhang ZG, Yu W, Lockwood FE, Grulke EA: Anomalous thermal conductivity enhancement in nano tube suspensions. *Appl Phys Lett* 2001, 78: 2252.
2. Li CH, Peterson GP: Experimental investigation of temperature and volume fraction variations on the effective thermal conductivity of nanoparticle suspensions (nanofluids). *J Appl Phys* 2006, 99: 1.
3. Lee S, Choi SUS, Li S, Eastman JA: Measuring thermal conductivity of fluids containing oxide nanoparticles. *J Heat Transfer* 1999, 121: 280. 10.1115/1.2825978
4. Kwek D, Crivoi A, Duan F: Effects of temperature and particle size on the thermal property measurements of Al2O3-water nanofluids. *J Chem Eng Data* 2010, 55: 5690. 10.1021/je1006407
5. Timofeeva EV, Gavrilov AN, McCloskey JM, Tolmachev YV: Thermal conductivity and particle agglomeration in alumina nanofluids: experiment and theory. *Phys Rev E* 2007, 76: 061203.
6. Sefiane K: On the role of structural disjoining pressure and contact line pinning in critical heat flux enhancement during boiling of nanofluids. *Appl Phys Lett* 2006, 89: 044106. 10.1063/1.2222283
7. Xuan Y, Roetzel W: Conceptions for heat transfer correlation of nanofluids. *Int J Heat Mass Transfer* 2000, 43: 3701. 10.1016/S0017-9310(99)00369-5
8. Tseng WJ, Lin K-C: Rheology and colloidal structure of aqueous TiO2 nanoparticle suspensions. *Mater Sci Eng A* 2003, 355: 186. 10.1016/S0921-5093(03)00063-7
9. Yang Y, Grulke EA, Zhang ZG, Wu G: Rheological behavior of carbon nanotube and graphite nanoparticle dispersions. *J Nanosci Nanotech* 2005, 5: 571. 10.1166/jnn.2005.079
10. Ding Y, Alias H, Wen D, Williams RA: Heat transfer of aqueous suspensions of carbon nanotubes (CNT nanofluids). *Int J Heat Mass Transfer* 2006, 49: 240. 10.1016/j.ijheatmasstransfer.2005.07.009
11. Duan F, Kwek D, Crivoi A: Viscosity affected by nanoparticle aggregation in Al2O3-water nanofluids. *Nanoscale Res Lett* 2011, 6: 248. 10.1186/1556-276X-6-248
12. Crivoi A, Duan F: Evaporation-induced formation of fractal-like structures from nanofluids. *Phys Chem Chem Phys* 2012, 14: 1449.
13. Cross MM: Rheology of non-Newtonian fluids: A new flow equation for

pseudoplastic systems. *J Colloid Sci* 1965, 20: 417. 10.1016/0095-8522(65)90022-X

14. Kim B, Park H, Sigmund WM: Rheological behavior of multiwall carbon nanotubes with polyelectrolyte dispersants. *Colloid Surface Physicochem Eng Aspect* 2005, 256: 123. 10.1016/j.colsurfa.2004.12.063
15. Phuoc TX, Massoudi M, Chen RH: Experimental investigation of temperature and volume fraction variations on the effective thermal conductivity of nanoparticle suspensions (nanofluids). *Int J Therm Sci* 2011, 50: 12. 10.1016/j.ijthermalsci.2010.09.008
16. Krieger IM, Dougherty TJ: A mechanism for non-newtonian flow in suspensions of rigid spheres. *J Rheol* 1959, 3: 137. 10.1122/1.548848
17. Rubio-hernandez FJ, Ayucar-Rubio MF, Velazquez-Navarro JF, Galindo-Rosales FJ: Intrinsic viscosity of SiO2, Al2O3 and TiO2 aqueous suspensions. *J Colloid Interface Sci*2006, 298: 967. 10.1016/j.jcis.2006.01.009
18. Jiang L, Gao L, Sun J: Production of aqueous colloidal dispersions of carbon nanotubes. *J Colloid Interface Sci* 2003, 260: 89. 10.1016/S0021-9797(02)00176-5
19. Tuinstra F, Koenig JL: Raman spectrum of graphite. *J Chem Phys* 1970, 53: 1126. 10.1063/1.1674108
20. Baby TT, Ramaprabhu S: Experimental investigation of the thermal transport properties of a carbon nanohybrid dispersed nanofluid. *Nanoscale* 2011, 3: 2208. 10.1039/c0nr01024c
21. Tamor MA, Vassell WC: Raman fingerprinting of amorphous carbon films. *J Appl Phys*1994, 76: 3823. 10.1063/1.357385
22. 2Ferrari AC, Robertson J: Interpretation of Raman spectra of disordered and amorphous carbon. *Phys Rev B* 2000, 61: 14095. 10.1103/PhysRevB.61.14095

9
CHAPTER

DIRECT MEASUREMENT OF ELECTROSTATIC FIELDS USING SINGLE TEFLON NANOPARTICLE ATTACHED TO AFM TIP

Joe-Ming Chang, Wei-Yu Chang, Fu-Rong Chen and Fan-Gang Tseng

Keywords: Electrostatic, Teflon, Nanoparticle, Atomic force microscopy

BACKGROUND

Measuring the electrical properties of devices at the micro/nanoscale is an important issue in the semiconductor industry and in materials science [1–4]. Electrical modes in scanning probe microscopy (SPM) [5] have become an essential tool in characterizing the electrical properties at the surface of samples, providing spatial resolution and sensitivity at the micro/nanoscale. Several methods have been developed for the measurement of surface electrical properties and local surface potential, such as electrostatic force microscopy [2, 3] and Kelvin probe force microscopy [6, 7]. The basic principle behind these techniques [5] is applying a direct current (DC) bias between the conductive probe and the sample to facilitate the recording of variations in the electrostatic force between the probe and sample. These signals are then analyzed in order to interpret the associated surface electrical properties. Jenke et al. [8] used a Pt-coated Si tip with a radius of about 380 nm to probe the electrostatic force generated above embedded nanoelectrodes in the vertical (*Z*) direction. The electrostatic force acting on a grounded conductive tip within an electrostatic field can also be characterized. In this approach, the electrostatic force acting on the atomic force microscopy (AFM) tip comprises Coulombic, induced charge, and image charge forces [9–11]. However, only the Coulombic force is capable of directly revealing the electrical properties of the sample because the two other terms are the result of the AFM tip effect. Kwek et al. [10] glued a charged microparticle to

Citation: Chang et al.: Direct measurement of electrostatic fields using single Teflon nanoparticle attached to AFM tip. Nanoscale, DOI: 10.1186/1556-276X-8-519.

an AFM cantilever to investigate the relative contributions of the Coulombic, induced charge, and image charge forces in the electrostatic force acting on the charged particle; however, the diameter of the charged particle was approximately 105 to 150 μm, which is unsuitable for measurement at the nanoscale.

This paper presents a novel microscopy probe for the direct measurement of electrostatic field (mainly Coulombic force) beside the top electrode of the parallel plate, at a spatial resolution of 250 nm and force resolution of 50 pN(Figure 1). The proposed probe comprises a single 210-nm Teflon nanoparticle (sTNP) attached to the vertex of an insulated Si_3N_4 AFM tip (sTNP tip) with charge deposited on the sTNP as an electret via contact electrification [2, 12–14]. The parallel plate condenser was fabricated by sputtering layers of Au (30-nm thick) and Ti (20-nm thick) on the top and bottom sides of a 1×1 cm glass slide (181 ± 0.25 μm thick). Au was used as the electrode surface and Ti as an adhesion layer. The glass slide was used as the dielectric material. The sTNP tip can be considered a point charge with which to probe the electrostatic force field beside the top electrode of the parallel plate condenser. The electrostatic force acting on the sTNP tip provides direct information related to the local electrostatic field generated in the sample. This technique provides the following advantages: (1) direct measurement of the electrostatic field of a sample is without complex operations or the need for an analytic model, and (2) minimum measurable electrostatic force is within 50 pN with a lateral/vertical spatial resolution of 250/100 nm, respectively.

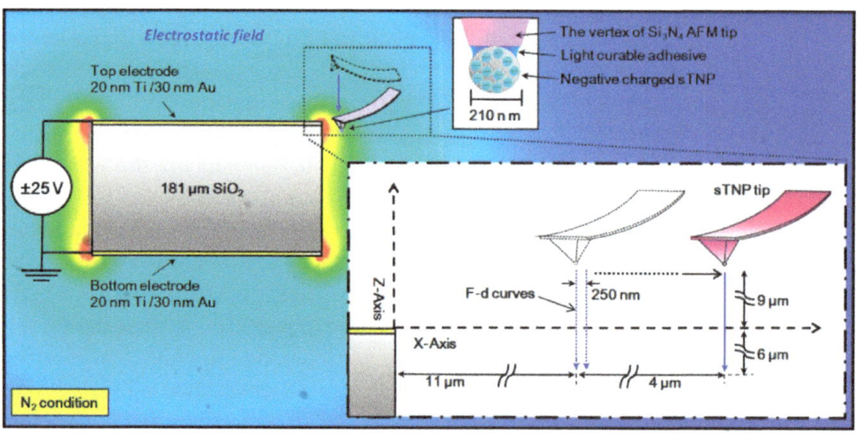

Figure 1: Schematic of experimental setup for the measurement of electrostatic field of a parallel plate condenser.

METHODS

The Process of Fabricating the sTNP tip

Figure 2 presents a schematic diagram illustrating the fabrication process of sTNP tip. To obtain insulating Si_3N_4 tips for accommodating sTNP, commercial Si_3N_4 AFM tips (OMCL-RC800PSA-1, Olympus, Tokyo, Japan) were immersed in gold etchant

(Transene, Danvers, MA, USA; 1:1 (*v/v*) in H_2O) for 15 min and in chromium etchant (Cyantek, Fremont, CA, USA; 1:3 (*v/v*) in H_2O) for 40 min to remove the reflective layer of gold (Au) and chromium (Cr) coating the back side of the cantilevers (Figure 2b), respectively. The normal spring constant of the insulating Si_3N_4 AFM tip was measured at 0.053 N/m using the thermal noise method [15] with JPK software (JPK Instrument, Berlin, Germany). In order to attach the 210-nm sTNPs, a flat square area with edge length of 300 nm at the vertex of the tip (Figure 2e) was fabricated by scanning a polished silicon nitride wafer (Mustek, Hsinchu, Taiwan) under a large contact loading force of 12 nN at a fast scanning speed of 80 µm/s (Figure 2c). The flattened Si_3N_4 AFM tip was cleaned by immersion in a heated (90°C) piranha solution (a 7:3 (*v/v*) of 95.5% $H_2SO_4$and 30% H_2O_2) for 30 min. Small droplets of light-curable adhesive (Loctite 3751, Henkel Corp., Way Rocky Hill, CT, USA) several microns in size were spread over the glass slide using a needle. In the application of light-curable adhesive, we employed an inverted optical microscope (IX 71, Olympus) to ensure uniformity in the size of droplets (approximately 5 µm) on the scale of the base length (approximately 4.5 µm) of the pyramidal AFM tip.

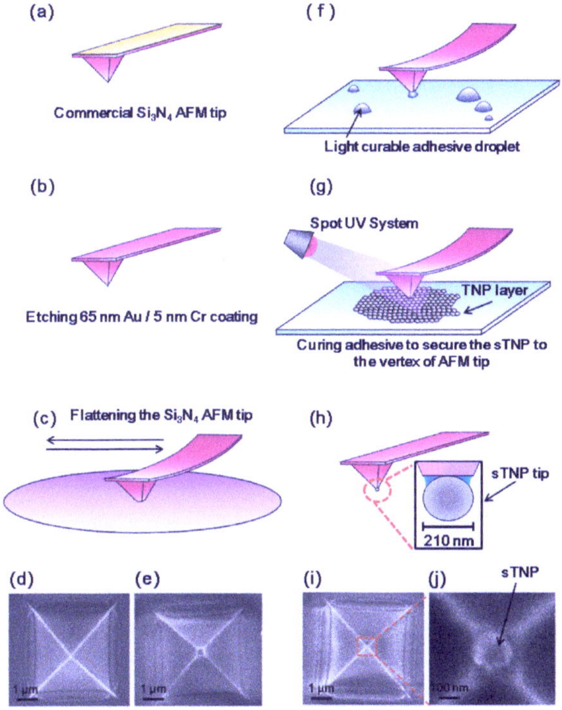

Figure 2: **Schematic diagram showing the process of fabricating the sTNP tip. (a, b)**Etching process for reflective metal layer on Olympus RC-800 Si_3N_4 tip. **(c)** The vertex of the tip was flattened by scanning the tip across a polished Si_3N_4 wafer. **(d, e)** Present SEM images of the Si_3N_4 AFM tip before and after the scanning process, respectively. **(f)**A small quantity of adhesive was applied to the flat top of the AFM tip. **(g)** Attached sTNP to the vertex of the flattened tip with

adhesive followed by curing. **(h)** Schematic diagram of fabricated sTNP tip. **(i, j)** SEM images of the sTNP tip.

The cleaned Si_3N_4 AFM tip was then mounted on the NanoWizard AFM scanner (JPK Instrument) and brought into contact with the adhesive droplet (Figure 2f). This allowed the placement of a small quantity of adhesive on the flat top of the AFM tip. The tip was then put into contact with the TNP layer deposited on the glass slide (Figure 2g). The TNP layer was prepared by drying a 30-μl droplet (200 nm in diameter) of 5% polytetrafluoroethylene (PTFE) aqueous dispersion (Teflon PTFE TE-3893, DuPont, Wilmington, DE, USA) on the glass slide. PTFE has been shown to possess excellent performance characteristics with regard to charge storage and is widely used in electret applications [16]. The adhesive was cured by exposure to UV radiation illuminated from a spot UV system (Aicure ANUP 5252 L, Panasonic, Osaka, Japan) at 3,000 mW/cm^2 for 3 min to secure the sTNP. Figure 2d,e presents typical images from a scanning electron microscope (SEM) showing the top views of the Si_3N_4 AFM tip before and after the flattening procedure. Figure 2i presents an SEM image of the sTNP tip. The diameter of the sTNP attached to the vertex of the Si3N4 AFM tip was measured by SEM at 210 nm (Figure 2j).

The Experimental Setup of the Deposition of Charge to the sTNP tip

The experimental setup used for the deposition of charge to the sTNP tip is presented in Figure 3. The back side of the sTNP tip was affixed to the 30-nm Au/ 20-nm Ti-coated glass slide using conductive copper tape (3 M, St. Paul, MN, USA).

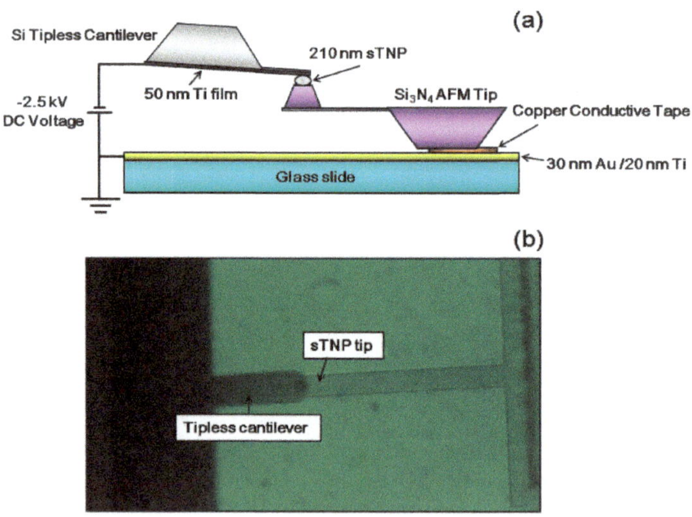

Figure 3: Schematic diagram of experimental setup for the deposition of charge to the sTNP tip. (a) Schematic diagram of experimental setup for the deposition of charge to the sTNP at the vertex of the Si_3N_4 AFM tip and **(b)** × 40 optical microscope image of the charging setup.

A 50-nm Ti-coated tipless cantilever (CSC12, MikroMasch, Tallinn, Estonia) was mounted on the JPK AFM scanner as the top electrode. The end of the tipless cantilever was positioned precisely on the sTNP at the vertex of the Si_3N_4 tip by aligning the JPK AFM scanner under an inverted optical microscope (IX 71, Olympus; Figure 3b). DC voltage (−2.5 kV) was applied to the tipless cantilever for 90 s under air, and the 30-nm Au/20-nm Ti-coated glass slide was used as the ground for the deposition of the negative charge to the sTNP tip. The force-distance (f-d) curves of the sTNP tip on the grounded gold surface were used to verify whether the charge was deposited [17].

Measurement of the Electrostatic Fields

The charged sTNP tip was then used for the measurement of f-d curves to determine the electrostatic field beside the top electrode of the parallel plate condenser (Figure 1). The sTNP tip is located slightly inward at the end of the AFM cantilever; therefore, the end of the AFM cantilever is susceptible to striking the edge of the top electrode when the distance between the AFM tip and the electrode is within 10 μm. To overcome this situation, 21 spots spaced at 0.25 μm along the X-axis at a distance of 10 to 15 μm are selected for the measurement of the f-d curves in order to derive the electrostatic field. As shown in Figure 1, the edge center of the condenser was plotted as the origin of the X- and Z-axes. DC voltage (V_{app}) of ±25 V was applied on the top electrode, and the bottom electrode was left grounded. Each curve measurement was conducted for distances of 15 μm along the Z-axis, from 6 μm below to 9 μm above the top electrode. The ramp rate and the ramp size of each f-d curve were 2 Hz and 15 μm, respectively. The sTNP tip did not come into contact with the substrate during the measurement of electrostatic fields. The measurement of f-d curves was conducted using the force mapping function in the JPK SPM software.

Simulation of the Electrostatic Field

The electric field was simulated using finite element method in Ansoft Maxwell simulation software [18] to estimate the electrostatic field. The current model deals only with the electric field in the Z direction from −10 to approximately 10 μm. After designing the model, the maximum length of elements was set at 0.4 μm; this was sufficient to provide accurate solutions to model at that scale. The Maxwell program automatically fits the mesh to estimate the electrostatic field.

RESULTS AND DISCUSSION

Figure 4a presents the f-d curves for tips before and after the charging process. A long-range attractive force [19] was observed between the charged sTNP tip and the grounded gold surface, mainly due to the electrostatic force. No attractive force was observed on the uncharged sTNP tip. The attractive force acting on the charged sTNP tip gradually increased as the tip was moved closer to the gold-coated surface. As shown in Figure 4a, the form of the f-d curve acting on the grounded metal surface using a charged sTNP is similar to that observed in a previous study involving the measurement of electrostatic force between a charged particle and a metal surface using the modified image charge method [17].

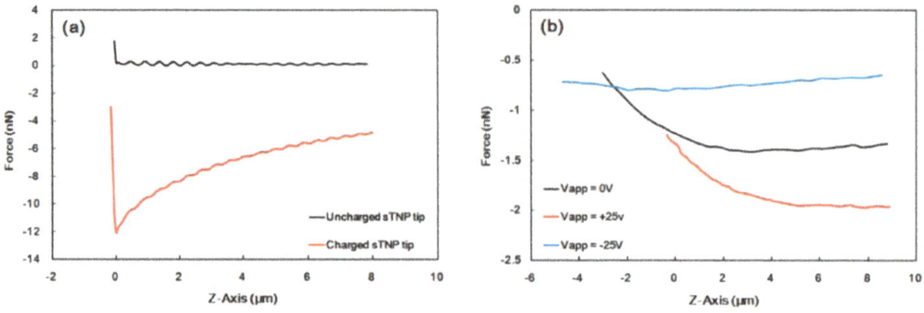

Figure 4: **Schematic diagram of f-d curves conducted using sTNP tip. (a)** f-d Curves obtained from a grounded metal surface using charged/uncharged sTNP tip. **(b)**Electrostatic force acting on charged sTNP tip when V_{app}=+25, 0, and −25 V in the *Z*direction at X=11 μm.

According to previous studies [9–11], the net electrostatic force (F_E) acting on a charged dielectric particle in an applied electric field that can be written as follows:

$$F_E = F_C + F_{image} + F_{pol}, \quad (1)$$

where F_C is the Coulombic force that resulted from the external field acting on the charged particle, F_{image} is the image force caused by the attraction of the particle to its net charge image, and F_{pol} is the force created by the attraction between the field-induced dipolar charge (polarization) in a particle in an electrostatic field and its dipole image in the electrode.

In this study, F_{pol} acting on the sTNP was due mainly to the thin layer of water adsorbed on the surface of the tip due to the large dielectric constant of water (ϵ_{water} = 80). To eliminate the influence of the water layer, the measurement of the electrostatic field was conducted under N_2conditions (RH<5%), such that F_{pol} acting on the sTNP could be disregarded; a plastic O-ring was placed between the scanner and sample to allow the injection of N_2 into the O-ring. Charges deposited on the sTNP under N_2 conditions can last (variation smaller than 5%) for over 90 min, and the measurement process can be completed within 10 min. In this study, the dissipation or generation of charge did not occur during charge verification, despite the fact that the charged sTNP tip touched the grounded metal surface under N_2 conditions. Figure 4b presents the three f-d curves at X=11 μm under N_2 conditions when V_{app}=+25, 0, and −25 V were applied to the top electrode, and the bottom electrode remained grounded. The *Z*-axis component of F_E acting on the sTNP tip can be revealed in the measured f-d curves (Figure 4b), expressed as $F_E(V_{app})$. F_E(0 V) acting on the sTNP tip is due mainly to F_{image}, which is always attractive to the top electrode of the condenser. The F_C(+25 V) is the attractive force acting on the negative-charged sTNP tip, such that F_E(+25 V) is smaller than F_E(0 V) above Z=0 μm.F_C(+25 V) always attracts the negative-charged sTNP tip, regardless of whether the sTNP tip is above or below the top electrode at Z=0 μm. This results in the charged sTNP tip being trapped at Z=0 μm, preventing it from moving forward during the measurement of the f-d curves, as shown in Figure 4b. F_C(−25 V) is a repulsive force acting on the negative-charged sTNP tip, such that F_E(−25 V) is larger than F_E(0 V) above Z=−2.6 μm; however, it is smaller below Z=−2.6 μm due to the attractive force induced from the bottom electrode.

Thus, $F_C(V_{app})$ acting on the negative-charged sTNP tip can be estimated according to the following formula: $F\,C(V_{app})=F_E(V_{app})-F_E(0\text{ V})$. The coulombic force acting on the positive charged sTNP produced by the electrostatic field of the parallel plate condenser is equal to $-F_C(V_{app})$, expressed as $F_{ele}(V_{app})$, which represents the electrostatic force field of the condenser. Figure 5a,c respectively presents the F_{ele}(+25 V) and F_{ele}(−25 V) distribution along the X-axis (0.25-μm spacing from 10 to 15 μm) and the Z-axis. As mention in previous discussion, F_{ele}(+25 V) below Z=0 μm cannot be measured but can be acquired through polynomial extrapolation. In this study, charge was deposited on the sTNP, a small portion of which was transferred to the edge of the pyramid shaped Si_3N_4 tip. As a result, the total charge on the sTNP was assumed to be a point charge located 2 μm above the vertex of the Si_3N_4 tip. The Z-axis in Figure 5a,c reveals the distance between the point charge and the top electrode in the Z direction. Figure 5b,d presents the results of Ansoft Maxwell simulation of electrostatic field distribution under V_{app}=+25 and −25 V, with trends similar to those in Figure 5a,c, respectively. The charge on the charged sTNP tip was approximately -1.7×10^{-14}C, as estimated through simulation. F_{ele}(−25 V) is the attractive force above Z=0 μm; however, this was converted into a repulsive force between Z=0 and −2 μm.

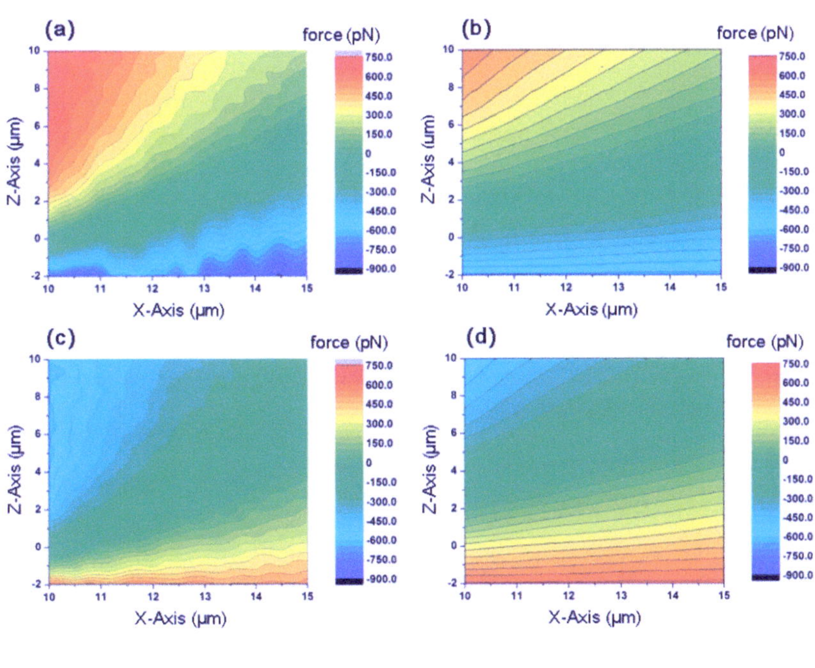

Figure 5: **Experimental results vs. Ansoft Maxwell simulation. (a, c)** The F_{ele}(+25 V) and-F_{ele}(−25 V) distribution along the X-axis (0.25-μm spacing from 10 to 15 μm) and the Z-axis. **(b, d)** The results of Ansoft Maxwell simulation of electrostatic field distribution under V_{app}=+25 and −25 V, respectively.

F_{ele}(+25 V) and F_{ele}(−25 V) are symmetrical about the *Z*-axis, revealing the inverse direction of the electrostatic field distribution. As shown in Figure 5a,c, the minimum F_{ele} that can be measured is less than 50 pN. The difference between the simulated and experimental results can be explained by the fact that a residual charge is transferred to the edge of the pyramid-shaped Si_3N_4 tip during charge deposition process; when the charged sTNP tip is brought closer to the top electrode (X=0 μm), the difference between the simulated and experimental results increases. The ripples shown in Figure 5a,c were caused by laser diffraction on the insulating Si_3N_4 cantilever (for more details, see Additional file 1). In the future, the pyramidal shape of the Si_3N_4 tip could be modified using a focused ion beam system to create a cylindrical shape in order to avoid the possibility of experimental fluctuations resulting from the shape of the tip. This probe could be employed to scan surface topographies by mapping f-d curves, and the interaction force between the charged Teflon particle and sample would give a direct indication of the local electric field and properties of the sample.

CONCLUSIONS

In summary, this paper reported the direct measurement of the electrostatic field beside a parallel plate condenser using a charged sTNP on an AFM tip. Experimental results were then compared with those obtained through simulation. A sTNP tip was fabricated by attaching a single 210-nm Teflon nanoparticle at the vertex of a Si_3N_4 AFM tip and was charged via contact electrification. The lateral/vertical resolution of the electrostatic force measurement is 250/100 nm, respectively. The minimum F_{ele} that can be measured using this method is less than 50 pN. This technique provides a novel means of studying the electric properties of electrical devices. The AFM tip is able to hold a single charged nanoparticle, making it possible to directly quantify the local electric/magnetic field, charge distribution, and electrostatic force of a sample surface using an AFM system. The charged sTNP tip could find a wide application in electrical research at the nanoscale.

Acknowledgements

This work was supported by grants from the National Science Council of Taiwan under the programs NSC102-2627-M-007-002, NSC100-2120-M-007-006, NSC 99-2120-M-007-009, NSC100-2627-M-007-013, and NSC 99-2627-M-007-002.

SUPPLEMENTARY MATERIAL

The Consistent of the charged sTNP Tip:

In this study, f-d curves of the sTNP tip on the grounded gold surface were used to verify whether the charge was deposited. An attractive force was observed to be acting on the sTNP tip (as shown on Fig.1), indicating the the sTNP tip was charged. The attractive force (F_{image}) acting on the charged sTNP tip before and after measurement was -10.82 nN and -10.97 nN, respectively. These results indicate that the deposited charge remained consistent throughout the measurement of the electrostatic field.

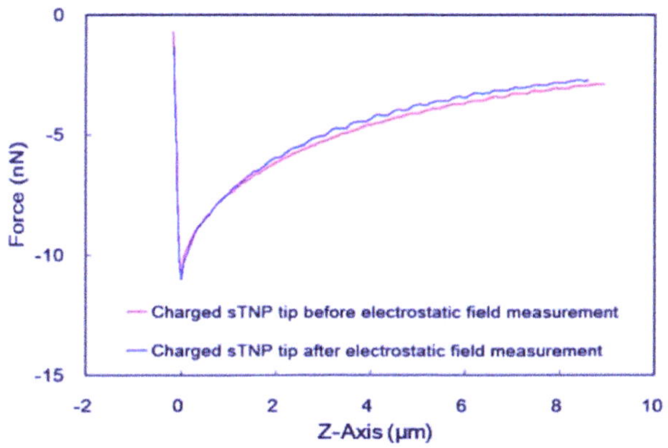

Figure 1: F-d curves obtained from a grounded metal surface before and after the measurement of the electrostatic field

The charge deposited on the sTNP exceeded charge dissipation due to moisture during the charging process. Charging the sTNP tip under ambient conditions for a period of 90 s was sufficient for the measurements in this study (Fig.1).

The Duration Time of The Charged sTNP tip Under N_2 Condition:

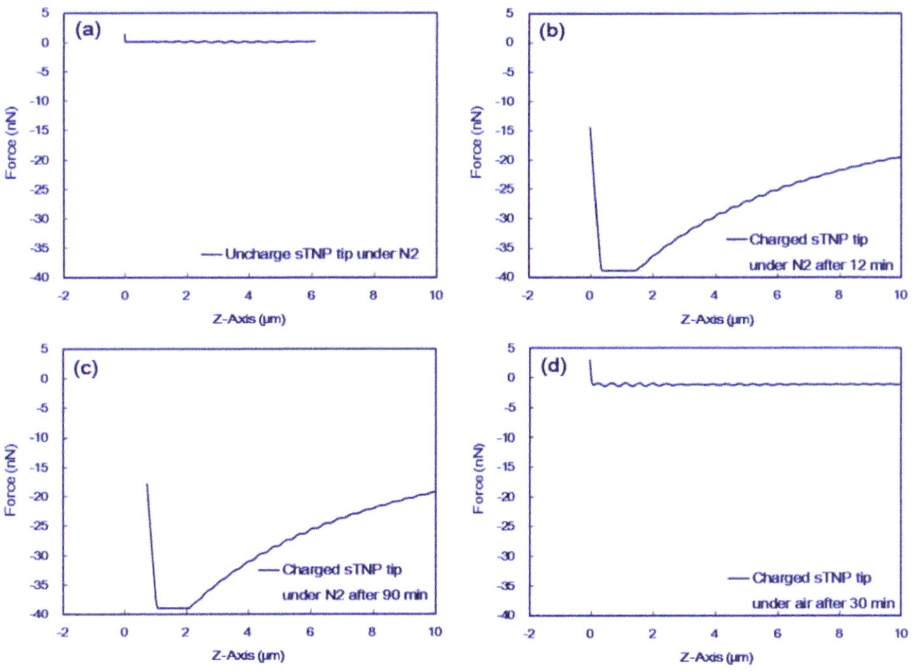

Figure 2: The duration time of the charged sTNP tip under N_2 condition. (a) F-d curves obtained

from a grounded metal surface using an uncharged sTNP tip under N2; (b) and (c) f-d curves obtained from a charged sTNP under N_2 after 12 and 90 min, respectively; (d) f-d curves obtained from charged sTNP under air after 30 min.

Fig. 2 presents the charging durations of the sTNP tip under N2. Fig. 2(a) presents the F-d curves obtained from a grounded metal surface using an uncharged sTNP tip under N2. No attractive force (F_{image}) was observed on the uncharged sTNP tip. Figs. 2(b) and (c) present the f-d curves obtained from charged sTNP tips under N2 after 12 and 90 min, respectively. The F_{image} acting on the sTNP tip is consistent with the results presented in Figs. 2(b) and (c), demonstrating that the charge deposited on a sTNP tip can last for over 90 min. A small portion of the f-d curves was outside the detection range of the photodetector.

Fig. 2(d) presents the f-d curves obtained from a charged sTNP under air after 30 min. The deposited charge dissipated through exposure to moisture and no F_{image} was observed on the charged sTNP tip.

The Measurement of the F_{pol}:

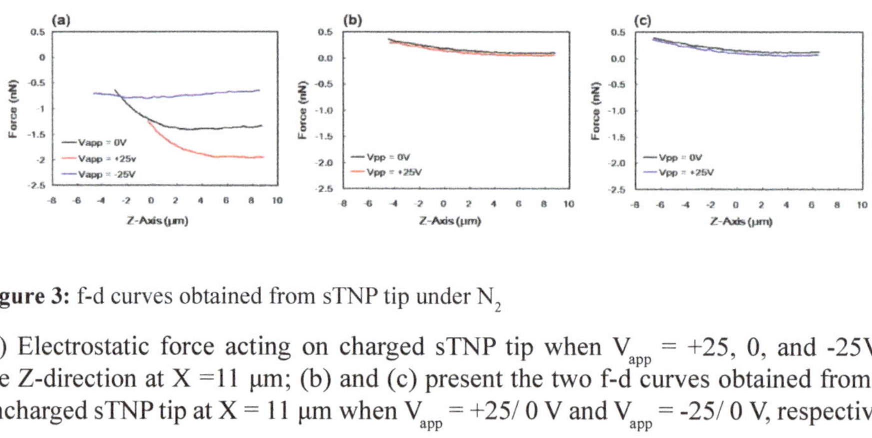

Figure 3: f-d curves obtained from sTNP tip under N_2

(a) Electrostatic force acting on charged sTNP tip when V_{app} = +25, 0, and -25V in the Z-direction at X =11 μm; (b) and (c) present the two f-d curves obtained from the uncharged sTNP tip at X = 11 μm when V_{app} = +25/ 0 V and V_{app} = -25/ 0 V, respectively.

Fig. 3(a) presents the electrostatic force acting on a charged sTNP tip when V_{app} = +25, 0 and -25 V in the Z-direction at X =11 μm. Figs. 3(b) and (c) respectively present the two f-d curves obtained from an uncharged sTNP tip at X =11 μm when V_{app} = +25/ 0V and V_{app} = -25/ 0V under N2 conditions. F_E(+25V) and F_E(-25V) acting on the uncharged sTNP tip under N2 conditions are due mainly to polarized charges in the uncharged sTNP tip (F_{image} and Fc = 0). The force (Fpol) induced by polarized charges can be estimated as follow:

$$F_{pol}(+25V) = F_E(+25V) - F_E(0V) \quad (1)$$

$$F_{pol}(-25V) = F_E(-25V) - F_E(0V) \quad (2)$$

The F_{pol} (+25V) was -49.41 ± 6.2 pN and F_{pol} (-25V) was -51.41 ± 8.9 pN. The contribution of the polarized charges (~ -50 pN) to the net electrostatic force FE (~500 pN) in Fig. 3(a)) was quite small (~10%).

The Differences between the Experimental Result and Ansoft Maxwell Simulation in the in the Region X = 10 to 11 Micrometers, Z = 4 to 10 Micrometers:

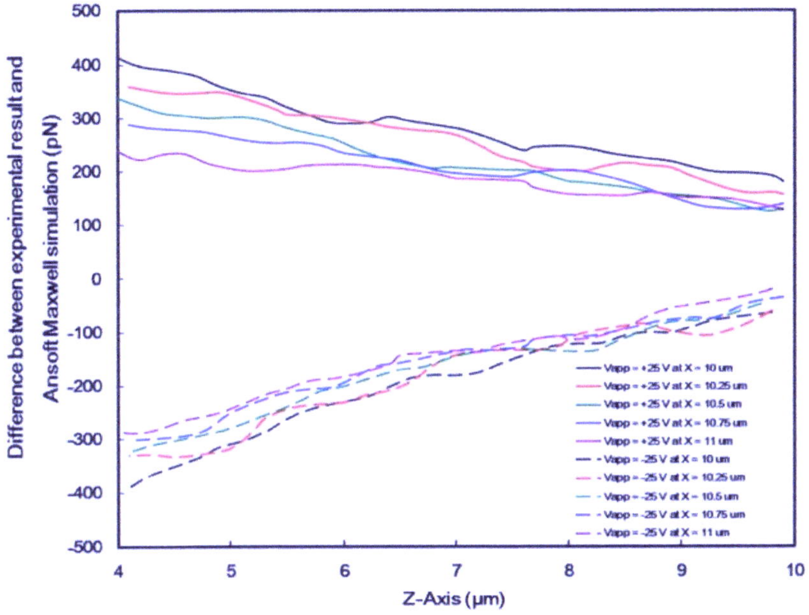

Vapp = +25 V	AVG of difference (pN)	STDV
At Z = 10 micrometers	147.04	22.12
At Z = 4 micrometers	327.99	67.06

Vapp = -25 V	AVG of difference (pN)	STDV
At Z = 10 micrometers	-45.22	18.08
At Z = 4 micrometers	-325.64	39.49

Figure 4: Schematic diagram of differences between experimental result and Ansoft Maxwell simulation. (Difference = F_{ele} measured by EXP - F_{ele} simulated by Ansoft Maxwell)

Fig. 4 presents the differences (Difference = F_{ele} measured by EXP - F_{ele} simulated by Ansoft Maxwell) between experimental result and Ansoft Maxwell simulation in the region X = 10 to 11 micrometers, Z = 4 to 10 micrometers when V_{app} = +25 and -25 V. As mentioned in the manuscript, when the charged sTNP tip was brought closer to the top electrode, the difference between experimental and simulation results increased.

Competing Interests

The authors declare that they have no competing interests.

REFERENCES

1. Martin Y, Williams CCHK, Wickramasinghe HK: Atomic force microscope-force

mapping and profiling on a sub 100-A scale. *J Appl Phys* 1987, 61: 4723–4729. 10.1063/1.338807

2. Stern JE, Terris BD, Mamin HJ, Rugar D: Deposition and imaging of localized charge on insulator surfaces using a force microscope. *Appl Phys Lett* 1988, 53: 2717–2719. 10.1063/1.100162
3. Terris BD, Sterna JE, Rugar D, Mamin HJ: Localized charge force microscopy. *J Vac Sci Technol* 1990, A8: 374–377.
4. Berger R, Butt HJ, Retschke MB, Weber SAL: Electrical modes in scanning probe microscopy. *Macromol Rapid Commun* 2009, 30: 1167–1178. 10.1002/marc.200900220
5. Bonnell DA: Electrostatic and magnetic force microscopy. In *Scanning Probe Microscopy and Spectroscopy*. New York: Wiley; 2001:207–210.
6. Nonnenmacher M, O'Boyle MP, Wickramasinghe HK: Kelvin probe force microscopy. *Appl Phys Lett* 1991, 58: 2921–2923. 10.1063/1.105227
7. Palermo V, Palma M, Samori P: Electronic characterization of organic thin films by Kelvin probe force microscopy. *Adv Mater* 2006, 18: 145–164. 10.1002/adma.200501394
8. Jenke MG, Santschi C, Hoffmann P: Two-dimensional electrostatic force field measurements with simultaneous topography measurement on embedded interdigitated nanoelectrodes using a force distance curve based method. *Appl Phys Lett* 2008, 92: 063113. 10.1063/1.2844882
9. Feng JQ, Hays DA: Relative importance of electrostatic forces on powder particles. *Powder Technol* 2003, 135–136: 65–75.
10. Kwek JW, Vakararelski IU, Ng WK, Heng JYY, Tan RBH: Novel parallel plate condenser for single particle electrostatic force measurements in atomic force microscope. *Colloids Surf Physicochem Eng Aspects* 2011, 385: 206–212. 10.1016/j.colsurfa.2011.06.008
11. Harris B: The electric field. In *University Physics*. New York: John Wily & Sons, Inc; 1995:455–475.
12. Terris BD, Stern JE, Rugar D, Mamin HJ: Contact electrification using force microscopy.*Phys Rev Lett* 1989, 63: 2669–2672. 10.1103/PhysRevLett.63.2669
13. Mesquida P, Stemmer A: Attaching silica nanoparticles from suspension onto surface charge patterns generated by a conductive atomic force microscope tip. *Adv Mater* 2001, 13: 1395–1398. 10.1002/1521-4095(200109)13:18<1395::AID-ADMA1395>3.0.CO;2-0
14. Mesquida P, Knapp HF, Stemmer A: Charge writing on the nanometre scale in a fluorocarbon film. *Surf Interface Anal* 2002, 33: 159–162. 10.1002/sia.1181
15. Hutter JL, Bechhoefer J: Calibration of atomic force microscope tips. *Rev Sci Instrum*1993, 64: 1868–1873. 10.1063/1.1143970
16. Kestelman VN, Pinchuk LS, Goldade VA: *Electrets Engineering: Fundamentals and Applications*. Boston: Kluwer Academic Publishers; 2000.

17. Matsuyama T, Ohtsuka M, Yamamoto H: Measurement of force curve due to electrostatic charge on a single particle using atomic force microscope. *KONA Powder Particle J* 2008, 26: 238–245.

18. ANSYS, Inc: ANSYS Maxwell.http://www.ansys.com/Products/Simulation+Technology/Electromagnetics/Electromechanical/ANSYS+Maxwell

19. Israelachvili JN: Contrasts between intermolecular, interparticle and intersurface forces. In*Intermolecular and Surface Forces*. San Diego: Academic; 1991:152–155.

10
CHAPTER

EXPERIMENTAL AND THEORETICAL STUDIES OF NANOFLUID THERMAL CONDUCTIVITY ENHANCEMENT

Clement Kleinstreuer, Yu Feng

INTRODUCTION

A nanofluid is a dilute suspension of nanometer-size particles and fibers dispersed in a liquid. As a result, when compared to the base fluid, changes in physical properties of such mixtures occur, *e.g.*, viscosity, density, and thermal conductivity. Of all the physical properties of nanofluids, the thermal conductivity (k_{nf}) is the most complex and for many applications the most important one. Interestingly, experimental findings have been controversial and theories do not fully explain the mechanisms of elevated thermal conductivity. In this paper, experimental and theoretical studies are reviewed for nanofluid thermal conductivity and convection heat transfer enhancement. Specifically, comparisons between thermal measurement techniques (*e.g.*, transient hot-wire (THW) method) and optical measurement techniques (*e.g.*, forced Rayleigh scattering (FRS) method) are discussed. Recent theoretical models for nanofluid thermal conductivity are presented and compared, including the authors' model assuming well-dispersed spherical nanoparticles subject to micro-mixing effects due to Brownian motion. Concerning theories/correlations which try to explain thermal conductivity enhancement for all nanofluids, not a single model can predict a wide range of experimental data. However, many experimental data sets may fit between the lower and upper mean-field bounds originally proposed by Maxwell where the static nanoparticle configurations may range from a dispersed phase to a pseudo-continuous phase. Dynamic k_{nf} models, assuming non-interacting metallic nano-spheres, postulate an enhancement above the classical Maxwell theory and thereby provide potentially additional physical insight. Clearly, it will be necessary to consider not only one possible mechanism but combine several mechanisms and compare predictive results to new benchmark experimental data sets.

Citation: Kleinstreuer and Feng: Experimental and theoretical studies of nanofluid thermal conductivity enhancement, DOI: 10.1186/1556-276X-6-229.

EXPERIMENTAL STUDIES

Nanofluids are a new class of heat transfer fluids by dispersing nanometer-size particles, *e.g.*, metal-oxide spheres or carbon nanotubes, with typical diameter scales of 1 to 100 nm in traditional heat transfer fluids. Such colloidal dispersions may be uniform or somewhat aggregated. Earlier experimental studies reported greater enhancement of thermal conductivity,k_{nf}, than predicted by the classical model of Maxwell [1], known as the mean-field or effective medium theory. For example, Masuda [2] showed that different nanofluids (*i.e.*, Al_2O_3-water, SiO_2-water, and TiO_2-water combinations) generated a k_{nf} increase of up to 30% at volume fractions of less than 4.3%. Such an enhancement phenomenon was also reported by Eastman and Choi [3] for CuO-water, Al_2O_3-water and Cu-Oil nanofluids, using the THW method. In the following decades, it was established that nanofluid thermal conductivity is a function of several parameters [4, 5], *i.e.*, nanoparticle material, volume fraction, spatial distribution, size, and shape, as well as base-fluid type, temperature, and pH value. In contrast, other experimentalists [6–9], reported that no correlation was observed between k_{nf} and nanofluid temperature *T*. Furthermore, no k_{nf} enhancement above predictions based on Maxwell's effective medium theory for non-interacting spherical nanoparticles was obtained [5]. Clearly, this poses the question if nanofluids can provide greater heat transfer performance, as it would be most desirable for cooling of microsystems. Some scientists argued that the anomalous k_{nf} enhancement data are caused by inaccuracies of thermal measurement methods, *i.e.*, mainly intrusive vs. non-intrusive techniques. However, some researchers [10, 11], relying on both optical and thermal measurements, reported k_{nf} enhancements well above classical model predictions. When comparing different measurement methods, error sources may result from the preparation of nanofluids, heating process, measurement process, cleanliness of apparatus, and if the nanoparticles stay uniformly dispersed in the base fluid or aggregate [12]. Thus, the controversy is still not over because of those uncertainties.

EXPERIMENTAL MEASUREMENT METHODS

The most common techniques for measuring the thermal conductivity of nanofluids are the transient hot-wire method [9, 12–15], temperature oscillation method [16, 17], and 3-ω method [18, 19]. As an example of a non-intrusive (optical) technique, forced Rayleigh scattering is discussed as well.

TRANSIENT HOT-WIRE METHOD

THW method is the most widely used static, linear source experimental technique for measuring the thermal conductivity of fluids. A hot wire is placed in the fluid, which functions as both a heat source and a thermometer [20, 21]. Based on Fourier's law, when heating the wire, a higher thermal conductivity of the fluid corresponds to a lower temperature rise. Das [22] claimed that during the short measurement interval of 2 to 8 s, natural convection will not influence the accuracy of the results.

The relationship between thermal conductivity k_{nf} and measured temperature *T* using the THW method is summarized as follows [20]. Assuming a thin, infinitely long line source dissipating heat into a fluid reservoir, the energy equation in cylindrical coordinates can be written as:

$$\frac{1}{\alpha_{nf}}\frac{\partial T}{\partial t} = \frac{1}{r}\frac{\partial}{\partial r}\left(r\frac{\partial T}{\partial r}\right) \tag{1}$$

with initial condition and boundary conditions

$$T(t = 0) = T_0 \tag{2a}$$

and

$$\lim_{r\to 0}\left(r\frac{\partial T}{\partial r}\right) = \frac{q}{2\pi k_{nf}} \text{ and } \left.\frac{\partial T}{\partial r}\right|_{r=\infty} = 0 \tag{2b-c}$$

The analytic solution reads:

$$T(r,t) = T_0 + \frac{q}{4\pi k_{nf}}\left\{-\gamma + \ln\left(\frac{4\alpha_{nf}t}{r^2}\right) + \left[\frac{\left(\frac{r^2}{4\alpha_{nf}t}\right)}{1\cdot 1!} - \frac{\left(\frac{r^2}{4\alpha_{nf}t}\right)^2}{2\cdot 2!} + -..+ ......\right]\right\} \tag{3}$$

where $\gamma = 0.5772$ is Euler›s constant. Hence, if the temperature of the hot wire at time t_1 and t_2 are T_1 and T_2, then by neglecting higher-order terms the thermal conductivity can be approximated as:

$$k_{nf} = \frac{q}{4\pi}\frac{\ln(t_1/t_2)}{T_1 - T_2} \tag{4}$$

For the experimental procedure, the wire is heated via a constant electric power supply at step time t. A temperature increase of the wire is determined from its change in resistance which can be measured in time using a Wheatstone-bridge circuit. Then the thermal conductivity is determined from Eq. 4, knowing the heating power (or heat flux q) and the slope of the curve $\ln(t)$ versus T.

The advantages of THW method are low cost and easy implementation. However, the assumptions of an infinite wire-length and the ambient acting like a reservoir (see Eqs. 1 and 2c) may introduce errors. In addition, nanoparticle interactions, sedimentation and/or aggregation as well as natural convection during extended measurement times may also increase experimental uncertainties [19, 23].

OTHER THERMAL MEASUREMENT METHODS

A number of improved hot-wire methods and experimental designs have been proposed. For example, Zhang [24] used a short-hot-wire method (see also Woodfield [25]) which can take into account boundary effects. Mintsa [26] inserted a mixer into his THW experimental devices in order to avoid nanoparticle aggregation/deposition in the suspensions. Ali et al. [27] combined a laser beam displacement method with the THW method to separate the detector and heater to avoid interference.

Alternative static experimental methods include the temperature oscillation method [16, 17,28], micro-hot-strip method [29], steady-state cut-bar method [30], 3-ω method [18, 31, 32], radial heat-flow method [33], photo-thermal radiometry method [34], and thermal comparator method [19, 35].

It is worth mentioning that most of the thermal measurement techniques are static or so called "bulk" methods (see Eq. 4). However, nanofluids could be used as coolants in forced convection, requiring convective measurement methods to obtain thermal conductivity data. Some experimental results of convective nanofluid heat transfer characteristics are listed in Table 1. For example, Lee [36] fabricated a microchannel, D_h = 200 μm, to measure the nanofluid thermal conductivity with a modest enhancement when compared to the result obtained by the THW method. Also, Kolade et al. [37] considered 2% Al_2O_3-water and 0.2% multi-wall carbon nano-tube (MWCNT)-silicone oil nanofluids. By measuring the thermal conductivities of nanofluids in a convective environment, Kolade et al. [37] obtained 6% enhancement for Al_2O_3-water nanofluid and 10% enhancement for MWCNT-silicone oil nanofluid. Such enhancements are very modest compared to the experimental data obtained by THW methods.

Table 1: Summary of experimental studies on convective heat transfer properties of nanofluids

Reference	**Nanofluids**	**Flow nature**	**Findings**
Pak and Cho [91]	d_p = 13 nm spherical Al_2O_3-water d_p = 27 nm spherical TiO_2-water	Tube/turbulent	Nu is 30% larger than conventional base fluid and larger than Dittus-Boelter prediction
Li and Xuan [92]	d_p < 100 nm spherical Cu-water	Tube/turbulent	Nu is larger than Dittus-Boelter prediction when volume fraction φ > 0.5%
Wen and Ding [93]	d_p = 27-56 nm spherical Al_2O_3-water	Tube/laminar	Nu > 4.36 for fully-developed pipe flow with constant wall heat flux
Ding [94]	d_p > 100 nm rodlike carbon nanotube-water	Tube/laminar	Nu increase more than 300% at Re = 800
Heris [95]	d_p = 20 nm spherical Al_2O_3-water	Tube/laminar	Nu measured is larger than Nu of pure water
Williams [49]	d_p = 46 nm spherical Al_2O_3-water d_p = 60 nm spherical ZrO_2-water	Tube/turbulent	Nu of nanofluids can be predicted by traditional correlations and models. No abnormal heat transfer enhancement was observed.

Kolade [37]	d_p = 40-50 nm spherical Al_2O_3-water rodlike carbon nanotube-oil	Tube/laminar	Nu is apparently larger than pure based fluid
Duangthongsuk [14]	d_p = 21 nm spherical TiO_2-water	Tube/turbulent	Pak and Cho (1998) correlation show better agreement to experimental data of Nu than Xuan and Li (2002) correlation
Rea [96]	d_p = 50 nm spherical Al_2O_3-water d_p = 50 nm spherical ZrO_2-water	Tube/laminar	Nu of Al_2O_3-water nanofluid show up to 27% more than pure water, ZrO_2-water displays much lower enhancement.
Jung [90]	d_p = 170 nm spherical Al_2O_3-water d_p = 170 nm spherical Al_2O_3-ethylene glycol	Rectangular microchannel/ laminar	Nu increases with increasing the Reynolds number in laminar flow regime, appreciable enhancement of Nu is measured
Heris [97]	spherical Al_2O_3-water	Tube/laminar	Nu increases with increasing the Peclet number and φ, Brownian motion may play role in convective heat transfer enhancement

Actually, "convective" k_{nf} values are not directly measured. Instead, wall temperature T_w and bulk temperature T_b are obtained and the heat transfer coefficient is then calculated as $h = q_w/(T_w - T_b)$. From the definition of the Nusselt number, $k_{nf} = hD/$ Nu where generally D is the hydraulic diameter. With h being basically measured and D known, either an analytic solution or an iterative numerical evaluation of Nu is required to calculate k_{nf}. Clearly, the accuracy of the "convective measurement method" largely depends on the degree of uncertainties related to the measured wall and bulk temperatures as well as the computed Nusselt number.

OPTICAL MEASUREMENT METHODS

In recent years, optical measurement methods have been proposed as non-invasive techniques for thermal conductivity measurements to improve accuracy [6–9, 13, 11, 27, 37]. Indeed, because the "hot wire" is a combination of heater and thermometer, interference is unavoidable. However, in optical techniques, detector and heater are always separated from each other, providing potentially more accurate data. Additionally, measurements are completed within several microseconds, *i.e.*, much shorter than reported THW-measurement times of 2 to 8 s, so that natural convection effects are avoided.

For example, Rusconi [6, 38] proposed a thermal-lensing (TL) measurement method to obtaink_{nf} data. The nanofluid sample was heated by a laser-diode module and the temperature difference was measured by photodiode as optical signals. After post-processing, the thermal conductivity values were generated, which did not exceed mean-field theory results. Similar to the TL method, FRS have been used to investigate the thermal conductivity of well-dispersed nanofluids [8, 39]. Again, their results did not show any anomalous enhancement either for Au-or Al_2O_3-nanofluids. Also, based on their data, no enhancement of thermal conductivity with temperature was observed. In contrast, Buongiorno et al. [9] presented data agreement when using both the THW method and FRS method. Another optical technique for thermal conductivity measurements of nanofluids is optical beam deflection [7, 40]. The nanofluid is heated by two parallel lines using a square current. The temperature change of nanofluids can be transformed to light signals captured by dual photodiodes. For Au-nanofluids, Putnam [7] reported significantly lower k_{nf} enhancement than the data collected with the THW method.

However, other papers based on optical measurement techniques showed similar enhancement trends for nanofluid thermal conductivities as obtained with the thermal measurement methods. For example, Shaikh et al. [10] used the modern light flash technique (LFA 447) and measured the thermal conductivity of three types of nanofluids. They reported a maximum enhancement of 161% for the thermal conductivity of carbon nanotube (CNT)-polyalphaolefin (PAO) suspensions. Such an enhancement is well above the prediction of the classical model by Hamilton and Crosser [41]. Also, Schmidt et al. [13] compared experimental data for Al_2O_3-PAO and $C_{10}H_{22}$-PAO nanofluids obtained via the Transient Optical Grating method and THW method. In both cases, the thermal conductivities were greater than expected from classical models. Additionally, Bazan [11] executed measurements by three different methods, *i.e.*, laser flash (LF), transient plane source, and THW for PAO-based nanofluids. They concluded that the THW method is the most accurate one while the LF method lacks precision when measuring nanofluids with low thermal conductivities. Also, no correlation between thermal conductivity and temperature was observed. Clearly, materials and experimental methods employed differ from study to study, where some of the new measurement methods were not verified repeatedly [6, 7]. Thus, it will be necessary for scientists to use different experimental techniques for the same nanofluids in order to achieve high comparable accuracy and prove reproducibility of the experimental results.

EXPERIMENTAL OBSERVATIONS

Nearly all experimental results before 2005 indicate an anomalous enhancement of nanofluid thermal conductivity, assuming well-dispersed nanoparticles. However, more recent efforts with refined transient hot-wire and optical methods spawned a controversy on whether the anomalous enhancement beyond the mean-field theory is real or not. Eapen et al. [5] suggested a solution, arguing that even for dilute nanoparticle suspensions k_{nf} enhancement is a function of the aggregation state and hence connectivity of the particles; specifically, almost all experimental k_{nf} data published fall between lower and upper bounds predicted by classical theories.

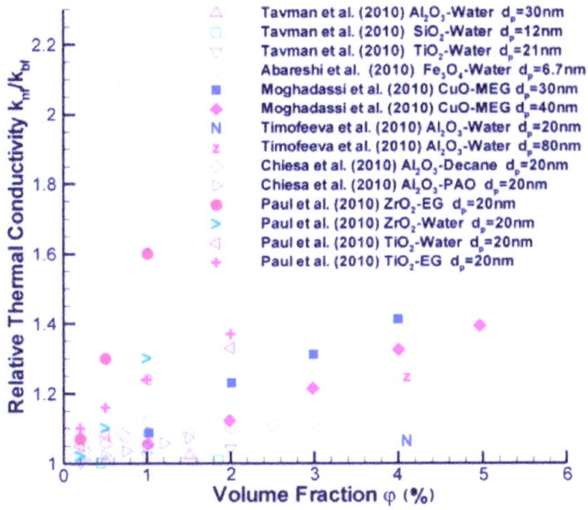

(a)

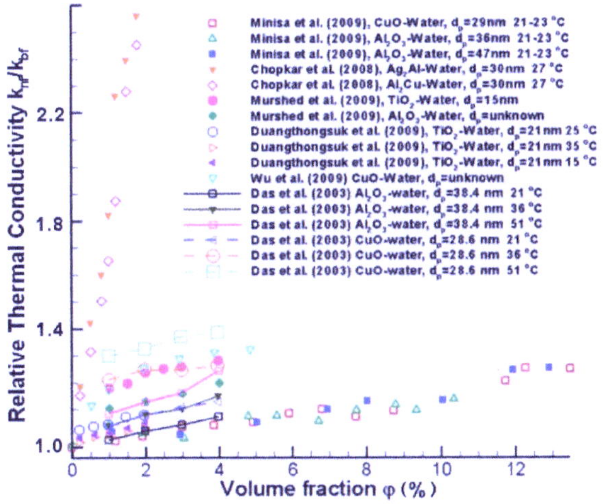

(b)

Figure 1: Experimental data for the relationship between k_{nf} and volume fraction. See refs. [14, 16, 19, 23, 26, 32, 46–48, 53, 87, 88].

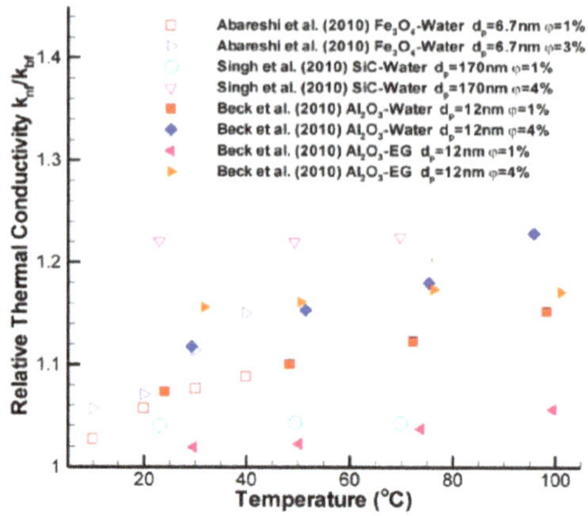

(a)

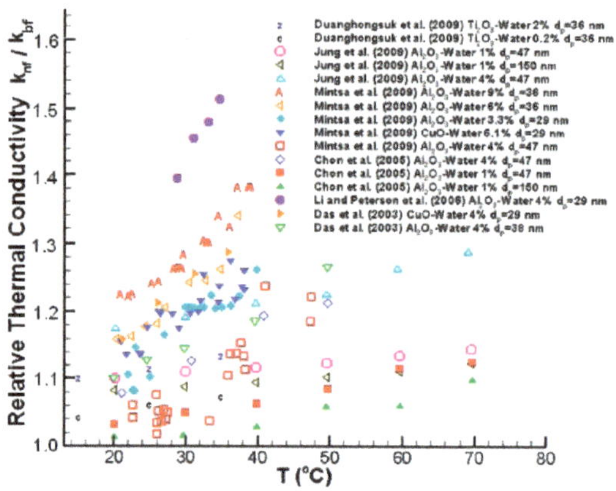

(b)

Figure 2: Experimental data for the relationship between k_{nf} and temperature. See refs. [14 , 16, 26, 44, 48, 57, 63, 89, 90].

In order to provide some physical insight, benchmark experimental data sets obtained in 2010 as well as before 2010 are displayed in Figures 1 and 2. Specifically, Figure 1a,b demonstrate that k_{nf} increases with nanoparticle volume fraction. This is because of a number of interactive mechanisms, where Brownian-motion-induced micro-mixing is arguably the most important one when uniformly distributed nanoparticles can be assumed. Figure 2a,b indicate that k_{nf} also increases with nanofluid bulk temperature. Such a relationship can be derived based on kinetics theory as outlined in Theoretical studies section. The impact of nanoparticle diameter on k_{nf} is given in Figures 1 and 2 as well. Compared to older benchmark data sets [16–19], new experimental results shown in Figures 1 and 2 indicate a smaller enhancement of nanofluid thermal conductivity, perhaps because of lower experimental uncertainties.

Nevertheless, discrepancies between the data sets provided by different research groups remain.

In summary, k_{nf} is likely to improve with nanoparticle volume fraction and temperature as well as particle diameter, conductivity, and degree of aggregation, as further demonstrated in subsequent sections.

Thermal Conductivity k_{nf} vs. Volume Fraction φ

Most experimental observations of nanofluids with just small nanoparticle volume fractions showed that k_{nf} will significantly increase when compared to the base fluid. For example, Lee and Choi [42] investigated CuO-water/ethylene glycol nanofluids with particle diameters 18.6 and 23.6 nm as well as Al_2O_3-water/ethylene glycol nanofluids with particle diameters 24.4 and 38.4 nm and discovered a 20% thermal conductivity increase at a volume fraction of 4%. Wang [43] measured a 12% increase in k_{nf} for 28-nm-diameter Al_2O_3-water and 23 nm CuO-water nanofluids with 3% volume fraction. Li and Peterson [44] provided thermal conductivity expressions in terms of temperature (T) and volume fraction (φ) by using curve fitting for CuO-water and Al_2O_3-water nanofluids. For non-metallic particles, *i.e.*, SiC-water nanofluids, Xie [45] showed a k_{nf} enhancement effect. Recently, Mintsa [26] provided new thermal conductivity expressions for Al_2O_3-water and CuO-water nanofluids with particle sizes of 47, 36, and 29 nm by curve fitting their in-house experimental data obtained by the THW method. Murshed [46] measured a 27% increase in 4% TiO_2-water nanofluids with particle size 15 nm and 20% increase for Al_2O_3-water nanofluids. However, Duangthongsuk [14] reported a more moderate increase of about 14% for TiO_2-water nanofluids. Quite surprising, Moghadassi [47] observed a 50% increment of thermal conductivity for 5% CuO-monoethylene glycol (MEG) and CuO-paraffin nanofluids.

Thermal Conductivity k_{nf} vs. Temperature T

Das [16] systematically discussed the relationship between thermal conductivity and temperature for nanofluids, noting significant increases of k_{nf} (T). More recently, Abareshi et al. [48] measured the thermal conductivity of Fe_3O_4-water with the THW method and asserted that k_{nf} increases with temperature T. Indeed, from a theoretical (*i.e.*, kinetics) view-point, with the increment of the nanofluid's bulk temperature T, molecules and nanoparticles are more active and able to transfer more energy from one location to another per unit time.

In contrast, many scientists using optical measurement techniques found no anomalous effective thermal conductivity enhancement when increasing the mixture temperature [[6–9,29, 30, 37, 49], etc.]. Additionally, Tavman et al. [32] measured SiO_2-water, TiO_2-water, and Al_2O_3-water by the 3-ω method and claimed, without showing actual data points, that there is no anomalous thermal conductivity enhancement with increment of both volume fraction and temperature. Whether anomalous enhancement relationship between k_{nf} and temperature T exist or not is still open for debate.

Dependence of k_{nf} on Other Parameters

Potentially influential parameters on thermal conductivity, other than volume fraction and temperature, include pH value, type of base fluid, nanoparticle shape, degree of nanoparticle dispersion/interaction, and various additives. For example, Zhu et al. [50] showed that the pH of a nanofluid strongly affects the thermal conductivity of suspensions. Indeed, pH value influence the stability of nanoparticle suspensions and the charges of the particle surface thereby affect the nanofluid thermal conductivity. For pH equal to 8.0-9.0, the thermal conductivity of nanofluid is higher than other situations [50] Of the most common base fluids, water exhibits a higher thermal conductivity when compared to ethylene glycol (EG) for the same nanoparticle volume fraction [43, 44, 51–53]. However, thermal conductivity enhancement of EG-based nanofluids is stronger than for water-based nanofluids [42, 43]. Different particle shapes may also influence the thermal conductivity of nanofluids. Nanoparticles with high aspect ratios seem to enhance the thermal conductivity further. For example, spherical particles show slightly less enhancement than those containing nanorods [54], while the thermal conductivity of CuO-water-based nanofluids containing shuttle-like-shaped CuO nanoparticles is larger than those for CuO nanofluids containing nearly spherical CuO nanoparticles [55]. Another parameter influencing nanofluid thermal conductivity is particle diameter. Das [16], Patel [56] and Chon [57] showed the inverse dependence of particle size on thermal conductivity enhancement, considering three sizes of alumina nanoparticles suspended in water. Beck et al. [58] and Moghadassi et al. [47] reported that the thermal conductivity will increase with the decrease of nanoparticle diameters. However, Timofeeva et al. [53] reported that k_{nf} increases with the increment of nanoparticle diameter for SiC-water nanofluids without publishing any data. Other factors which may influence the thermal conductivity of nanofluids are sonification time [32] and/or surfactant mass fraction [32] to obtain well-dispersed nanoparticles.

For other new experimental data, Wei X. et al. [59] reported nonlinear correlation between k_{nf} and synthesis parameters of nanoparticles as well as temperature T. Li and Peterson [60] showed natural convection deterioration with increase in nanoparticle volume fraction. This may be because the nanoparticle's Brownian motion smoothen the temperature gradient leading to the delay of the onset of natural convection. Also, higher viscosity of nanofluids can also induce such an effect. Wei et al. [61] claimed that the measured apparent thermal conductivity show time-dependent characteristics within 15 min when using the THW method. They suggested that measurements should be made after 15 min in order to obtain accurate data. Chiesa et al. [23] investigated the impact of the THW apparatus orientation on thermal conductivity measurements; however, that aspect was found not to be significant. Shalkevich et al. [62] reported no abnormal thermal conductivity enhancement for 0.11% and 0.00055% of gold nanoparticle suspensions, which are rather low volume fractions. Beck et al. [63] and Teng et al. [15] provided curve-fitted results based on their in-house experimental data, reflecting correlations between k_{nf} and several parameters, *i.e.*, volume fraction, bulk temperature and particle size. Both models are easy to use for certain types of nanofluids. Ali et al. [27] proposed hot wire-laser probe beam method to measure nanofluid thermal conductivity and confirmed that particle clustering has a significant effect on thermal conductivity enhancement.

THEORETICAL STUDIES

Significant differences among published experimental data sets clearly indicate that some findings were inaccurate. Theoretical analyses, mathematical models, and associated computer simulations may provide additional physical insight which helps to explain possibly anomalous enhancement of the thermal conductivity of nanofluids.

CLASSICAL MODELS

The static model of Maxwell [1] has been used to determine the effective electrical or thermal conductivity of liquid-solid suspensions of monodisperse, low-volume-fraction mixtures of spherical particles. Hamilton and Crosser [41] extended Maxwell's theory to non-spherical particles. For other classical models, please refer to Jeffery [64], Davis [65] and Bruggeman [66] as summarized in Table 2. The classical models originated from continuum formulations which typically involve only the particle size/shape and volume fraction and assume diffusive heat transfer in both fluid and solid phases [67]. Although they can give good predictions for micrometer or larger-size multiphase systems, the classical models usually underestimate the enhancement of thermal conductivity increase of *nanofluids* as a function of volume fraction. Nevertheless, stressing that nanoparticle aggregation is the major cause of k_{nf} enhancement, Eapen et al. [5] revived Maxwell's lower and upper bounds for the thermal conductivities of dilute suspensions (see also the derivation by Hashin and Shtrikman [68]). While for the lower bound, it is assumed that heat conducts through the mixture path where the nanoparticles are well dispersed, the upper bound is valid when connected/interacting nanoparticles are the dominant heat conduction pathway. The effect of particle contact in liquids was analyzed by Koo et al. [69], *i.e.*, actually for CNTs, and successfully compared to various experimental data sets. Their stochastic model considered the CNT-length as well as the number of contacts per CNT to explain the nonlinear behavior of k_{nf} with volume fraction.

Table 2: Classical models for effective thermal conductivity of mixtures

Models	Expressions	Remarks
Maxwell	$\frac{k_{nf}}{k_{bf}} = 1 + \frac{3(k_p/k_{bf} - 1)\varphi}{(k_p/k_{bf} + 2) - (k_p/k_{bf} - 1)\varphi}$	Spherical particles
Hamilton-Crosser	$\frac{k_{nf}}{k_{bf}} = 1 + \frac{k_p/k_{bf} + (n-1) - (n-1)(1 - k_p/k_{bf})\varphi}{k_p/k_{bf} + (n-1) + (1 - k_p/k_{bf})\varphi}$	$n = 3$ for spheres $n = 6$ for cylinders
Jeffrey	$\frac{k_{nf}}{k_{bf}} = 1 + 3\left(\frac{k_p/k_{bf} - 1}{k_p/k_{bf} + 2}\right)\varphi + \left(3\left(\frac{k_p/k_{bf} - 1}{k_p/k_{bf} + 2}\right)^2 + \frac{3}{4}\left(\frac{k_p/k_{bf} - 1}{k_p/k_{bf} + 2}\right)^2 + \frac{9}{16}\left(\frac{k_p/k_{bf} - 1}{k_p/k_{bf} + 2}\right)^3\left(\frac{k_p/k_{bf} + 2}{2k_p/k_{bf} + 3}\right)\cdots\right)\varphi^2$	Spherical particles
Davis	$\frac{k_{nf}}{k_{bf}} = 1 + \frac{3(k_p/k_{bf} - 1)\varphi}{(k_p/k_{bf} + 2) - (k_p/k_{bf} - 1)\varphi}(\varphi + f(k_p/k_{bf})\varphi^2 + O(\varphi^3))$	High-order terms represent pair interaction of randomly dispersed sphere
Lu-Lin	$\frac{k_{nf}}{k_{bf}} = 1 + (k_p/k_{bf})\varphi + b\varphi^2$	Spherical and non-spherical particles

DYNAMICAL MODELS AND COMPARISONS WITH EXPERIMENTAL DATA

When using the classical models, it is implied that the nanoparticles are stationary to the base fluid. In contrast, dynamic models are taking the effect of the nanoparticles' random motion into account, leading to a "micro-mixing" effect [70]. In general, anomalous thermal conductivity enhancement of nanofluids may be due to:

- Brownian-motion-induced micro-mixing;
- heat-resistance lowering liquid-molecule layering at the particle surface;
- higher heat conduction in metallic nanoparticles;
- preferred conduction pathway as a function of nanoparticle shape, *e.g.*, for carbon nanotubes;
- augmented conduction due to nanoparticle clustering.

Up front, while the impact of micro-scale mixing due to Brownian motion is still being debated, the effects of nanoparticle clustering and preferred conduction pathways also require further studies.

Oezerinc et al. [71] systematically reviewed existing heat transfer mechanisms which can be categorized into conduction, nano-scale convection and/or near-field radiation [22], thermal waves propagation [67, 72], quantum mechanics [73], and local thermal non-equilibrium [74].

For a better understanding of the micro-mixing effect due to Brownian motion, the works by Leal [75] and Gupte [76] are of interest. Starting with the paper by Koo and Kleinstreuer [70], several models stressing the Brownian motion effect have been published [22]. Nevertheless, that effect leading to micro-mixing was dismissed by several authors. For example, Wang [43] compared Brownian particle diffusion time scale and heat transfer time scale and declared that the effective thermal conductivity enhancement due to Brownian motion (including particle rotation) is unimportant. Keblinski [77] concluded that the heat transferred by nanoparticle diffusion contributes little to thermal conductivity enhancement. However, Wang [43] and Keblinski [77] failed to consider the surrounding fluid motion induced by the Brownian particles.

Incorporating indirectly the Brownian-motion effect, Jang and Choi [78] proposed four modes of energy transport where random nanoparticle motion produces a convection-like effect at the nano-scale. Their effective thermal conductivity is written as:

$$k_{\mathrm{nf}} = k_{\mathrm{bf}}(1-\varphi) + k_{\mathrm{p}}\varphi + 3C_1 \frac{d_{\mathrm{bf}}}{d_{\mathrm{p}}} k_{\mathrm{bf}} \mathrm{Re}_{d_{\mathrm{p}}} \Pr \varphi \qquad (5)$$

where C_1 is an empirical constant and d_{bf} is the base fluid molecule diameter. Re_{dp} is the Reynolds number, defined as:

$$\mathrm{Re}_{d_{\mathrm{p}}} = \frac{\bar{V}_{\mathrm{p}} \cdot d_{\mathrm{p}}}{\upsilon_{\mathrm{bf}}} \qquad (6)$$

with

$$\bar{V}'_{p} = \frac{D}{\lambda_{bf}} = \frac{\kappa_{Boltzmann} T}{3\pi \mu_{bf} d_{p}} \quad (7)$$

where D is the nanoparticle diffusion coefficient, $\kappa_{Boltzmann}$ = 1.3807e-23 J/K is the Boltzmann constant, $\bar{V}'_p$ is the root mean square velocity of particles and λ_{bf} is the base fluid molecular mean free path. The definition of $\bar{V}'_p$ (see Eq. 7b) is different from Jang and Choi's 2006 model [79]. The arbitrary definitions of the coefficient "random motion velocity" brought questions about the model's generality [78]. Considering the model by Jang and Choi [78], Kleinstreuer and Li [80] examined thermal conductivities of nanofluids subject to different definitions of "random motion velocity". The results heavily deviated from benchmark experimental data (see Figure3a,b), because there is no accepted way for calculating the random motion velocity. Clearly, such a rather arbitrary parameter is not physically sound, leading to questions about the model's generality [80].

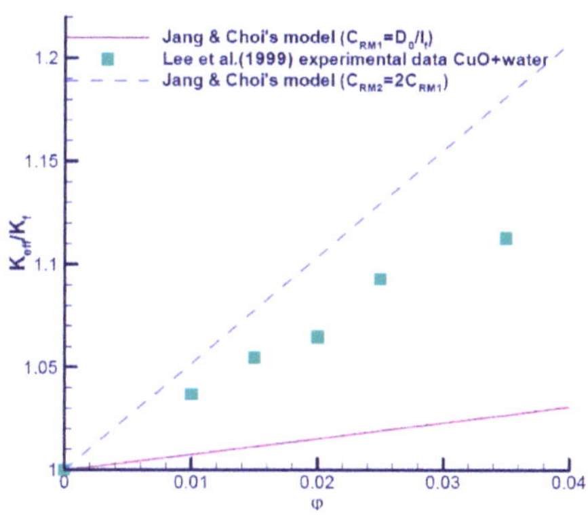

(a)

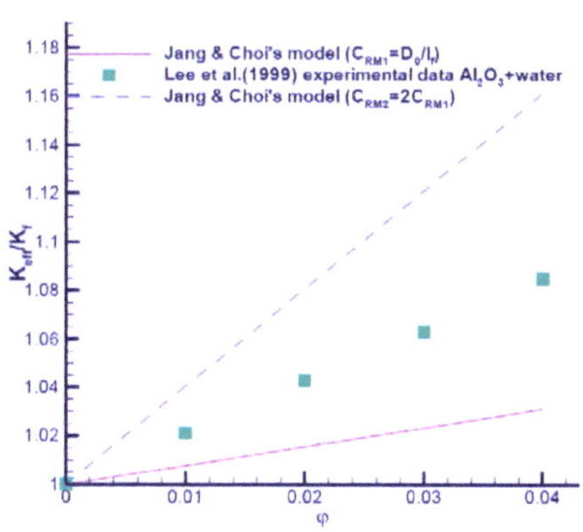

(b)

Figure 3: Comparison of experimental data. (**a**) Comparison of the experimental data for CuO-water nanofluids with Jang and Choi's model [78] for different random motion velocity definitions [80]. (**b**) Comparison of the experimental data for Al_2O_3-water nanofluids with Jang and Choi's model [78] for different random motion velocity definitions [80].

Prasher [81] incorporated semi-empirically the random particle motion effect in a multi-sphere Brownian (MSB) model which reads:

$$\frac{k_{nf}}{k_{bf}} = \left(1 + A\mathrm{Re}^{m}\mathrm{Pr}^{0.333}\varphi\right) \times \left(\frac{\left[k_p(1+2\alpha)+2k_m\right]+2\varphi\left[k_p(1-2\alpha)-k_m\right]}{\left[k_p(1+2\alpha)+2k_m\right]-\varphi\left[k_p(1-2\alpha)-k_m\right]}\right) \tag{8}$$

Here, Re is defined by Eq. 7a, $\alpha = 2R_b k_m/d_p$ is the nanoparticle Biot number, and $R_b = 0.77 \times 10^{-8}$ Km²/W for water-based nanofluids which is the so-called thermal interface resistance, while A and m are empirical constants. As mentioned by Li [82] and Kleinstreuer and Li [80], the MSB model fails to predict the thermal conductivity enhancement trend when the particle are too small or too large. Also, because of the need for curve-fitting parameters A and m, Prasher's model lacks generality (Figure 4).

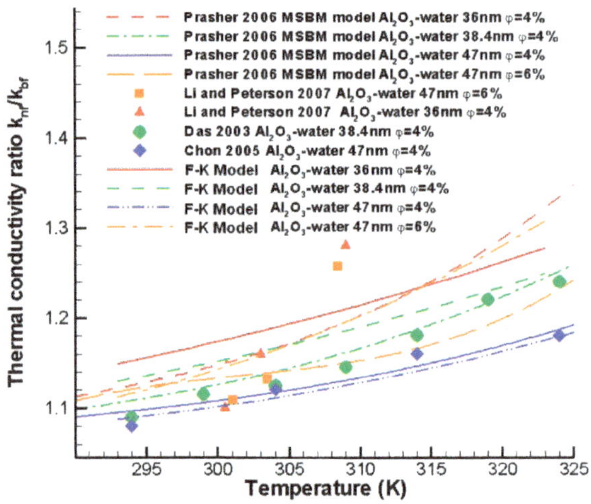

Figure 4: **Comparisons between Prasher's model** [81], **the F-K model** [86], **and benchmark experimental data** [16, 44, 57].

Kumar [83] proposed a "moving nanoparticle" model, where the effective thermal conductivity relates to the average particle velocity which is determined by the mixture temperature. However, the solid-fluid interaction effect was not taken into account.

Koo and Kleinstreuer [70] considered the effective thermal conductivity to be composed of two parts:

$$k_{nf} = k_{static} + k_{Brownian} \tag{9}$$

where k_{static} is the static thermal conductivity after Maxwell [1], *i.e.*,

$$\frac{k_{static}}{k_{bf}} = 1 + \frac{3\left(\frac{k_p}{k_{bf}} - 1\right)\cdot\varphi}{\left(\frac{k_d}{k_{bf}} + 2\right) - \left(\frac{k_d}{k_{bf}} - 1\right)\varphi} \tag{10}$$

Now, $k_{Brownian}$ is the enhanced thermal conductivity part generated by midro-scale convective heat transfer of a particle's Brownian motion and affected ambient fluid motion, obtained as Stokes flow around a sphere. By introducing two empirical functions β and f, Koo [84] combined the interaction between nanoparticles as well as temperature effect into the model and produced:

$$k_{\text{Brownian}} = 5 \times 10^4 \beta\varphi(\rho c_p)_{bf} \times \sqrt{\frac{\kappa_B T}{\rho_p d_p}} f(T,\varphi) \tag{11}$$

Li [82] revisited the model of Koo and Kleinstreuer (2004) and replaced the functions β and$f(T,\varphi)$ with a new g-function which captures the influences of particle diameter, temperature and volume fraction. The empirical g-function depends on the type of nanofluid [82]. Also, by introducing a thermal interfacial resistance R_f = 4e - 8 km^2/W the original k_p in Eq. 10 was replaced by a new $k_{p,eff}$ in the form:

$$R_f + \frac{d_p}{k_p} = \frac{d_p}{k_{p,eff}} \tag{12}$$

Finally, the KKL (Koo-Kleinstreuer-Li) correlation is written as:

$$k_{\text{Brownian}} = 5 \times 10^4 \varphi(\rho c_p)_{bf} \times \sqrt{\frac{\kappa_B T}{\rho_p d_p}} g(T,\varphi,d_p) \tag{13}$$

where $g(T,\varphi,d_p)$ is:

$$g(T,\varphi,d_p) = \left(a + b\ln(d_p) + c\ln(\varphi) + d\ln(\varphi)\ln(d_p) + e\ln(d_p)^2\right)\ln(T) + \left(g + h\ln(d_p) + i\ln(\varphi) + j\ln(\varphi)\ln(d_p) + k\ln(d_p)^2\right) \tag{14}$$

The coefficients *a*-*k* are based on the type of particle-liquid pairing [82]. The comparison between KKL model and benchmark experimental data are shown in Figure 5.

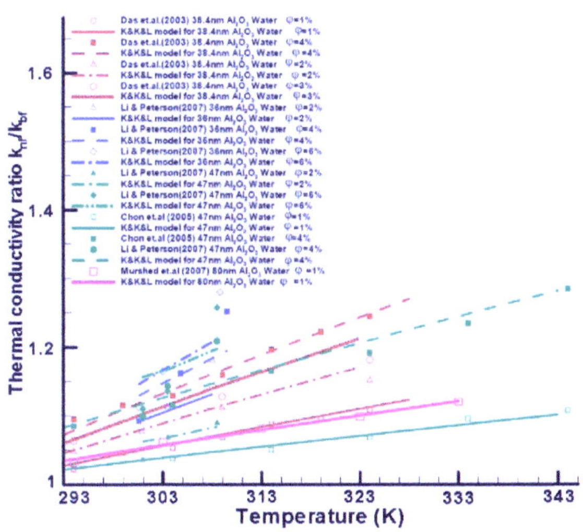

Figure 5: Comparisons between KKL model and benchmark experimental data [82].

In a more recent paper dealing with the Brownian motion effect, Bao [85] also considered the effective thermal conductivity to consist of a static part and a Brownian motion part. In a deviation from the KKL model, he assumed the velocity of the nanoparticles to be constant, and hence treated the ambient fluid around nanoparticle as steady flow. Considering convective heat transfer through the boundary of the ambient fluid, which follows the same concept as in the KKL model, Bao [85] provided an expression for Brownian motion thermal conductivity as a function of volume fraction φ, particle Brownian motion velocity v_p and Brownian motion time interval τ. Bao asserted that the fluctuating particle velocity v_p can be measured and τ can be expressed via a velocity correlation function based on the stochastic process describing Brownian motion. Unfortunately, he did not consider nanoparticle interaction, and the physical interpretation of $R(t)$ is not clear. The comparisons between Bao's model and experimental data are shown in Figure 6. For certain sets of experimental data, Bao's model shows good agreement; however, it is necessary to select a proper value of a matching constant M which is not discussed in Bao [85].

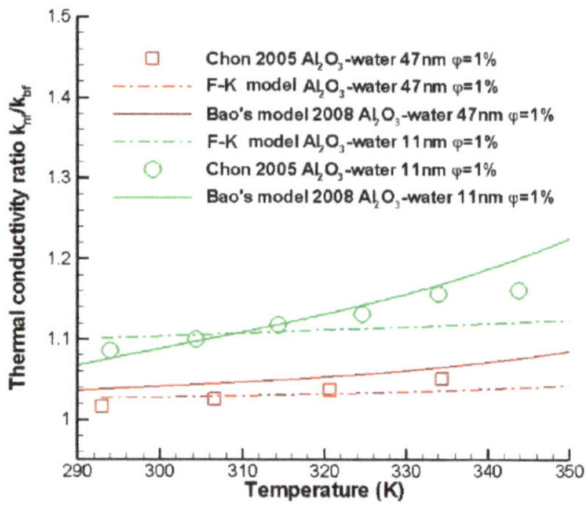

Figure 6: Comparisons between Bao's model, F-K model and benchmark experimental data.

Feng and Kleinstreuer [86] proposed a new thermal conductivity model (labeled the F-K model for convenience). Enlightened by the turbulence concept, *i.e.*, just random quantity fluctuations which can cause additional fluid mixing and not turbulence structures such as diverse eddies, an analogy was made between random Brownian-motion-generated fluid-cell fluctuations and turbulence. The extended Langevin equation was employed to take into account the inter-particle potentials, Stokes force, and random force.

$$m_p \frac{d\vec{v}_p}{dt} = -\nabla \left(\Psi_{LD} + \Psi_{Rep}\right) - \vec{F}_{\text{Stokes}} + \vec{F}_B(t) \tag{15}$$

Combining the continuity equation, momentum equations and energy equation with Reynolds decompositions of parameters, *i.e.*, velocity and temperature, the F-K model can be expressed as:

$$k_{nf} = k_{static} + k_{mm} \quad (16)$$

The static part is given by Maxwell's model [1], while the micro-mixing part is given by:

$$k_{mm} = 49500 \cdot \frac{\kappa_B \tau_p}{2m_p} \cdot C_c \cdot (\rho c_p)_{nf} \cdot \varphi^2 \cdot (T \ln T - T) \cdot$$
$$\exp(-\zeta \omega_n \tau_p) \sinh\left(\sqrt{\frac{(3\pi\mu_{bf}d_p)^2}{4m_p^2} - \frac{K_{p\_p}}{m_p}}\,\frac{m_p}{3\pi\mu_{bf}d_p}\right) \Bigg/ \left(\tau_p\sqrt{\frac{(3\pi\mu_{bf}d_p)^2}{4m_p^2} - \frac{K_{p\_p}}{m_p}}\right) \quad (17)$$

The comparisons between the F-K model and benchmark experimental data are shown in Figures 4, 6, 7a,b. Figure 7a also provides comparisons between F-K model predictions and two sets of newer experimental data [26, 32]. The F-K model indicates higher k_{nf} trends when compared to data by Tavman and Turgut [32], but it shows a good agreement with measurements by Mintsa et al. [26]. The reason may be that the volume fraction of the nanofluid used by Tavman and Turgut [32] was too small, *i.e.*, less than 1.5%. Overall, the F-K model is suitable for several types of metal-oxide nanoparticles ($20 < d_p < 50$ nm) in water with volume fractions up to 5%, and mixture temperatures below 350 K.

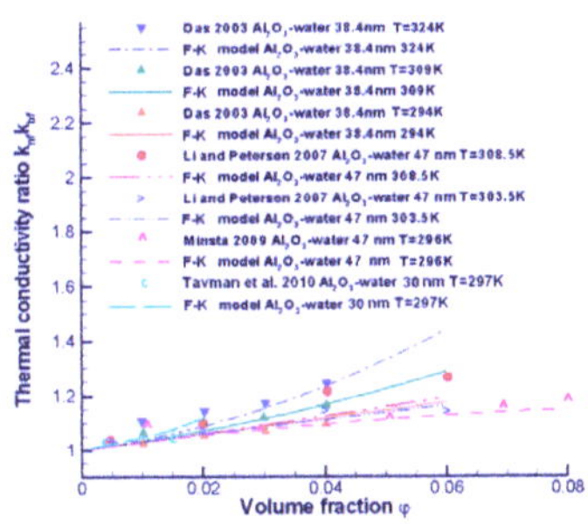

(a)

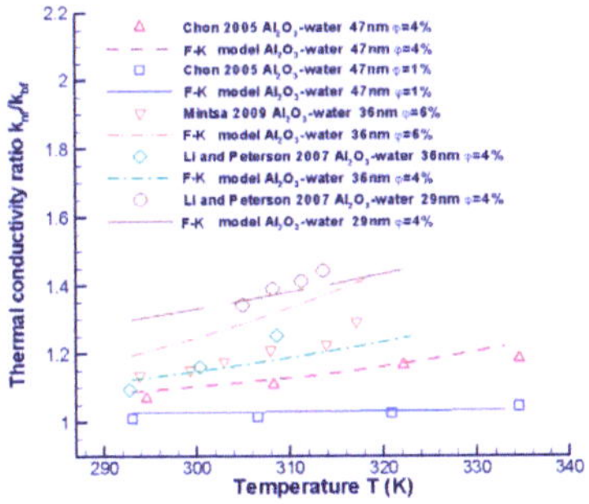

(b)

Figure 7: Comparisons between the F-K model and benchmark experimental data.

SUMMARY AND FUTURE WORK

Nanofluids, *i.e.*, well-dispersed metallic nanoparticles at low volume fractions in liquids, enhance the mixture's thermal conductivity over the base-fluid values. Thus, they are potentially useful for advanced cooling of micro-systems. Still, key questions linger concerning the best nanoparticle-and-liquid pairing and conditioning, reliable measurements of achievablek_{nf} values, and easy-to-use, physically sound computer models which fully describe the particle dynamics and heat transfer of nanofluids. At present, experimental data and measurement methods are lacking consistency. In fact, debates are still going on whether the anomalous enhancement is real or not, and what are repeatable correlations between k_{nf} and temperature, nanoparticle size/shape, and aggregation state. Clearly, additional benchmark experiments are needed, using the same nanofluids subject to different measurement methods as well as variations in nanofluid characteristics. This would validate new, minimally intrusive techniques and verify the reproducibility of experimental results.

Concerning theories/correlations which try to explain thermal conductivity enhancement for all nanofluids, not a single model can predict a wide range of experimental observations. However, many experimental data sets may fit between the lower and upper mean-field bounds originally proposed by Maxwell [1], where the static nanoparticle configurations may range between the two extremes of a dispersed phase to a continuous phase. Dynamic k_{nf} models postulate an enhancement above the classic Maxwell theory and thereby provide additional physical insight. Clearly, it will be necessary to consider not only one possible mechanism but combine several mechanisms and compare predictive results to new benchmark experimental data sets.

Competing Interests

The authors declare that they have no competing interests.

Authors' Contributions

YF conducted the extensive literature review and CK wrote the article. Both authors read and approved the final manuscript.

REFERENCES

1. Maxwell JC: *A Treatise on Electricity and Magnetism*. Oxford: Clarendon; 1891.
2. Masuda H, Ebata A, Teramea K, Hishinuma N: Alteration of thermal conductivity and viscosity of liquid by dispersing ultra-fine particles. *Netsu Bussei* 1993, 4: 227–233.
3. Eastman JA, Choi US, Li S, Thompson LJ, Lee S: Enhanced thermal conductivity through the development of nanofluids. In *Nanophase and Nanocomposite Materials II*. Edited by: Komarneni S, Parker JC, Wollenberger HJ. Pittsburg: Materials Research Society; 1997:3–11.
4. Yu WH, France DM, Routbort JL, Choi SUS: Review and comparison of nanofluid thermal conductivity and heat transfer enhancements. *Heat Transfer*

Engineering 2009, 29: 432–460. 10.1080/01457630701850851

5. Eapen J, Rusconi R, Piazza R, Yip S: The classical nature of thermal conduction in nanofluids. *Journal of Heat Transfer* 2010., 132: 102402–1-102402–14. 102402-1-102402-14. 10.1115/1.4001304

6. Rusconi R, Rodari E, Piazza R: Optical measurements of the thermal properties of nanofluids. *Applied Physics Letters* 2006., 89: 261916–1-2619163. 261916-1-2619163. 10.1063/1.2425015

7. Putnam SA, Cahill DG, Braun PV: Thermal conductivity of nanoparticle suspensions.*Journal of Applied Physics* 2006., 99: 084308–1-084308–6. 084308-1-084308-6. 10.1063/1.2189933

8. Venerus DC, Kabadi MS, Lee S, Perez-Luna V: Study of thermal transport in nanoparticle suspensions using forced Rayleigh scattering. *Journal of Applied Physics* 2006., 100: 094310–1-094310–5. 094310-1-094310-5. 10.1063/1.2360378

9. Buongiorno J, Venerus DC, Prabhat N: A benchmark study on the thermal conductivity of nanofluids. *Journal of Applied Physics* 2009., 106: 094312–1-094312–14. 094312-1-094312-14. 10.1063/1.3245330

10. Shaikh S, Lafdi K: Thermal conductivity improvement in carbon nanoparticle doped PAO oil: An Experimental Study. *Journal of Applied Physics* 101: 064302–1-064302–7. 064302-1-064302-7.

11. Bazan JAN: *Thermal conductivity of poly-aelpha-olefin (PAO)-based nanofluids. Ph.D. Thesis, University of Dayton, Dayton, OH, USA*. 2010.

12. Yoo DH, Hong KS, Yang HS: Study of thermal conductivity of nanofluids for the application of heat transfer fluids. *Thermochim Acta* 2007, 455.

13. Schmidt AJ, Chiesa M, Torchinsky DH, Johnson JA, Nelson KA, Chen G: Thermal conductivity of nanoparticle suspension in insulating media measured with a transient optical grating and a hotwire. *Journal of Applied Physics* 2008., 103: 083529–1-083529–5. 083529-1-083529-5.

14. Duangthongsuk W, Wongwises S: Measurement of temperature-dependent thermal conductivity and viscosity of TiO_2**-water nanofluids.** *Experimental Thermal and Fluid Science* 2009, 33: 706–714. 10.1016/j.expthermflusci.2009.01.005

15. Teng TP, Hung YH, Teng TC, Mo HE, Hsu HG: The effect of alumina/ water nanofluid particle size on thermal conductivity. *Applied Thermal Engineering* 2010, 30: 2213–2218. 10.1016/j.applthermaleng.2010.05.036

16. Das SK, Putra N, Theisen P, Roetzel W: Temperature dependence of thermal conductivity enhancement for nanofluids. *Journal of Heat Transfer* 2003, 125: 567–574. 10.1115/1.1571080

17. Czarnetzki W, Roetzel W: Temperature oscillation techniques for simultaneous measurement of thermal diffusivitiy and conductivity. *International Journal of Thermophysics* 1995, 16: 413–422. 10.1007/BF01441907

18. Choi TY, Maneshian MH: Measurement of the thermal conductivity of a water-based single-wall carbon nanotube colloidal suspension with a modified 3-w

method.*Nanotechnology* 2009., 20: 315706–1-315706–6. 315706-1-315706-6.

19. Paul G, Chopkar M, Manna I, Das PK: Techniques for measuring the thermal conductivity of nanofluids: a review. *Renewable and Sustainable Energy Reviews* 2010, 14: 1913–1924. 10.1016/j.rser.2010.03.017
20. Feng Y: *A new thermal conductivity model for nanofluids with convection heat transfer application. MS thesis, North Carolina State University, Raleigh, NC, USA*. 2010.
21. Vadasz P: Rendering the transient hot wire experimental method for thermal conductivity estimation to two-phase systems-theoretical leading order results. *Journal of Heat Transfer* 2010., 132: 081601–1-081601–7. 081601-1-081601-7.
22. Das SK, Choi SUS, Yu W, Pradeep T: *Nanofluids: Science and Technology*. New Jersey: Wiley; 2008.
23. Chiesa M, Simonsen AJ: The importance of suspension stability for hot-wire measurements of thermal conductivity of colloidal suspensions. *16th Australasian Fluid Mechanics Conference, Gold Coast, Australia* 2010.
24. Zhang X, Gu H, Fujii M: Experimental study on the effective thermal conductivity and thermal diffusivity of nanofluid. *AIAA Journal* 2006, 41: 831–840. 10.2514/2.2044
25. Woodfield PL: A two-dimensional analytical solution for the transient short-hot-wire method. *International Journal of Thermophysics* 2008, 29: 1278–1298. 10.1007/s10765-008-0469-y
26. Mintsa HA, Roy G, Nguyen CT, Doucet D: New temperature dependent thermal conductivity data for water-based nanofluids. *International Journal of Thermal Sciences*2009, 48: 363–371. 10.1016/j.ijthermalsci.2008.03.009
27. Ali FM, Yunus WMM, Moksin MM, Talib ZA: The effect of volume fraction concentration on the thermal conductivity and thermal diffusivity of nanofluids: numerical and experimental. *Review of Scientific Instruments* 2010., 81: 074901–1-074901–9. 074901-1-074901-9. 10.1063/1.3458011
28. Patel HE, Sundararajan T, Das SK: An experimental investigation into the thermal conductivity enhancement in oxide and metallic nanofluids. *Journal of Nanoparticle Research* 2010, 12: 1015–1031. 10.1007/s11051-009-9658-2
29. Ju YS, Kim J, Hung MT: Experimental study of heat conduction in aqueous suspensions of aluminum oxide nanoparticles. *Journal of Heat Transfer* 2008., 130: 092403–1-092403–6. 092403-1-092403-6.
30. Li CH, Williams W: Transient and steady-state experimental comparison study of effective thermal conductivity of Al_2O_3**-water nanofluids.** *Journal of Heat Transfer* 2008., 130: 042407–1-042407–7. 042407-1-042407-7.
31. Turgut A, Tavman I, Chirtoc M, Schuchmann HP, Sauter C, Tavman S: Thermal conductivity and viscosity measurements of water-based TiO2 nanofluids. *International Journal of Thermophysics* 2009, 30: 1213–1226. 10.1007/s10765-009-0594-2

32. Tavman I, Turgut A: An investigation on thermal conductivity and viscosity of water based nanofluids. *Microfluidics Based Microsystems* 2010, 0: 139–162. full_text

33. Iygengar AS, Abramson AR: Comparative radial heat flow method for thermal conductivity measurement of liquids. *Journal of Heat Transfer* 2009., 131: 064502–1-064502–3. 064502-1-064502-3.

34. Kusiak A, Pradere C, Battaglia JL: Measuring the thermal conductivity of liquids using photo-thermal radiometry. *Measurement Science and Technology* 2010., 21: 015403–1-015403–6. 015403-1-015403-6. 10.1088/0957-0233/21/1/015403

35. Rousan AA, Roy DM: A thermal comparator method for measuring thermal conductivity of cementitious materials. *Industrial and Engineering Chemistry Product Research and Development* 1983, 22: 349–351. 10.1021/i300010a035

36. Lee JH: *Convection Performance of Nanofluids for Electronics Cooling, Ph. D. Dissertation*. Stanford University, CA, USA; 2009.

37. Kolade B, Goodson KE, Eaton JK: Convective performance of nanofluids in a laminar thermally developing tube flow. *Journal of Heat Transfer* 2009., 131: 052402–1-052402–8. 052402-1-052402-8. 10.1115/1.3013831

38. Rusconi R, Isa L, Piazza R: Thermal-lensing measurement of particle thermophoresis in aqueous dispersions. *Journal of Optical Society of America* 2004, 21(3):605–616. 10.1364/JOSAB.21.000605

39. Venerus DC, Schieber JD, Iddir H, Guzman JD, Broerman AW: Measurement of thermal diffusivity in polymer melts using forced rayleigh light scattering. *Journal of Polymer Science: Part B: Polymer Physics* 1999, 37: 1069–1078. 10.1002/(SICI)1099-0488(19990601)37:11<1069::AID-POLB3>3.0.CO;2-U

40. 4Putnam SA, Cahill DG: Micron-scale apparatus for measurements of thermodiffusion in liquids. *Review of Scientific Instruments* 2004, 75(7):2368–2372. 10.1063/1.1765761

41. Hamilton RL, Crosser OK: Thermal conductivity of heterogeneous two-component systems. *Industrial Engineering and Chemistry Fundamentals* 1962, 1(3):187–191. 10.1021/i160003a005

42. Lee S, Choi SUS: Measuring thermal conductivity of fluids containing oxide nanoparticles.*Journal of Heat Transfer* 1999, 121: 280–289. 10.1115/1.2825978

43. Wang XW, Xu XF, Choi SUS: Thermal conductivity of nanoparticle-fluid mixture. *Journal of Thermalphysics and Heat Transfer* 1999, 13: 474–480. 10.2514/2.6486

44. Li CH, Peterson GP: Experimental investigation of temperature and volume fraction variations on the effective thermal conductivity of nanoparticle suspensions (nanofluids).*Journal of Applied Physics* 2006, 99(8):084314. 10.1063/1.2191571

45. Xie H, Wang J, Xi T, Liu Y: Thermal conductivity of suspensions containing nanosized SiC particles. *International Journal of Thermophysics* 2002, 23(2):571–580. 10.1023/A:1015121805842

46. Murshed SMS, Leong KC, Yang C: A combined model for the effective thermal conductivity of nanofluids. *Applied Thermal Engineering* 2009, 29: 2477–2483. 10.1016/j.applthermaleng.2008.12.018

47. Moghadassi AR, Hosseini SM, Henneke DE: Effect of CuO nanoparticles in enhancing the thermal conductivities of monoethylene glycol and paraffin fluids. *Industrial Engineering and Chemistry Research* 2010, 49: 1900–1904. 10.1021/ie901060e

48. Abareshi M, Goharshiadi EK, Zebarjad SM, Fadafan HK, Youssefi A: Fabrication, characterization and measurement of thermal conductivity of Fe_3O_4nanofluids. *Journal of Magnetism and Magnetic Materials* 2010, 322(24):3895–3901. 10.1016/j.jmmm.2010.08.016

49. Williams W, Buongiorno J, Hu LW: Experimental investigation of turbulent convective heat transfer and pressure loss of alumina/water and zirconia/water nanoparticle colloids (nanofluids) in horizontal tubes. *Journal of Heat Transfer* 2008., 130: 042412–1-042412–7. 042412-1-042412-7. 10.1115/1.2818775

50. Zhu D, Li X, Wang N, Wang X, Gao J, Li H: Dispersion behavior and thermal conductivity characteristics of Al_2O_3-**H_2O** nanofluids. *Current Applied Physics* 2009, 9: 131–139. 10.1016/j.cap.2007.12.008

51. Jang SP, Choi SUS: Role of Brownian motion in the enhanced thermal conductivity of nanofluids. *Applied Physics Letters* 2004, 84: 4316–4318. 10.1063/1.1756684

52. Timofeeva EV, Gavrilov AN, McCloskey JM, Tolmachev YV: Thermal conductivity and particle agglomeration in alumina nanofluids: experiment and theory. *Physical Review E*2007., 76: 061203–1-061203–16. 061203-1-061203-16. 10.1103/PhysRevE.76.061203

53. Timofeeva EV, Smith DS, Yu W, France DM, Singh D, Routbort JL: Particle size and interfacial effects on thermo-physical and heat transfer characteristics of water-based alpha-SiC nanofluids. *Nanotechnology* 2010., 21: 215703–1-215703–10. 215703-1-215703-10. 10.1088/0957-4484/21/21/215703

54. Murshed SMS, Leong KC, Yang C: Enhanced thermal conductivity of TiO2-water based nanofluids. *International Journal of Thermal Sciences* 2005, 44: 367. 10.1016/j.ijthermalsci.2004.12.005

55. Zhu HT, Zhang CY, Tang YM, Wang JX: Novel synthesis and thermal conductivity of CuO nanofluid. *Journal of Physical Chemistry C* 2007, 111: 1646. 10.1021/jp065926t

56. Patel HE, Das SK, Sundararajan T, Sreekumanran NA, George B, Pradeep T: Thermal conductivities of naked and monolayer protected metal nanoparticle based nanofluids: manifestation of anomalous enhancement and chemical effects. *Applied Physics Letters*2003, 83: 2931–2933. 10.1063/1.1602578

57. Chon CH, Kihm KD, Lee SP, Choi SUS: Empirical correlation finding the role of temperature and particle size for nanofluid (Al_2O_3) **thermal conductivity enhancement.** *Applied Physics Letters* 2005., 87: 153107–1-153107–3. 153107-1-153107-3. 10.1063/1.2093936

58. Beck MP, Yuan Y, Warrier P, Teja AS: The effect of particle size on the thermal conductivity of alumina nanofluids. *Journal of Nanoparticle Research* 2009, 11: 1129–1136. 10.1007/s11051-008-9500-2

59. Wei X, Kong T, Zhu H, Wang L: CuS/Cu_2S nanofluids: synthesis and thermal conductivity.*International of Heat and Mass Transfer* 2010, 53: 1841–1843. 10.1016/j.ijheatmasstransfer.2010.01.006

60. Li CH, Peterson GP: Experimental studies of natural convection heat transfer of Al_2O_3**/DI water nanoparticle suspensions (nanofluids).** *Advances in Mechanical Engineering*2010., 2010: 742739–1-742739–10. 742739-1-742739-10.

61. Wei Y, Xie H, Chen L, Li Y: Investigation on the thermal transport properties of ethylene glycol-based nanofluids containing copper nanoparticles. *Powder Technology* 2010, 197: 218–221. 10.1016/j.powtec.2009.09.016

62. Shalkevich N, Escher W, Buergi T, Michel B, Ahmed L, Poulikakos D: On the thermal conductivity of gold nanoparticle colloid. *Langmuir* 2010, 26(2):663–670. 10.1021/la9022757

63. Beck MP, Yuan Y, Warrier P, Teja AS: The thermal conductivity of alumina nanofluids in water, ethylene glycol, and ethylene glycol + water mixture. *Journal of Nanoparticles Research* 2010, 12: 1469–1477. 10.1007/s11051-009-9716-9

64. Jeffrey DJ: Conduction through a random suspension of spheres. *Proceedings of Royal Society, A* 1973, 335: 355–367. 10.1098/rspa.1973.0130

65. Davis RH: The effective thermal conductivity of a composite material with spherical inclusions. *International Journal of Thermophysics* 1986, 7: 609–620. 10.1007/BF00502394

66. Bruggeman DAG: Berechnung verschiedener physikalischer konstanten von heterogenen substanzen, I-Dielektrizitatskonstanten und leitfahigkeiten der mischkorper aus isotropen substanzen. *Annalen der Physik, Leipzig* 1935, 24: 636–679. 10.1002/andp.19354160705

67. Wang LQ, Zhou XS, Wei XH: *Heat conduction mathematical models and analytical solutions*. Berlin: Springer-Verlag; 2008.

68. Hashin Z, Shtrikman S: Conductivity of polycrystals. *Physical Review* 1963, 130: 129–133. 10.1103/PhysRev.130.129

69. Koo J, Kang Y, Kleinstreuer C: A nonlinear effective thermal conductivity model for carbon nanotube and nanofiber suspensions. *Nanotechnology* 2008., 19: 375705–1-375705–7. 375705-1-375705-7.

70. Koo J, Kleinstreuer C: A new thermal conductivity model for nanofluids. *Journal of Nanoparticle Research* 2004, 6: 577–588. 10.1007/s11051-004-3170-5

71. Oezerinc S, Kakac S, Yazicioglu AG: Enhanced thermal conductivity of nanofluids: A state-of-the-art review. *Microfluid Nanofluid* 2010, 8: 145–170. 10.1007/s10404-009-0524-4

72. Wang LQ, Fan J: Nanofluids research: key issues. *Nanoscale Research Letter* 2010, 5: 1241–1252. 10.1007/s11671-010-9638-6

73. Prevenslik T: Nanoscale Heat Transfer by Quantum Mechanics. *Fifth International*

*Conference on Thermal Engineering: Theory and Applications, Marrakesh, Morocco*2010.

74. Nield DA, Kuznetsov AV: The effect of local thermal nonequilibrium on the onset of convection in a nanofluid. *Journal of Heat Transfer* 2010., 132: 052405–1-052405–7 052405-1-052405-7
75. Leal LG: On the effective conductivity of a dilute suspension of spherical drops in the limit of low particle Peclet number. *Chem Eng Commun* 1973, 1: 21–31. 10.1080/00986447308960412
76. Gupte SK, Advani SG: Role of micro-convection due to non-affine motion of particles in a mono-disperse suspension. *Int J Heat Mass Transfer* 1995, 38(16):2945–2958. 10.1016/0017-9310(95)00060-M
77. Keblinski P, Phillpot SR, Choi SUS, Eastman JA: Mechanisms of heat flow in suspensions of nanos-sezed particles (nanofluids). *International Journal of Heat and Mass Transfer*2002, 45: 855–863. 10.1016/S0017-9310(01)00175-2
78. Jang SP, Choi SUS: Effects of various parameters on nanofluid thermal conductivity. *ASME J Heat Transfer* 2007, 129: 617–623. 10.1115/1.2712475
79. Jang SP, Choi SUS: Cooling performance of a microchannel heat sink with nanofluids. *Appl Therm Eng* 2006, 26: 2457–2463. 10.1016/j.applthermaleng.2006.02.036
80. Kleinstreuer C, Li J: Discussion: effects of various parameters on nanofluid thermal conductivity. *ASME Journal of Heat Transfer* 2008., 130: 025501–1-025501–3. 025501-1-025501-3. 10.1115/1.2812307
81. Prasher R: Brownian-motion-based convective-conductive model for the effective thermal conductivity of nanofluids. *Journal of Heat Transfer* 2006, 128: 588–595. 10.1115/1.2188509
82. Li J: *Computational analysis of nanofluid flow in microchannels with applications to micro-heat sinks and bio-MEMS, PhD Thesis*. NC State University, Raleigh, NC, the United States; 2008.
83. Kumar DH, Patel HE, Kumar VRR, Sundararajan T, Pradeep T, Das SK: Model for heat conduction in nanofluids. *Physical Review Letters* 2004., 93: 144301–1-144301–4. 144301-1-144301-4. 10.1103/PhysRevLett.93.144301
84. Koo JM: *Computational nanofluid flow and heat transfer analyses applied to micro-systems, Ph.D Thesis*. NC State University, Raleigh, NC, USA; 2005.
85. Bao Y: Thermal conductivity equations based on Brownian motion in suspensions of nanoparticles (nanofluids). *Journal of Heat Transfer* 130: 042408–1-042408–5. 042408-1-042408-5.
86. Feng Y, Kleinstreuer C: Nanofluid convective heat transfer in a parallel-disk system.*International Journal of Heat and Mass Transfer* 2010, 53: 4619–4628. 10.1016/j.ijheatmasstransfer.2010.06.031
87. Chopkar M, Sudarshan S, Das PK, Manna I: Effect of particle size on thermal conductivity of nanofluid. *Metallurgical and Materials Transactions A* 2008, 39A: 1535–1542. 10.1007/s11661-007-9444-7

88. Wu D, Zhu H, Wang L, Liu L: Critical issues in nanofluids preparation, characterization and thermal conductivity. *Current Nanoscience* 2009, 5: 103–112. 10.2174/157341309787314548

89. Singh D, Timofeeva E, Yu W, Routbort J, France D, Smith D, Lopez-Cepero JM: An investigation of silicon carbide-water nanofluid for heat transfer applications. *Journal of Applied Physics* 2009., 15: 064306–1-064306–6. 064306-1-064306-6.

90. Jung YJ, Jung YY: Thermal conductivity enhancement of nanofluids in conjunction with electrical double layer (EDL). *International Journal of Heat and Mass Transfer* 2009, 52: 525–528. 10.1016/j.ijheatmasstransfer.2008.07.016

91. Pak BC, Cho YI: Hydrodynamic and heat transfer study of dispersed fluids with submicron metallic oxide particles. *Exp Heat Transfer* 1998, 11: 151–170. 10.1080/08916159808946559

92. Li Q, Xuan Y: Convective heat transfer and flow characteristics of Cu-water nanofluid.*Science in China (Series E)* 2002, 45: 408–416.

93. Wen D, Ding Y: Experimental investigation into convective heat transfer of nanofluids at the entrance region under laminar flow conditions. *International Journal of Heat and Mass Transfer* 2004, 47: 5181–5188. 10.1016/j.ijheatmasstransfer.2004.07.012

94. Ding Y, Alias H, Wen D, Williams RA: Heat transfer of aqueous suspensions of carbon nanotubes (CNT nanofluids). *International Journal of Heat and Mass Transfer* 2006, 49: 240–250. 10.1016/j.ijheatmasstransfer.2005.07.009

95. Heris SZ, Esfahany MN, Etemad SG: Experimental investigation of convective heat transfer of Al_2O_3**/water nanofluid in circular tube.** *International Journal of Heat and Fluid Flow* 2007, 28: 203–210. 10.1016/j.ijheatfluidflow.2006.05.001

96. Rea U, McKrell T, Hu L, Buongiorno J: Laminar convective heat transfer and viscous pressure loss of alumina-water and zirconia-water nanofluids. *International Journal of Heat and Mass Transfer* 2009, 52: 2042–2048. 10.1016/j.ijheatmasstransfer.2008.10.025

97. Heris SZ, Etemad SGh, Esfahany MN: Convective heat transfer of a Cu/water nanofluid flowing through a circular tube. *Experimental Heat Transfer* 2009, 22: 217–227. 10.1080/08916150902950145

11
CHAPTER

PENETRATION LOSS MEASUREMENT AND MODELING FOR HAP MOBILE SYSTEMS IN URBAN ENVIRONMENT

Jaroslav Holis, Pavel Pechac

INTRODUCTION

Urban areas are covered by a variety of mobile wireless systems. One disadvantage of these systems is that they are vulnerable to disasters—either natural or manmade disasters such as terrorist attacks. To avoid this, terrestrial networks could potentially be complemented with high altitude platforms (HAPs) situated in the stratosphere at an altitude of about 20 km [1]. HAPs can be promptly deployed and easily change their locations. In addition, HAPs can be located in multiple deployments [2]. A great benefit of HAPs against terrestrial mobile networks is the absence of shadowing for high elevation angles. HAP stations located in the stratosphere offer noticeably lower free space loss than satellites. Another benefit when compared to satellites is the HAP position maintenance, with deviation of only 0.5 km [3]. Based on these facts, it is obvious that HAPs can be successfully used for urban outdoor coverage, but there is a question of whether HAP can also provide mobile or wireless services in general inside the buildings. Prospective mobile systems for HAPs seem to be the third-generation universal mobile telecommunication system (UMTS) operating in a frequency band of about 2.0 GHz, which is also allocated to HAPs [4], and the emerging mobile WiMAX systems [5] operating in the frequency band between 2–6 GHz. The problems of penetration loss inside the buildings is a challenging area for HAP systems. Currently, an HAP elevation-based empirical model has been published to predict penetration loss into buildings at a frequency of 2.0 GHz for high elevation angles only [6, 7]. This model applies to scenarios where the coverage of urban/suburban areas is achieved at elevation angles ranging between 55–90 degrees and the HAP is positioned above the city center. In [8], the authors studied the impact of high elevation angles in the

Citation: Jaroslav Holis, Pavel Pechac, Penetration Loss Measurement and Modeling for HAP Mobile Systems in Urban Environment, doi:10.1155/2008/543290.

propagation mechanisms; it was shown that in urban coverage scenarios by HAPs most of the buildings have at least one of the walls and the rooftop directly illuminated.

The propagation prediction model for terrestrial systems is not defined as a function of the elevation angle [9, 10], which is the crucial parameter in the case of the HAP systems. The aim of this paper is to introduce a novel empirical model of penetration loss inside buildings in urban areas as a function of the elevation angle developed based on empirical data obtained from trials using a remotecontrolled airship. The remote-controlled airship was used to measure penetration loss inside the buildings because of its perfect flight control possibilities (so that a whole range of elevation angles can be observed). The measurements were carried out in different types of buildings, at different positions, at different heights, and at frequencies of 2.0 GHz and 3.5 GHz. The ray tracing approach was not used here since the scenarios are very complicated to describe, and the aim of this paper is to develop a universal model applicable to urban areas rather than a detailed description of a single concrete situation.

Figure 1: Remote-controlled airship.

The second section of this paper describes the measurement campaign, introduces details about the airship used, the developed payload, the antennas, and the considered scenarios. The third section deals with the processing of the measured data. A novel, universal empirical model is introduced as a function of the elevation angle. During the measurements, the airship was flying over a building in various directions in order to obtain results statistically independent on the azimuth angle and so the final model is not a function of the azimuth angle. The final section discusses the applicability of a universal model at frequencies of 2.0 and 3.5 GHz. In addition, this model can be successfully used for planning of prospective HAP mobile systems in S band.

MEASUREMENT CAMPAIGN

The measurement campaign was planned in different types of buildings and using selected scenarios in the center of the city of Prague. A statistical analysis of the collected data was then performed as a function of the elevation angle.

The Remote-Controlled Airship

A remote-controlled airship [11] was utilized to carry the transmitters. This airship (see Figure 1) is designed to be low altitude, but for the trial we have presented here it is not necessary to reach the stratosphere. The key point is to obtain a sufficient range of elevation angles during the trial. This requirement excellently suits the remote-controlled airship, which offers the great benefit of flexible manoeuvrability. The main parameters of the airship used for the experiment are as follows:

(i) length 9 m,

(ii) maximum diameter 2.3 m,

(iii) payload 7 kg,

(iv) hull filling helium

The altitude is limited by visual contact with the operator at this stage of airship development. The altitude fluctuated by hundreds of metres during the experiments. Another significant benefit of the airship for the trial is the possibility it offers of fast and simple transportation.

During the measurements, the airship was flying over a building in various directions in order to obtain results statistically independent on the azimuth angle. Small movements of airship were compensated using a special stabilization platform to equalize inclinations caused by wind gusts.

Penetration Loss Measurement

The measurement campaign was planned and executed in the following way. The signal was transmitted from the airship and received inside a building. The special payload designed for this trial was mounted in the airship gondola. Continuous wave (CW) generators with an output power of 26 dBm and rectangular patch antennas were used to transmit the signal. The carrier frequencies were equal to 2.0 GHz and 3.5 GHz. These frequencies represent the central value of the carrier frequency for the third-generation mobile services UMTS and mobile WiMAX systems.

The receiver station was composed of a spectrum analyzer and a spiral broadband antenna (circular polarization) receiving the signal. The receiving antenna was situated 1.5 m above the floor and 2 m in front of a wall. The wide beamwidth antennas were utilized for this trial in order to minimize the effect of antenna radiation patterns on the interpretation of propagation effects. The possible error caused by antennas was lower than 3 dB. The position of the airship was recorded using the global positioning system (GPS), and the altitude of the airship was measured based on a barometer, which provides more accurate information concerning the altitude. The airship position determines the distance between the receiver station and the transmitter antennas. The penetration loss can then be calculated from the measured received signal level separating the impact of free space loss (FSL) and the measurement system (antenna gain, transmitted signal level, cable losses, etc.). The penetration loss is usually defined as the difference between the mean received signal strength in the surroundings of a building and the mean signal strength inside the building. In this paper, the penetration loss is defined as the additional path loss with respect to the FSL from the transmitter up to the building. Figure 2 presents an example of received signal level during the measurement at a frequency of 2.0 GHz in the ground floor of an office building during two flyovers of the airship above the building. Figure 3 illustrates the path loss in addition to FSL from the measured signal level. It is obvious from Figure 3 that the additional path loss lies in the range 10–60 dB depending on the airship's position (lower additional path loss was obtained for higher elevation angles and in a situation where the HAP directly illuminated the wall neighboring the receiver station). The very high additional path loss was measured for very low elevation angles, where there is a high probability that the signal is incoming across more walls inside the building, and other propagation effects such as shadowing and diffraction on surrounding buildings affect the final signal level.

Selected Scenarios

Measurements were carried out in four different types of buildings:

(i) an office building,

(ii) an office building with storeys higher than surrounding buildings,

(iii) a brick building,

(iv) a prefab residential building.

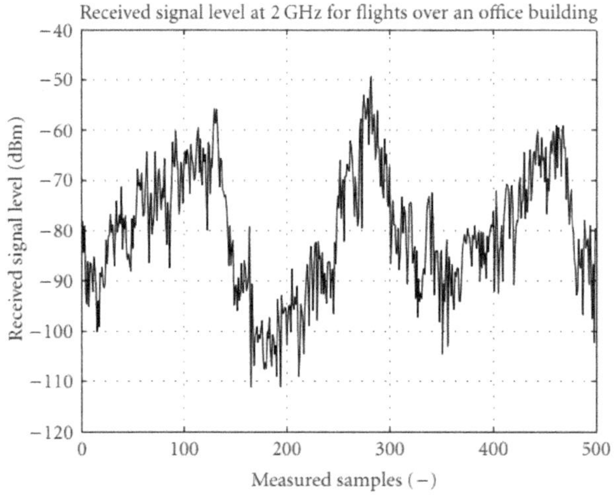

Figure 2: An example of measured received signal level at 2.0 GHz.

In addition, the receiver station was situated in different typical positions inside the buildings. It will be shown that the resulting penetration loss is crucially dependent on the receiver position within the building. One of the selected positions of the receiver station was, for example, 2 m in front of the external wall, and another in the middle of the building and distant from directly illuminated external walls, and so forth. The storey number influences the penetration loss as well, but the impact of the floor level was not as crucial as the position of the receiving station within the floor. More than 50 000 measured samples were used for further processing. About 5 samples per second were collected during measurements.

PENETRATION LOSS

This section described the processing of the measurement results. Figure 4 shows an example of the processed data measured on the second floor of an office building at 2.0 GHz. The measured samples were averaged in 1 degree step of the elevation angle. A histogram of measured data for the elevation angle of 33 degrees is shown inside Figure 4. The receiver station was situated 2 metres in front of an external wall. The dependence of the penetration loss in decibels (dB) on the elevation angle (θ) can be found by the following function:

$$L_{PL}(\theta) = \sqrt{a - b(\theta - c)^2}, \qquad (1)$$

where a, b, and c are empirical parameters.

The measurement was carried out in the same position three floors above the ground floor. All of these floors are below the rooftop level of surrounding buildings in the area.

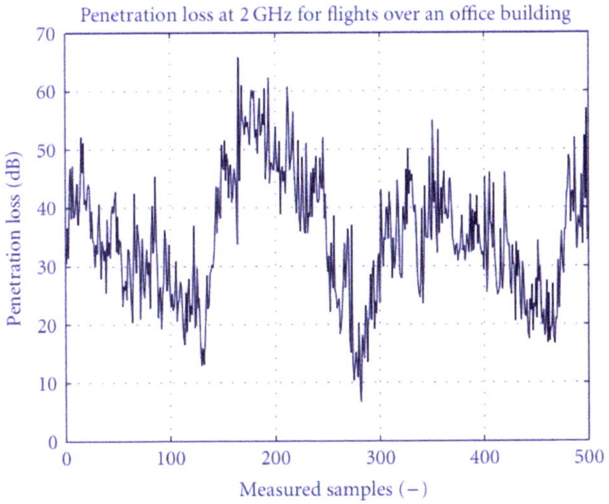

Figure 3: An example of extracted penetration loss into the ground floor of an office building.

The fitted measured data are presented in Figure 5. The very similar dependence of the penetration loss on the elevation angle is distinguishable from Figure 5.

Lower penetration loss was measured at the higher floors for some elevation angles. The multipath propagation plays an important role here, so that the lowest penetration loss for higher elevation angles was measured on the first floor. Nevertheless, differences in penetration loss were not as crucial in this measurement scenario. In order to obtain a universal model, all the data from this measurement scenario were processed together.

The final dependence of the penetration loss on the elevation angle in the office building for storeys situated below the rooftop level of surrounding buildings is depicted in Figure 6.

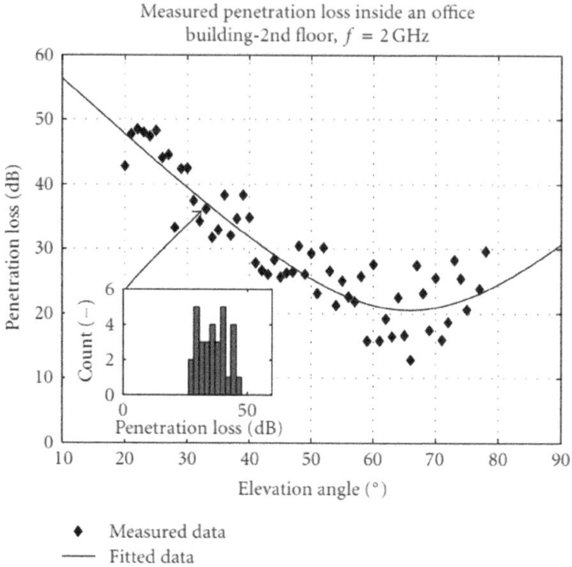

Figure 4: Penetration loss as a function of the elevation angle on the second floor of an office building at 2.0 GHz.

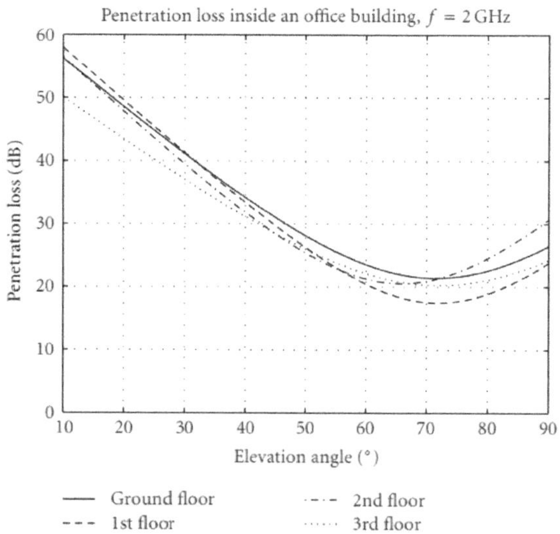

Figure 5: Penetration loss as a function of the elevation angle for range of floors in office building, f = 2.0 GHz.

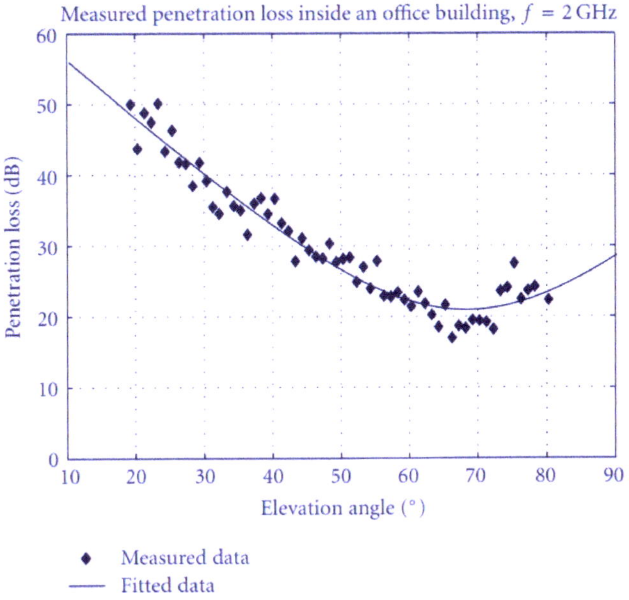

Figure 6: Mean penetration loss as a function of the elevation angle in office building at 2.0 GHz.

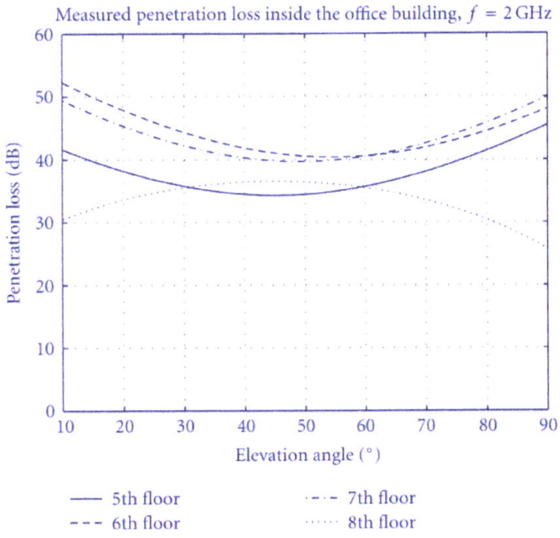

Figure 7: Penetration loss as a function of the elevation angle for a range of floors in an office building, f = 2.0 GHz.

The corresponding empirical parameters for model (1) are summarized in Table 1. This dependence can easily be used in system level simulation for the planning of a wireless system provided using HAP stations. In the second scenario, the receiver station was situated in the same office building, but in the floors above the rooftop level of surrounding buildings. In this case, the effect of multipath propagation is not so dominant as the signal cannot reflect from neighboring buildings and so forth. The measured data were again successfully approximated by (1). The results for the four floors are shown in Figure 7.

In the 6th and 7th floors, a higher penetration loss was measured than in the 5th floor because the reflections from surrounding buildings can be still dominant for the 5th floor. The atypical behavior in the 8th floor is given by the position of the receiver station (directly under the flat roof of the building being studied), and by the fact that the 8th floor is a superstructure with smaller floor projection than the 7th floor so that the signal can reflect from the 7th floor roof rim. In another scenario, the receiver station was situated in 7 floors of a prefab residential building. It was placed in the corridor in the middle of the building so that the rooms were symmetrically located around the receiver station, meaning that the receiver station was as far as possible from the external walls. For this measurement scenario, the penetration loss at 2.0 GHz was about 50 dB almost independently of the elevation angle. The mean measured value for the receiver station situated in the middle of the residential building is depicted in Figure 8. Figure 8 shows the impact of the type of building and the receiver position inside the building on the penetration loss.

The penetration loss into an older building built of bricks was also measured. The receiver station was situated 2 m from an external wall. The data measurement was divided to cover two situations. In the first, the external wall, in front of which the receiver station was located, was directly illuminated by the transmitter. In the second, the external wall was not directly illuminated, that is, the airship was situated on the opposite side of the building. The results of this study are presented in Figure 9. For the situation where the external wall was directly illuminated, the penetration loss decreased for higher elevation angles. On the other hand, the penetration loss increases for a decreasing angle of elevation in the case of external wall that was not directly illuminated. The empirical model for high elevation angles [6] assumes that the external wall is directly illuminated.

This assumption significantly simplifies the calculation. In any event, there is a high level of agreement between the data measured in the case of a directly illuminated wall and the empirical model [6] for the high elevation angle, but this model does not consider the scenario of a wall that is not directly illuminated, which has a crucial impact on the final signal level. The final dependence for all data measured in the brick building has similar shape of curve to Figure 6, for example.

Finally, the impact of the carrier frequency was also studied. The trial was accomplished at frequencies of 2.0 GHz and 3.5 GHz. It is obvious that the penetration loss is higher for the higher frequencies. The behavior of the penetration loss dependence on an elevation angle was found to be almost the same at 2.0 and 3.5 GHz. The increase of penetration loss at 3.5 GHz compared to 2.0 GHz depends on the type of building. For an office building where metal frames are used, the penetration loss at 3.5 GHz is about 5.0 dB higher than for a frequency of 2.0 GHz. In the residential building, a difference in penetration loss of 2.3 dB was found. This trend concords with the measurements carried out for terrestrial systems [9]. Figure 10 illustrates the comparison of penetration losses as a function of the elevation angle for both frequencies and two types of buildings. For more clarity, only a fitted curve is shown for 2.0 GHz.

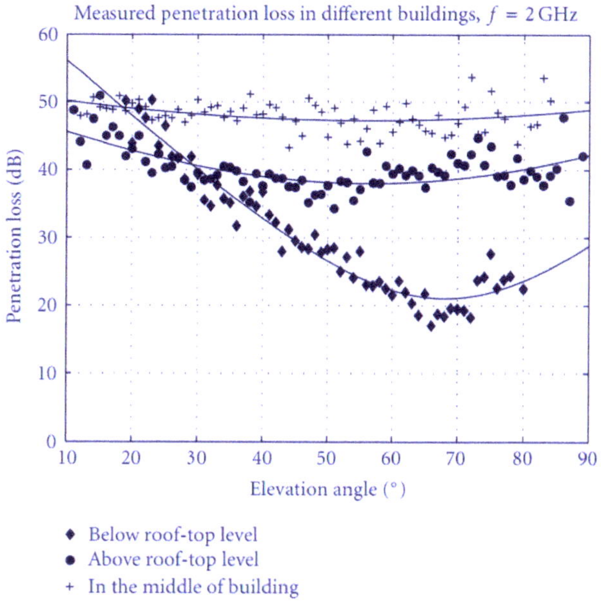

Figure 8: Penetration loss as a function of the elevation angle for different scenarios, f = 2.0 GHz.

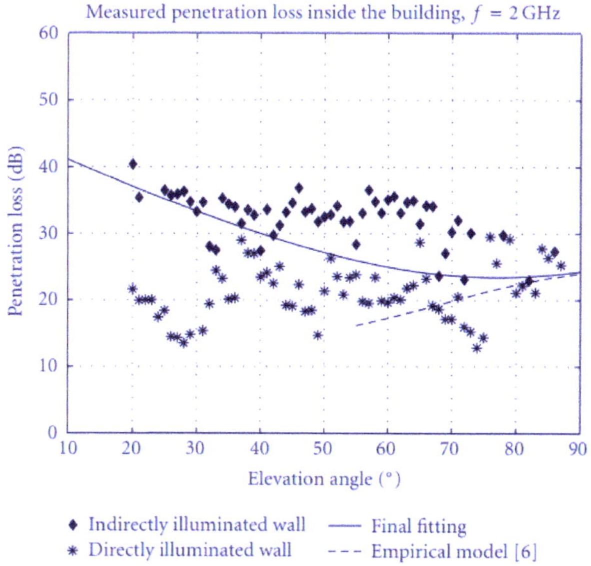

Figure 9: Penetration loss as a function of the elevation angle for different scenarios at 2.0 GHz.

The aim of this work was to explore the behavior of penetration loss into the building as a function of the elevation angle in urban areas for HAP systems. The different behavior of penetration loss dependence on the elevation angle was observed in different scenarios. This means that for precise modeling model (1) should be calibrated according to the specific scenarios. Anyway, the final universal model as a function of the elevation angle was determined based on the

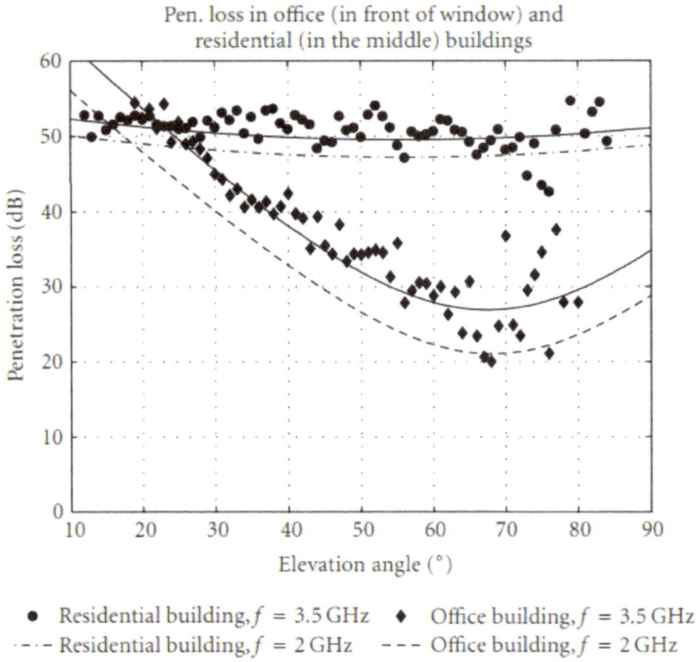

Figure 10: Comparison of penetration loss as a function of the elevation angle for different environments and carrier frequencies of 2.0 GHz and 3.5 GHz.

all measured data from the most common scenarios:

$$L_{\mathrm{PL}}(\theta) = \begin{cases} \sqrt{506 + 0.512(\theta - 70.4)^2}, & f = 2.0\,\mathrm{GHz}, \\ \sqrt{692 + 0.571(\theta - 70.2)^2}, & f = 3.5\,\mathrm{GHz}. \end{cases} \tag{2}$$

The empirical parameters for (1) are summarized for different environments in Table 1 including the final model.

Table 1: Parameters of the penetration loss model for different environments and frequencies in an urban area.

		Office building		Brick building	Final
		Below rooftop level	Above rooftop level		
2.0 GHz	a	444	1443	550	506
	b	−0.802	−0.296	−0.252	−0.512
	c	68.1	56.3	77.2	70.4
3.5 GHz	a	723	1858	664	692
	b	−0.960	−0.324	−0.251	−0.571
	c	67.5	54.8	79.0	70.2

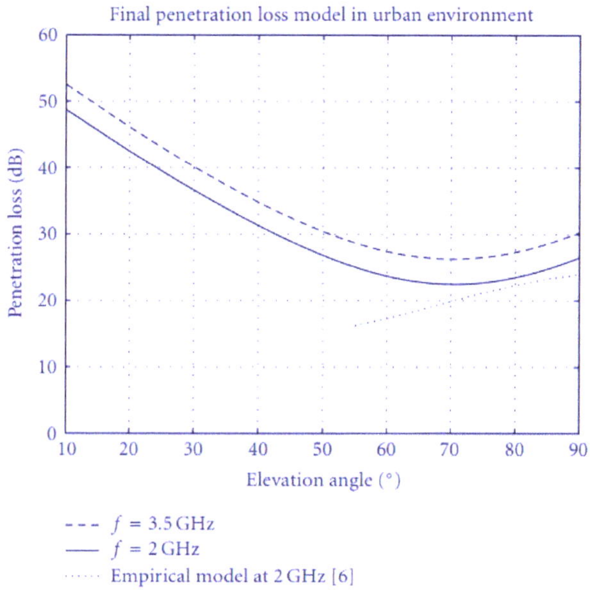

Figure 11: Comparison of the new penetration loss model with an empirical model developed for high elevation angles in [6].

DISCUSSION

The penetration loss as a function of the elevation angle was measured using a remote-controlled airship. The campaign was motivated by a lack of penetration models for HAP systems with the exception of an empirical model for building penetration loss at 2.0 GHz for high elevation angles [6]. This model was developed for a scenario, where the 2 walls of a room with the receiver are directly illuminated. This condition is very hard to achieve, especially for lower elevation angles below 60 degrees. As shown above, the penetration loss closely depends on the position of the HAP and the receiver position within a building. Based on the measurement data, a model was developed for typical scenarios in urban areas. The penetration loss calculated using (2) is shown in Figure 11 and compared with the empirical model for high elevation angles [6]. Difference in penetration loss at 2.0 GHz and 3.5 GHz is about 3.6 dB, but it depends on building material. Generally, it could be from 1 to 6 dB [12]. The total propagation loss for HAP mobile systems when the mobile terminal is situated inside a building can be determined based on the free space loss and the empirical penetration loss model (2):

$$L = L_{\text{FSL}} + L_{\text{PL}}, \tag{3}$$

where L_{FSL} is the free space loss in dB, and L_{PL} is the penetration loss in dB defined in (2). The free space loss is equal to

$$L_{\text{FSL}} = 20\log(d_{\text{km}}) + 20\log(f_{\text{GHz}}) + 92.4, \tag{4}$$

where f_{GHz} is the carrier frequency in GHz, and d_{km} is the distance between a platform and a user in km. For more realistic approach, an additional random lognormal fade margin should be added as a location variability. The standard deviation of the log-normal distribution for indoor environment is about 10 dB.

The multipath fading was eliminated thanks to the averaging of the narrowband measurement data. It is obvious that due to the measurement in real urban scenarios the multipath propagation effects and, especially in case of low elevation angles, the shadowing of surrounding buildings play an important role. We have not tried to distinguish the different components of the propagation phenomena since the goal was to model the average signal level for an indoor mobile station served from HAPs regardless the azimuth angle. This way the introduced penetration loss model can be, together with information such as the type of building and construction material used that collectively define a street model [13], successfully utilized for simple propagation predictions in urban areas for HAP scenarios. This model could be used for system level simulations of mobile services provided via HAPs.

CONCLUSION

An empirical propagation prediction model for calculation of penetration loss into the building as a function of the elevation angle is presented in this paper. The model was developed based on measured data obtained using a remotecontrolled airship. It is shown that the penetration loss closely relates to the position of the user within the building as well as to the type of building. The measured data were approximated by a simple function given by three empirical parameters depending on the type of building and the frequency. The universal model was derived from the statistical processing of data obtained in different scenarios based on measurements for four building types in the Czech Republic. Further calibration may be needed for the model to be applied in other specific scenarios. This model can be used for the radio network planning of mobile systems provided via high altitude platforms in the whole range of elevation angles.

ACKNOWLEDGMENT

This work was partly supported by the Czech Ministry of Education, Youth and Sports under projects OC092-COST Action 297 and MSM 6840770014.

REFERENCES

1. T. C. Tozer and D. Grace, "High-altitude platforms for wireless communications," Electronics & Communications Engineering Journal, vol. 13, no. 3, pp. 127–137, 2001.
2. D. Grace, J. Thornton, G. Chen, G. P. White, and T. C. Tozer, "Improving the system capacity of broadband services using multiple high-altitude platforms," IEEE Transactions on Wireless Communications, vol. 4, no. 2, pp. 700–709, 2005.
3. "Revised technical and operational parameters for typical IMT-2000 terrestrial systems using High Altitude Platforms Stations and CDMA radio transmission technologies," ITU-R Study Groups Document 8-1/307-E, ITU, March 1999.
4. "Minimum performance characteristics and operational conditions for high altitude platform stations providing IMT- 2000 in the Bands 1885–1980 MHz, 2010–2025 MHz and 2110–2170 MHz and in Regions 1 and 3 and 1885–1980 MHz and 2110–2160 MHz in Region 2," Recommendation ITU-R M.1456, ITU, 2000.
5. "Air interface for fixed and mobile broadband wireless access systems—Amendment for physical and medium access control layers for combined fixed and mobile operation in licensed band," IEEE Standard 802.16e-2005, 2005.
6. D. I. Axiotis and M. E. Theologou, "An empirical model for predicting building penetration loss at 2 GHz for high elevation angles," IEEE Antennas and Wireless Propagation Letters, vol. 2, no. 1, pp. 234–237, 2003.
7. D. I. Axiotis and M. E. Theologou, "Building penetration loss at 2 GHz for mobile communications at high elevation angles by HAPS," in Proceedings of the 5th IEEE International Symposium on Wireless Personal Multimedia Communications (WPMC '02), vol. 1, pp. 282–285, Honolulu, Hawaii, USA, October 2002.
8. D. I. Axiotis and M. E. Theologou, "2 GHz outdoor to indoor propagation at high elevation angles," in Proceedings of the 13th International IEEE Symposium on Personal, Indoor and Mobile Radio Communications (PIMRC '02), vol. 2, pp. 901– 905, Lisbon, Portugal, September 2002.
9. S. R. Saunders and A. Argo-Zavala, Antennas and Propagation for Wireless Communication Systems, John Wiley & Sons, New York, NY, USA, 2nd edition, 2007.
10. COST 231 Final report, "Digital Mobile Radio: COST 231 VIEW on the Evolution Towards 3rd Generation Systems," Commission of the European Communities and COST Telecommunications, Brussels, Belgium, 1999.
11. http://www.airshipclub.com/.
12. S. Aguirre, L. H. Loew, and Y. Lo, "Radio propagation into buildings at 912, 1920, and 5990 MHz using microcells," in Proceedings of the 3rd Annual International Conference on Universal Personal Communications (ICUPC '94), pp. 129–134, San Diego, Calif, USA, September-October 1994.

13. J. Holis and P. Pechac, "Elevation dependent shadowing model for mobile communications via high altitude platforms in built-up areas," IEEE Transactions on Antennas and Propagation, vol. 56, no. 4, pp. 1078–1084, 2008.

12
CHAPTER

THE USE OF ADVANCED OPTICAL MEASUREMENT METHODS FOR THE MECHANICAL ANALYSIS OF SHEAR DEFICIENT PRESTRESSED CONCRETE MEMBERS

K. De Wilder, G. De Roeck, L. Vandewalle

Keywords: experimental mechanics, prestressed concrete, coordinate measurement machine, stereo-vision digital image correlation

INTRODUCTION

Despite the long-established and worldwide research effort, a widely accepted theory for determining the shear capacity of a structural concrete member remains open for discussion (Balázs 2010; Collins 2010). This can be mainly attributed to the complex nature of shear in structural concrete. Indeed, after the occurrence of inclined cracking, various shear transfer mechanisms are activated (Jeong and Kim 2014). The aforementioned mechanisms are interrelated and highly susceptible to various parameters such as the geometry, the amount of prestressing, material properties, the amount and type of shear and longitudinal reinforcement and the loading conditions. As a consequence, numerous analytical modelling approaches can be found in literature. An overview of recent approaches to shear in structural concrete elements can be found in Fédération Internationale du Béton (fib) (2010).

The on-going debate on how to deal with shear in structural concrete members is also reflected by current codes of practice (European Committee for Standardization 2004; Canadian Standards Association 2004; American Concrete Institute 2011) which propose different design provisions resulting in varying design shear capacities and take factors affecting the shear capacity into account in a different way. Due to our incomplete knowledge and the brittle failure modes typically associated with shear,

Citation: K. De Wilder, G. De Roeck, L. Vandewalle, The Use of Advanced Optical Measurement Methods for the Mechanical Analysis of Shear Deficient Prestressed Concrete Members, DOI: 10.1007/s40069-016-0135-x.

current codes of practice generally propose highly conservative shear provisions, specifically in the case of prestressed concrete elements (Nakamura et al. 2013). For the design of structural concrete members, these design equations lead to excessive material usage and corresponding construction costs. Reversely, using current codes of practice to determine the shear capacity, existing concrete structures are often found to be unable to withstand the applied service loads whereas no structural problems are reported in reality (Valerio et al. 2011; Lantsoght et al. 2013).

Improving analytical modelling approaches for shear in structural concrete members thus remains of utmost importance to optimize the design and analysis of structural concrete elements. A prerequisite for developing suitable models is a clear understanding of the actual mechanical behaviour observed during experimental tests. Traditional measurement techniques, i.e. linear variable differential transformers (LVDTs) or demountable mechanical strain gauges (DEMEC), usually provide limited test data in one or two directions. If the actual mechanical behaviour is to be understood, more elaborate experimental data is required. This paper therefore investigates on the use of advanced optical measurement methods, i.e. 3D coordinate measurement machines (CMMs) and Stereo-vision digital image correlation (3D DIC), for the mechanical analysis of shear-deficient structural concrete elements. The main focus in this paper will be on prestressed concrete beams. Firstly, the experimental program is elaborated. Secondly, the adopted measurement methods are presented and their corresponding setup is described. Thirdly, the experimental results are presented including the results of the precision assessment of the measurements and the full-scale experimental results. Finally, the mechanical behaviour of the reported test beams is investigated based on the acquired measurement data and compared to current modelling approaches found in codes of practice.

EXPERIMENTAL RESEARCH

Geometry and Materials

Nine test specimens were constructed and labelled with the descriptive letter B followed by a number ranging from 101 to 109. All specimens were characterized by an I-shaped cross section, were 7000 mm long, 630 mm high. The flange width was equal to 240 mm whereas the web width was 70 mm wide. All specimens were prestressed using 8 or 4 seven-wire strands (nominal diameter 12.5 mm) at the bottom and two seven-wire strands at the top (nominal diameter 9.3 mm). The initial strain given to each prestressing strand was 7.5×10^{-3} mm/mm (σ_{p0}=1488 MPa) for specimens B101–B103 and B107–B109 whereas the stress level σ_{p0} in the strands of beams B104–B106 was equal to 750 MPa (initial strain equal to 3.8×10^{-3} mm/mm). Shear reinforcement was provided in six specimens consisting of open single-legged stirrups with a nominal diameter of 6 mm (center-to-center distance equal to 150 mm). To avoid splitting failure at the end of each specimen due to the gradual development of the prestressing force over the member's height, splitting reinforcement was provided in each specimen (nominal diameter 8 mm, center to center distance equal to 50 mm). An overview of the geometry of the reported test specimens is given in Figs. 1a to 1f.

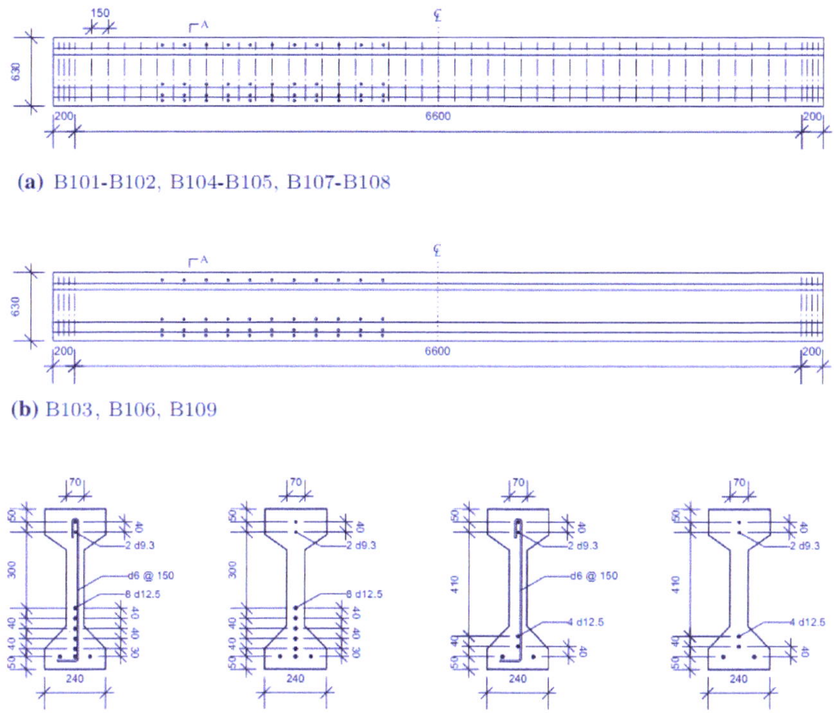

(a) B101-B102, B104-B105, B107-B108

(b) B103, B106, B109

(c) B101-B102, B104-B105 **(d)** B103 **(e)** B107-B108 **(f)** B109

Figure 1: Geometry and reinforcement layout of the reported test specimens: **a**, **b** Longitudinal drawings; **c**–**f** Cross sectional drawings.

A self-compacting concrete mixture, designed to have a characteristic cylindrical compressive strength f_{ck} equal to 50 MPa, was used to cast each specimen. The concrete mixture consisted of cement CEM I 52.5 R (16.0 wt%), limestone gravel 2/12 (43.8 wt%), sand 0/2 (26.5 wt%%), water (6.7 wt%), limestone filler (6.7 wt%%) and a high-range water reducer (0.3 wt%%). The expected density of this mixture was equal to 2378 kg/m33. Standardized tests were performed to determine the mean cylindrical compressive strength fcmfcm as well as the mean cube compressive strength $f_{cm,cube}$. Moreover, the mean flexural tensile strength $f_{ctm,fl}$ and the secant modulus of elasticity E_{cm} were determined for each specimen. The testing procedures can be found in Bureau for Standardisation NBN (2009a, b, 2014) The results are presented in Table 1. The age of each specimen at the day of testing is also indicated in the aforementioned Table 1.

Standard tensile tests were performed on the reinforcement bars to determine the mean yield stress f_{ym}, the modulus of elasticity E_s, the mean ultimate tensile strength f_{tm} and the corresponding strain at failure ϵ_{su}. The same characteristics were adopted from the technical information provided by the manufacturer of the prestressing strands. The results are shown in Table 2.

Table 1: Concrete material properties for reported specimens

Specimens	$f_{cm,cube}$ (MPa) (#*, s**)	f_{cm} (MPa) (#, s)	E_{cm} (GPa) (#, s)	$f_{ctm,fl}$ (MPa) (#, s)	Age (days)
B101-B103	87.1 (8, 6.8)	77.5 (5, 10.7)	43.4 (3, 2.3)	5.8 (6, 0.6)	28-233-393
B104-B106	82.8 (9, 10.5)	88.9 (6, 10.2)	43.5 (6, 8.0)	6.5 (6, 1.0)	412-404-407
B107-B109	74.6 (9, 9.6)	89.3 (6, 14.2)	42.2 (6, 4.6)	5.7 (6, 1.1)	428-424-412

* Number of tested specimens, ** Standard deviation.

Table 2: Reinforcement properties.

Reinf. type	Type	d_s d_p* (mm)	E_s E_p (GPa)	f_{ym} $f_{p0.1m}$ (MPa)	f_{tm} f_{pm} (MPa)	ϵ_{su} ϵ_{pu} (%)
Top prestress. reinf.	7-wire	9.3	198.0	1737	1930	5.20
Bot. prestress. reinf.	7-wire	12.5	198.0	1737	1930	5.20
Shear reinf.	Cold worked	6.0	210.0	608	636	2.73
Splitting reinf.	Cold worked	8.0	203.0	542	603	5.97

* Subscript p and s denote properties of prestressing and conventional reinforcement types, respectively.

Experimental Setup

All test specimens are subjected to a four-point bending test, schematically shown in Fig. 2. A force-controlled loading scheme was followed using a hydraulic press (Instron, maximum capacity 2.5 MN). The load exerted by the hydraulic press was transferred to two point loads using a steel transfer beam. The total load was applied at a rate equal to 0.250 kN/s (shear force rate V˙=0.125V˙=0.125 kN/s). The distance between both support points was equal to 5000 mm. The distance outside the support points was therefore equal to 1000 mm.

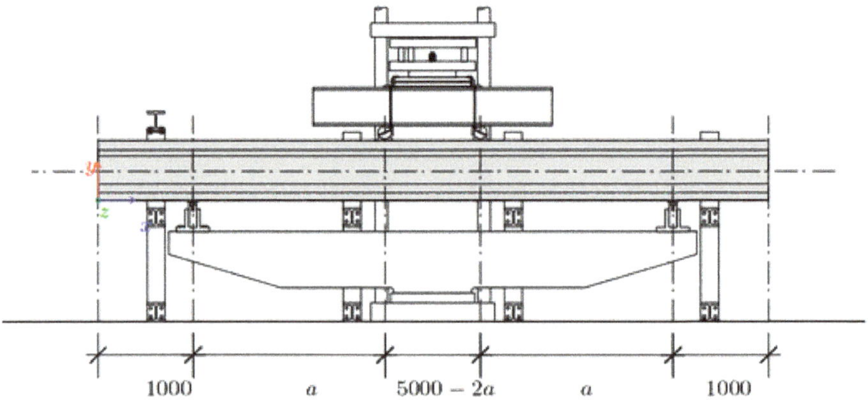

Figure 2: Schematic representation of the experimental setup (*note* (x, y, z) denote Cartesian coordinates).

Table 3: Overview of the investigated parameters per specimen

Specimen	Type	d* (mm)	σ_{p0} (MPa)	a (mm)	$\frac{a}{d}$ (–)	ρ_l^{**} (–)	ρ_w^{***} ($\times 10^{-3}$)
B101	I	511	1488	1600	3.13	0.0208	2.693
B102	I	511	1488	2000	3.91	0.0208	2.693
B103	I	511	1488	2000	3.91	0.0208	0
B104	I	511	750	1600	3.13	0.0208	2.693
B105	I	511	750	2000	3.91	0.0208	2.693
B106	I	511	750	2000	3.91	0.0208	0
B107	I	550	1488	1600	2.91	0.0097	2.693
B108	I	550	1488	2000	3.64	0.0097	2.693
B109	I	550	1488	2000	3.64	0.0097	0

* Effective depth.
** Longitudinal reinforcement ratio $\frac{A_{sl}}{b_w d}$ with A_{sl} area of longitudinal reinforcement and b_w web width.
*** Shear reinforcement ratio $\rho_w = \frac{A_{sw}}{b_w s}$ with $\frac{A_{sw}}{s}$ the area of shear reinforcement per unit length.

This allowed the authors to investigate on the mechanical behaviour of shear in the presented test beams outside the length needed for the prestressing force to gradually develop over the member's height. The shear span a, i.e. the distance between the support point and the load point, was equal to 1600 mm (specimens B101, B104 and B107) or 2000 mm (B102–B103, B105–B106, B108–B109). Given the geometry and reinforcement layout, this resulted in shear span-to-effective depth ratios (adad-ratio) between 2.91 and 3.91. An overview of the investigated parameters per specimen is given in Table 3.

Adopted Measurement Methods

CCD-LED Coordinate Measurement Machine (CMM)

As an optical motion tracking system, coordinate measurement machines [CMMs, also referred to as dynamic measurement machines (DMMs)] have proven to be an ideal tool for a wide range of applications, including biomechanical (Burg et al. 2013), automotive (Deconinck et al. 2004) and civil engineering applications (De Roeck et al. 2004; Sun 2007). The basic principle behind this technique is to determine the three-dimensional location of a discrete amount of points by means of triangulation. Here, two Krypton K600 CMM (Nikon Metrology, formerly known as Metris, Leuven, Belgium) systems were used for the experiment on specimen B101. The points of which the three-dimensional coordinates were to be determined were infra-red (IR) light emitting diodes (LEDs). In total 112 LEDs were used and each LED was placed onto an orthogonal grid covering both zones where a shear force occurred, refer to Fig. 3a, and was glued onto the concrete surface by using a thermoplastic adhesive. Since the LEDs were attached to the material underneath, the displacement of each LED, from which deformations can be derived, is the same as the material under investigation. The IR light emitted by each LED was seen by a camera unit consisting of three 2048 px charge-coupled device (CCD) line-element cameras, as shown in Fig. 3b. All LEDs were connected in series to a camera control unit which also acted as an interface between the cameras and the DAQ laptops.

Figure 3: Experimental setup for the adopted coordinate measurement machine (CMM): **a** IR LEDs (indicated with *red rectangle*); **b** Camera units mounted on tripod.

To assess the accuracy (closeness of the measurement to the true value, related to systematic errors) and precision (the degree to which repeated measurements under unchanged conditions show the same results, related to random errors) of the CCD-LED system, a reference measurement was performed. During the aforementioned reference measurement, each coordinate of each LED was measured during 50 s at a measurement frequency of 20 Hz. Typical results of the variation of each coordinate around its initial value u_i (with i=x,y,z) for one LED are shown in Figs. 4a to 4c. The noise on the measurement data can assumed to be approximately normally distributed. Therefore, Figs. 4d to 4f show the fitted normal probability density functions (PDF) of u_i (i=x,y,z). The mean value and corresponding standard deviation are also indicated in the aforementioned figures.

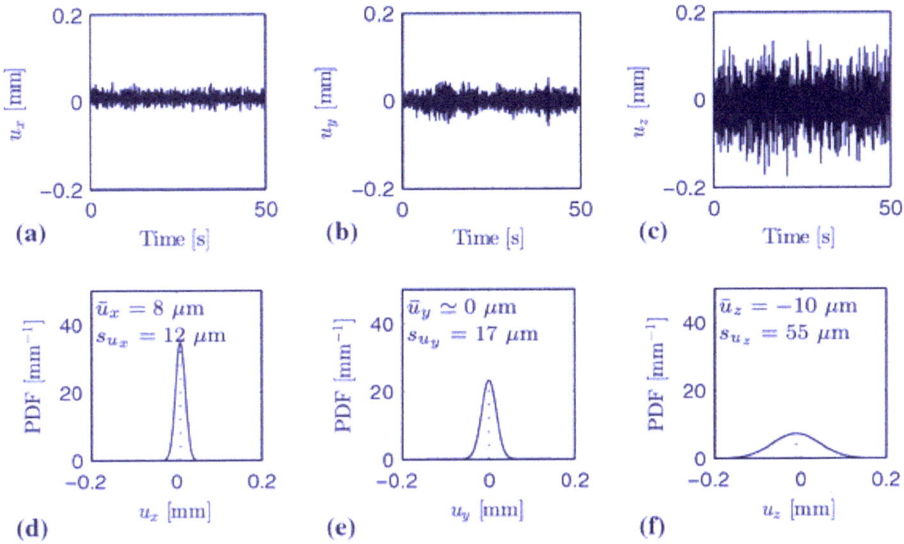

Figure 4: Accuracy and precision assessment of the Krypton K600 CMM for the *upper left* LED (refer to Fig. 3a): **a** horizontal in-plane displacement u_x; vertical in-plane displacement u_y; out-of-plane displacement u_z; fitted normal PDF for **d** u_x; **e** u_y; **f** u_z.

Based on the coordinate measurements obtained during the aforementioned reference measurement, in-plane strain values (both horizontal ϵ_x and vertical ϵ_y) can be derived between two consecutive LEDs. The standard deviation of the in-plane horizontal and vertical strain value, s_{ϵ_x} and s_{ϵ_y} can then be calculated as a function of the surface coordinates. The results are presented in Figs. 5a and 5b. From the aforementioned Fig. 5a, it can be seen that the value of the standard deviation of the horizontal strain s_{ϵ_x} is well below the typical strain values occurring for concrete in compression (maximum compressive strain $\epsilon_{cu} \simeq 35 \times 10^{-4}$ mm/mm). Figure 5b indicates that less precise measurements of the vertical strain are obtained near the top flange of the beam but good overall results are acquired in the web of the specimens. Indeed, the values of the vertical strain will primarily be used to assess the strain and thus stress levels in the shear reinforcement elements. Yielding of the shear reinforcement occurs at nearly 29×10^{-4} mm/mm, refer to Table 2 whereas rupture occurs at a strain value equal to 273×10^{-4}. Therefore, the expected vertical strains are well outside the obtained noise levels. The variation of the derived strain values as a function of the surface coordinate is primarily due to the different relative positions of each IR LED with respect to the optical centre of the camera control units.

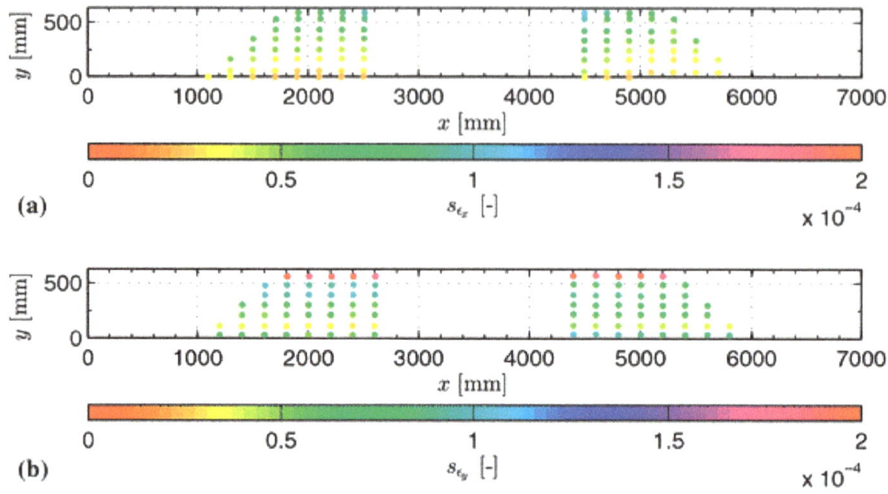

Figure 5: Precision assessment of the adopted Krypton K600 CMM for the derived strain measurements: **a** standard deviation of the horizontal strain s_{ϵ_x} as a function of the location on the specimen; **b** standard deviation of the vertical strain s_{ϵ_y} as a function of the location on the specimen.

Stereo-Vision Digital Image Correlation (3D-DIC)

As an optical-numerical full-field measurement method, digital image correlation (DIC) has proven to be an ideal tool for a wide range of applications, including the identification of the mechanical material behavior through inverse modeling (Cooreman et al. 2007, 2008), structural health monitoring (Sas et al. 2012) and the study of the deformation characteristics of a wide range of materials (Ivanov et al. 2009; Van Paepegem et al. 2009; Srikar et al. 2016). During the testing of specimens B103–B109, the stereo-vision digital image correlation technique (3D-DIC) has been adopted to assess the displacements and deformations during the loading procedure. Stereo-vision implies the use of two cameras to allow for the reconstruction of the three-dimensional geometry and measurement of the three-dimensional displacements opposed to single-camera measurements which yield only two-dimensional data. The basic principle behind this technique is to calculate the displacements on the surface of an object by taking images of a random speckle pattern in undeformed and deformed state. As the speckle pattern is attached to the material underneath, the displacement and deformation of the speckle pattern is the same as the surface material of the object under investigation. To follow the displacement and deformation evolution during the full-scale experiments, a series of images of the object is made and the speckle pattern displacement and deformation in the series of images can be followed. The grey-value images of the speckle pattern are captured with a CCD camera. Similar to the CMM measurements, both zones where a shear force occurs were investigated using two DIC systems, each consisting of two cameras. This setup allowed for stereo-vision measurements. Both cameras of each DIC system take simultaneously a picture of the investigated object, in the following denoted as a frame. Each stereo-vision DIC system consisted of two 8-bit CCD cameras (AVT F201 B; 1628 px × 1236 px resolution) with wide-angle lenses (focal length equal to 12 mm) mounted on a tripod. Each zone under investigation measured approximately 1500 mm × 630 mm. The cameras were located at a perpendicular distance of approximately 2700 mm from the web of the specimen. To ensure good lighting conditions and allow for small exposure times, two 500 W quartz iodine lamps were provided per investigated zone. The image acquisition rate of each camera was equal to 2 Hz. All images were transferred to a desktop computer and synchronized with the analogue data of the hydraulic press (i.e. applied load and corresponding displacement). The experimental setup is depicted in Figs. 6a and 6b.

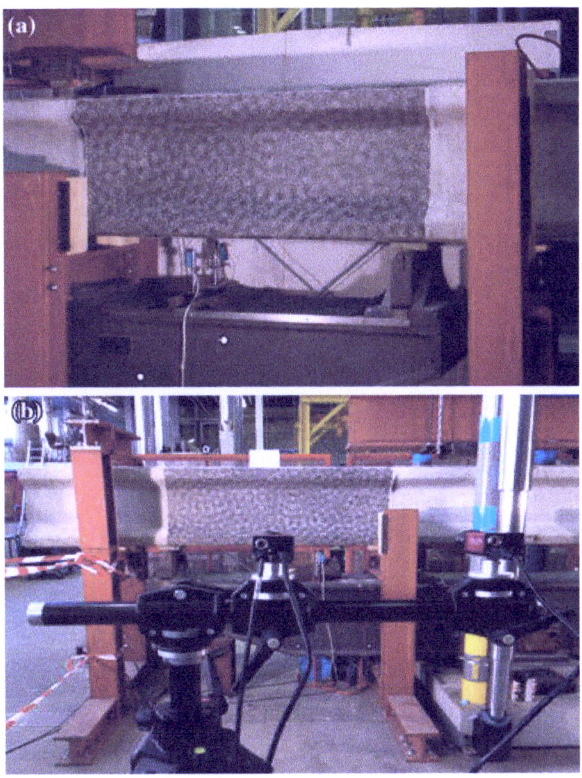

Figure 6: Experimental setup for DIC measurements: **a** numerically generated speckle pattern applied onto specimen B104; **b** 8-bit CCD cameras mounted on tripod located at a perpendicular distance of 2700 mm from the web of the test specimens.

The analysis of the frames taken during the loading was done with specialized software. In this work, the in-house code MatchID (KU Leuven, Campus Ghent) (Lava et al. 2009, 2010, 2011; Wang et al. 2011) was used. To reveal the displacements and deformations of an object during an experiment, typically a square subset of (2M+12M+1) pixels (px) from the undeformed image is taken and its location in the deformed image is traced. The principle of the stereo-vision DIC algorithm is clearly explained by Lava et al. (2011). Matching of two (2*M*+1) px subsets in the undeformed image $F(x_i, y_j)$ and deformed image $G^t(x_i, y_j)$ at a certain time *t* (i.e. load step) is performed by adopting an optimization routine for a degree of similarity expressed by a correlation criterion. Here, the Zero Normalized Sum of Squared Differences (ZNSSD) correlation criterion was adopted. This correlation criterion is independent of scale and offset in lighting (Sutton et al. 2009) and is therefore the most suitable correlation criterion to yield accurate results, especially in the zones where the lighting conditions are difficult to control, i.e. transition zones between web and flange of the I-shaped specimens. The mathematical formalism is clearly explained in the reference work by Sutton et al. (2009).

Lecompte (2007) states that *it has already been shown that the subset size is a critical parameter in the correlation process* (Knauss et al. 2003). *On the one hand it should be chosen small enough to allow for a reasonable linear approximation of the displacement field, within the region of the subset. On the other hand the subset size should not be chosen too small, to avoid correlation problems due to the non-uniqueness of the subset information content. This indicates the importance of an adequate speckle pattern for digital image correlation.* From a solely black or white pattern, no valuable displacement information can be gathered so that the subset size should be larger than the speckle size. Destrycker (2012) states that *to isolate the effect of the speckle pattern, there should be a way of applying the speckle pattern in a controlled way, e.g. controlling the speckle size, the speckle size distribution, the grey value distribution and the actual colour of the black and white paint.* Therefore, to be able to generate suitable DIC speckle patterns, a numerical technique recently proposed by Bossuyt (2012) was adopted. In his work, Bossuyt (2012) firstly defines two concepts which are used to assess the suitability of a speckle pattern for DIC measurements. Firstly, the autocorrelation peak sharpness radius of a pre-processed image of the considered pattern is proposed to quantitatively evaluate how a particular strain sensor pattern influences the sensitivity of a DIC measurement. Secondly, the autocorrelation margin is proposed to evaluate how that pattern influences the robustness of the DIC measurement. The former is related to the measurement precision whereas the latter is correlated to the measurement accuracy. Ideal patterns for DIC would combine a sharp autocorrelation peak with a well-defined autocorrelation margin. For simple patterns, these characteristics vary in direct proportion to each other. However, Bossuyt (2012) proposes a method based on morphological image processing and Fourier transform to synthesize a DIC pattern with wide autocorrelation margins even though the autocorrelation peaks are sharp. Such patterns are exceptionally well-suited for DIC measurements. A detail of the numerically generated speckle pattern is shown in Fig. 7a. The generated pattern is then applied onto each specimen where the DIC technique was used by adopting a heat-sensitive stencil printing technique which consists of three layers: a vinyl base layer, the inverse of the speckle pattern and a top protective heat-sensitive polypropylene layer. The printed speckles have a precalculated oversampling of at least five pixels in order to avoid aliasing effects in the obtained results due to the expected small magnitude of the displacement and deformation field. Given the camera sensor properties and the dimensions of the field of view, speckles with a diameter of nearly 5 mm were required. Obtaining large speckles is nearly impossible with traditional speckle techniques (i.e.

spray painting). However, the adopted numerical technique allows for the generation of a speckle pattern tailored to the needs of the experiment. Figure 7b shows the same detail as presented in Fig. 7a applied onto beam B105. Figure 7c shows the speckle pattern, area of interest and adopted subset size. Since full-field displacement data is readily available, Green-Lagrange strains can be easily derived from the aforementioned displacement data. Therefore, the displacement data is smoothed over a certain zone to damp out the effect of noise and local uncertainties. A bilinear plane can be fitted through the displacement values in the points around the center of the strain window.

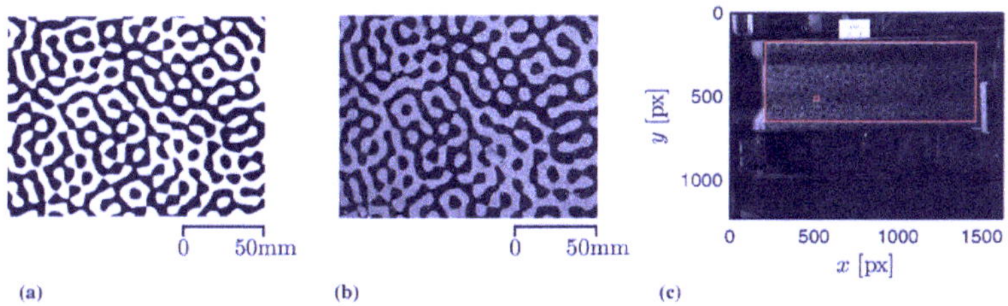

Figure 7: **a** Detail of numerically generated speckle pattern; **b** Detail of the speckle pattern shown in (**a**) applied onto specimen B105; **c** indication of area of interest (AOI) and subset size in *red* for the left measurement field of specimen B103.

To assess the precision of the 3D-DIC setup for the presented experiments, a number (N=50) of reference images were taken prior to each test under a zero-loading condition. The subset size was taken equal to 27 px whereas the stepsize was chosen equal to 3 px. Given the experimental setup, the physical dimension of 1 pixel approximated 1 mm. The subset was allowed to undergo an affine transformation thus taking into account translation, rotation, shear and normal straining. As the displacements from one frame to another may be smaller than one pixel, the subset in the image of the deformed state is not likely to fit on the pixel grid and an interpolation method between the pixels is needed. Therefore, a bicubic interpolation scheme has been adopted in the MatchID software during the analyses. Additionally, Gaussian prefiltering (5×5 px kernel size) of the subset information was adopted as a low pass-filter to attenuate high-frequency signals and allow for proper interpolation. Based on the measured displacements u_i (with i=x,y,z), Green-Lagrange strains can be determined by means of bilinear-quadrilateral smoothing of the displacement field. This method was adopted for the analyses of the frames during the loading procedure. For the purpose of assessing the precision, strains were derived directly from the displacement data with similar initial base lengths L_0 as the CMM and without smoothing of the displacement field to allow for a reasonable comparison between both adopted systems. The base length to determine the horizontal strains L_0^- was chosen approximately equal to 200 mm whereas the base length to determine vertical strains $L_0^|$ was taken roughly equal to 75 mm. An overview of the adopted DIC settings is presented in Table 4. Similar to the CMM system, Figs. 8a and 8b present the standard deviation of the horizontal and vertical strain s_{ϵ_i} (with i=x,y) based on the *N* pairs of reference images obtained from both DIC systems for specimen B103. Comparable results were found for the remaining specimens where the DIC technique has been adopted.

Table 4: Measurement settings for the DIC tests.

	Unit	B103	B104	B105	B106	B107	B108	B109
Subset	(px)	27	27	27	27	27	27	27
Step	(px)	3	3	3	3	3	3	3
Measurement points	(−)	68327	67482	67819	70290	68843	68868	75864
Temporal resolution	(fps)	2	2	2	2	2	2	2
Camera distance	(mm)	2700	2700	2700	2700	2700	2700	2700
Interpolation	(−)	b.c.s.†	b.c.s.	b.c.s.	b.c.s.	b.c.s.	b.c.s.	b.c.s.
				Displacement				
Spatial resolution[a]	(mm)	31.2	30.9	29.9	29.5	29.9	30.8	28.4
In-plane resolution†	(mm)	0.018	0.019	0.034	0.027	0.020	0.021	0.023
Out-of-plane resolution‡	(mm)	0.125	0.120	0.165	0.167	0.147	0.149	0.152

† Bicubic spline.
‡ Maximum mean value of the standard deviation of the in-plane horizontal and vertical displacement; mean value of the standard deviation of the out-of-plane displacement.
[a] Physical dimension of subset.

From the aforementioned Figs. 8a and 8b, it can be firstly seen that similar values for the standard deviation of both the horizontal and vertical strain are found if the data of the 3D DIC system is compared to the data obtained from the Krypton K600 system, refer to Figs. 5a and 5b. Secondly, it can be seen that less precise measurements of the horizontal strain ϵ_x occur near the edge of the area of interest, refer to Fig. 8a.

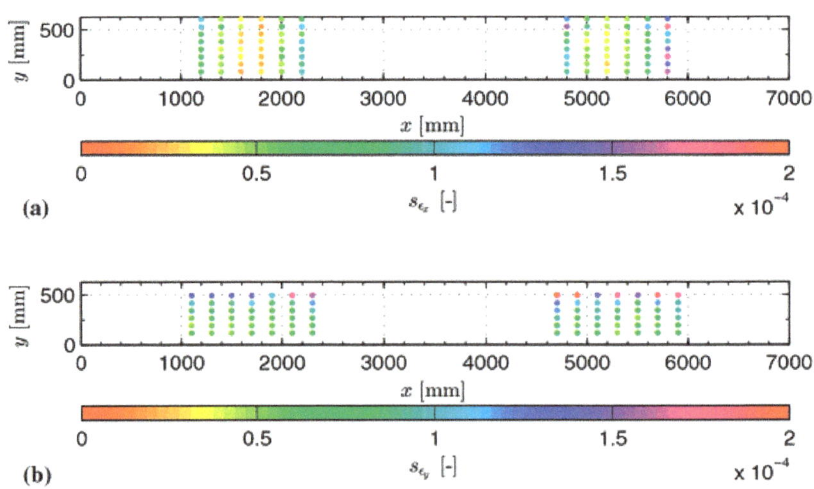

Figure 8: Precision assessment of the adopted 3D DIC system for the derived strain measurements for specimen B103: **a** standard deviation of the horizontal strain s_{ϵ_x} as a function of the location on the specimen; **b** standard deviation of the vertical strain s_{ϵ_y} as a function of the location on the specimen.

This can be attributed to optical aberration effects near the edge of the field of view. However, the value of the expected strains during the loading procedure well exceed the presented values of the standard deviation of the horizontal strain. Thirdly, due to the shaded area near the transition between the top flange and the web, refer to Fig. 7c, less precise measurements of the vertical displacement are obtained resulting in less accurate vertical strain measurements near the top flange. However, vertical strains will primarily be investigated in the web of the specimens to assess the strain and stress levels in the shear reinforcement elements.

RESULTS AND DISCUSSION

Figure 9a–c depict the experimentally observed load-displacement response curves of the presented test specimens. The onset of bending (∘) and diagonal (Δ) cracking is also indicated in the aforementioned Figs. 9a to 9c. All I-shaped test specimens but beams B107 and B108 failed due to shear in a very brittle manner. The aforementioned specimens exhibited severe diagonal web cracking leading to excessive yielding and sudden rupture of the shear reinforcement bars. Crushing of the diagonal compressive struts was not observed. Beams B107 and B108 exhibited a ductile bending failure mode leading to rupture of the prestressing strands. Typical experimentally observed failure modes of the presented test beams are shown in Figs. 10a to 10d.

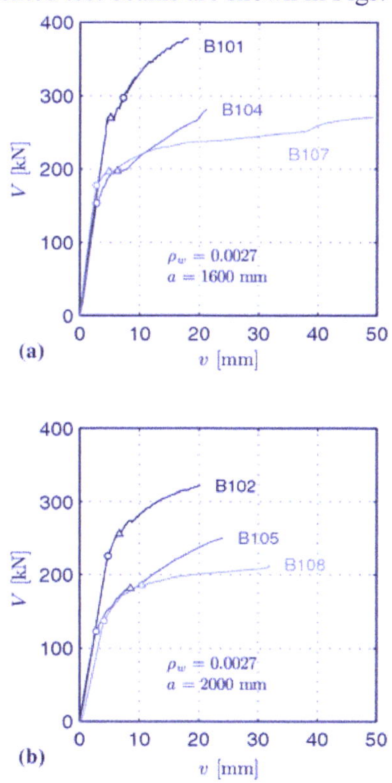

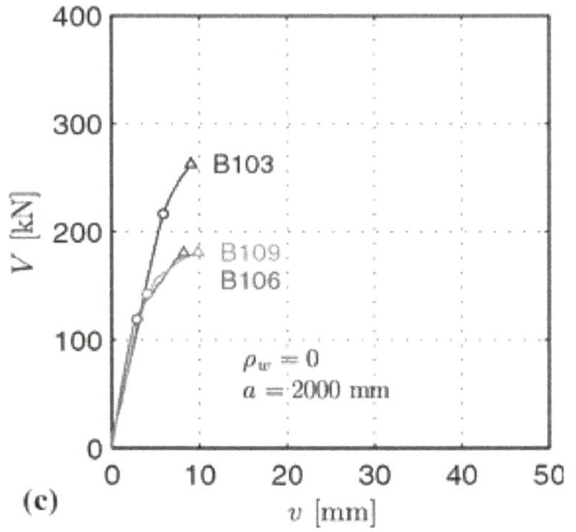

(c)

Figure 9: Experimentally determined load-displacement response curves measured at 1200 mm from the support point: **a** specimens B101, B104 and B107; **b** specimens B102, B105 and B108; **c** specimens B103, B106 and B109 (*note* onset of bending and diagonal cracking indicated with (*circle*) respectively (*triangle*).

(a) Diagonal tension (DT) failure B102 **(b)** Diagonal tension failure B104

(c) Bending failure mode B107 **(d)** Strand rupture B107

Figure 10: Typical experimentally observed failure modes for the reported test specimens.

A number of observations can be made based on Figs. 9a to 9c and 10a to 10d:

- Prior to the onset of cracking, the stiffness in the elastic regime is comparable for all specimens with the same shear span and overall height. Indeed, prior to the occurrence of cracking, the response of the test specimens to the applied load is governed by the bending stiffness *EI*. Due to the comparable secant modulus of elasticity, refer to Table 1, and the negligible influence of the area of longitudinal reinforcement on the second moment of inertia, it can be concluded that the bending stiffness is similar for the reported test specimens.
- The occurrence of cracks determines the transition between linear and nonlinear behaviour. All specimens exhibited both bending and web cracks. The load at which web cracks occurs, is function of the amount of prestressing and the concrete tensile strength respectively.
- Specimens where shear reinforcement was provided and which failed in shear (B101–B102, B104–B105) exhibited a significant post-cracking stiffness and post-cracking bearing capacity resulting in a brittle shear failure mode due to diagonal tension. Specimens B107 and B108 which failed in bending, show a highly ductile behaviour with a limited post-cracking bearing capacity. Finally, beams without shear reinforcement (B103, B106 and B109) failed immediately after the occurrence of the first inclined web crack.

The failure load and failure mode of each specimen is summarized and presented in Table 5. The experimentally determined failure load and failure mode can be compared to analytical predictions of the bearing capacity according to the shear design procedures outlined in Eurocode 2 (EC 2) (European Committee for Standardization 2004; Bureau for Standardisation NBN 2010). For the calculation of the shear capacity of a structural concrete member, a distinction is to be made between members with and without shear reinforcement. Structural concrete members with shear reinforcement are to be designed according to EC 2 using the variable angle truss model (VATM). This approach assumes that the behaviour of a structural concrete member near failure can be idealized by means of a parallel chord truss. The bottom and top flanges resist the applied bending moment whereas the combination of a compressive stress field with constant inclination and vertical tension bars resist the applied shear force. The design shear capacity V_{Rd} of a member with shear reinforcement is the minimum of the shear force required to obtain yielding of the shear reinforcement $V_{Rd,s}$ and the shear force which causes crushing of the compressive struts $V_{Rd,max}$ refer to Eq. (1).

$$V_{Rd} = \min\left\{V_{Rd,s}, V_{Rd,max}\right\} \tag{1}$$

The expressions for $V_{Rd,s}$ and VRd,max$V_{Rd,s}$ can be derived from equilibrium conditions of the adopted truss model where the angle between the horizontal and the inclined compressive stresses is denoted by θ, refer to Eqs. (2)–(3).

$$V_{Rd,s} = \frac{A_{sw}}{s} z f_{ywd} \cot\theta \tag{2}$$

$$V_{Rd,max} = \frac{\alpha_{cw} v_1 b_w z f_{cd}}{\cot\theta + \tan\theta} \tag{3}$$

In Eqs. (2)–(3), A_{sw}/s denotes the area of shear reinforcement per unit length, *z* is the internal lever arm equal to 0.9*d* whereas f_{ywd} is the design value of the yield strength of the shear reinforcement bars. Factors α_{cw} and v_1 take into account the stress distribution of the compressive chord respectively the effect of lateral tensile straining on the ultimate compressive strength. The width of the web is denoted by b_w. The inclination angle θ can be chosen freely between certain limits as presented in Eq. (4).

$$1 \leq \cot\theta \leq \cot\theta_{\max} \tag{4}$$

The maximum value of $\cot_\theta$, thus the minimum allowable value for the angle θ can be determined using Eq. (5) (Bureau for Standardisation NBN 2010).

$$\cot\theta_{\max} = 2 + \frac{0.15\sigma_{cp}b_w d}{\frac{A_{sw}}{s} z f_{ywd}} \leq 3 \tag{5}$$

Equation (5) clearly indicates that for highly prestressed members (average normal stress due to prestressing σ_{cp}), a lower angle of inclination is allowed with a minimul value of θ=18.4°.

In the case of structural concrete members without shear reinforcement, a distinction has to be made between members cracked respectively uncracked due to the acting bending moment, refer to Eqs. (6)–(7).

$$V_{Rd} = V_{Rd,c1} = \left[\frac{0.18}{\gamma_c}\left(1+\sqrt{\frac{200}{d}}\right)\left(100\frac{A_{sl}}{b_w d} f_{ck}\right)^{\frac{1}{3}} + 0.15\sigma_{cp}\right] b_w d \tag{6}$$

$$V_{Rd} = V_{Rd,c2} = \frac{I b_w}{S}\sqrt{f_{ctd}^2 + \alpha_l \sigma_{cp} f_{ctd}} \tag{7}$$

In the previous equations, γ_c is a partial safety factor (equal to 1.5), A_{sl} is the area of longitudinal tensile reinforcement, *I* is the second moment of area whereas *S* is the first moment of area. Finally, f_{ctd} is the design value of the characteristic uni-axial concrete tensile strength. If Eqs. (1)–(7) are used to estimate the actual failure load, partial safety factors should be omitted and average material properties are to be used rather than characteristic or design values. Eqs. (2)–(3) and Eqs. (5)–(7) are then transformed to Eqs. (8)–(12).

$$V_{R,s} = \frac{A_{sw}}{s} z f_{ym} \cot\theta_{\max} \tag{8}$$

$$V_{r,max} = \frac{\alpha_{cw} v_1 b_w z f_{cm}}{\cot\theta_{\max} + (\cot\theta_{\max})^{-1}} \tag{9}$$

$$\cot\theta_{\max} = 2 + \frac{0.15\sigma_{cp}b_w d}{\frac{A_{sw}}{s} z f_{ym}} \leq 3 \tag{10}$$

$$V_{R,c1} = \left[0.18\left(1+\sqrt{\frac{200}{d}}\right)\left(100\frac{A_{sl}}{b_w d}f_{cm}\right)^{\frac{1}{3}} + 0.15\sigma_{cp}\right]b_w d \tag{11}$$

$$V_{R,c2} = \frac{Ib_w}{S}\sqrt{(0.67f_{ctm,fl})^2 + \alpha_l\sigma_{cp}(0.67f_{ctm,fl})} \tag{12}$$

In Eq. (12), the mean value of the uni-axial tensile strength is written as a function of the experimentally determined mean flexural tensile strength. The adopted relation is given in Müller et al. (2013). The results of the aforementioned calculation using Eqs. (8)–(12) are also indicated in Table 5. The theoretical failure load required to obtain a bending failure mode, derived from a general plane section analysis $V_{u,bend}$, is also indicated in the aforementioned Table 5.

Table 5: Experimentally observed and analytically predicted failure load and failure mode.

Specimen	Experiment		Eurocode 2 (European Committee for Standardization 2004, Bureau for Standardisation NBN 2010)			
	$V_{u,exp}$ (kN)	Failure mode	$V_{u,pred}$ (kN)	$V_{u,bend}$ (kN)	Failure mode	$\frac{V_{u,exp}}{V_{u,pred}}$ (−)
B101	377.7	S-DT†	158.1	412.2	S	2.39
B102	321.6	S-DT	158.1	329.6	S	2.03
B103	262.8	S-DT	243.3‡	329.6	S	1.08
B104	281.8	S-DT	135.2	406.6	S	2.08
B105	251.2	S-DT	135.2	325.3	S	1.86
B106	179.7	S-DT	206.9‡	325.3	S	0.87
B107	271.3	B	147.9	236.5	S	1.83
B108	213.8	B	147.9	189.2	S	1.45
B109	181.0	S-DT	197.0*	189.2	B	0.92
					$\overline{\frac{V_{u,exp}}{V_{u,pred}}}$	1.61
					COV[a]	55.4 %

† Shear failure mode due to diagonal tension.
‡ Obtained using Eq. (12).
[a] Coefficient of Variation.

Based on the results presented in Table 5, the following observations and preliminary conclusions can be made:

- In general, a poor correlation is found between the experimental results and analytical calculations according to EC 2 for all specimens apart for beams B103, B106 and B109 without shear reinforcement. Even if all partial safety factors are omitted and average material strength properties are used rather than characteristic or design values, an average experimental-to-predicted failure load ratio of 1.61 is found with a coefficient of variation (COV) equal to 55.4 %%.
- The failure mode is correctly predicted for all specimens apart for beams B107–B109. Specimens B107 and B108 failed due to bending despite having a lower shear capacity in comparison to the corresponding bending strength. Indeed, the

experimental failure load correlates well with the theoretical load required to obtain the bending capacity for the aforementioned specimens. Moreover, the wrongly predicted failure mode for specimen B109 can be attributed to the small difference in the analytically calculated shear and bending capacity.

- As expected, increasing the shear reinforcement ratio, increases the shear capacity (B102–B103, B105–B106, B108–B109).
- Increasing the shear span-to-effective depth ratio while keeping all other investigated parameters constant, consistently decreases the experimentally observed shear failure load (B101–B102, B104–B105).
- Increasing the prestressing force while keeping the longitudinal reinforcement ratio approximately constant, increases the shear capacity of specimens with (B101–B104, B102–B105) and without shear reinforcement (B103–B106).
- Decreasing the longitudinal reinforcement ratio while keeping the prestressing force constant does not significantly influence the failure load (B104–B07, B105–B108 and B106–B109). However, the failure mode shifts from a brittle shear induced failure mode towards a more ductile bending induced failure mode for specimens with shear reinforcement.

The found discrepancy between the experimentally observed and analytically calculated failure load in the case of prestressed concrete members with shear reinforcement is certainly to be further investigated. Therefore, the experimentally observed mechanical behaviour is assessed based on the acquired data of the optical (-numerical) measurement methods. Figure 11a, b present the experimentally determined angle $|\theta_{\epsilon 2}|$ between the principal compressive strain $\epsilon_2$with respect to the horizontal at mid-depth determined from the Krypton K600 CMMs respectively 3D-DIC as a function of the location along the *x*-axis and the applied load. The measured value for | $|\theta_{\epsilon 2}|$ is compared to the adopted value for θ used for the presented strength calculations, refer to Table 5. Similar results were obtained for the remaining test specimens.

Figures 11a and 11b indicate that the value of the experimentally determined angle $|\theta_{\epsilon 2}|$ in the middle of the shear span (x=1800 and 5200 mm) correlates well with the adopted value of the angle θ in the presented shear strength calculations, refer to Table 5. However, the value of $|\theta_{\epsilon 2}|$ is not constant along the longitudinal *x*-axis of the specimen contrary to the assumption made by the VATM approach. Instead, a parabolic course of the angle $|\theta_{\epsilon 2}|$ is observed which does not tend to change significantly if the loading is furthermore increased. This observation if clarified by Figs. 12a and 12b which presents the full-field magnitude and direction of the principal compressive strain field ϵ_2 for specimen B104 as a function of the surface coordinates and the applied load. Similar results were observed for the remaining specimens where the DIC technique has been adopted.

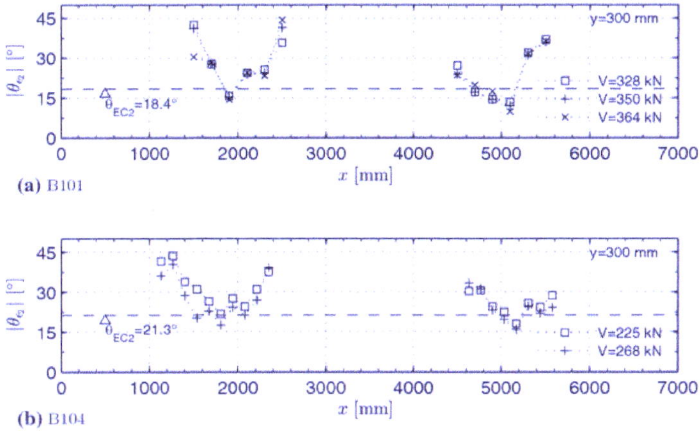

Figure 11: Experimentally measured angle $|\theta\epsilon 2||\theta\epsilon 2|$ between the direction of the principal compressive strain $\epsilon 2\epsilon 2$ as a function of the location along the *x*-axis and applied load for specimens **a**B101 (derived from Krypton K600 CMMs) and **b** B104 (derived from 3D-DIC).

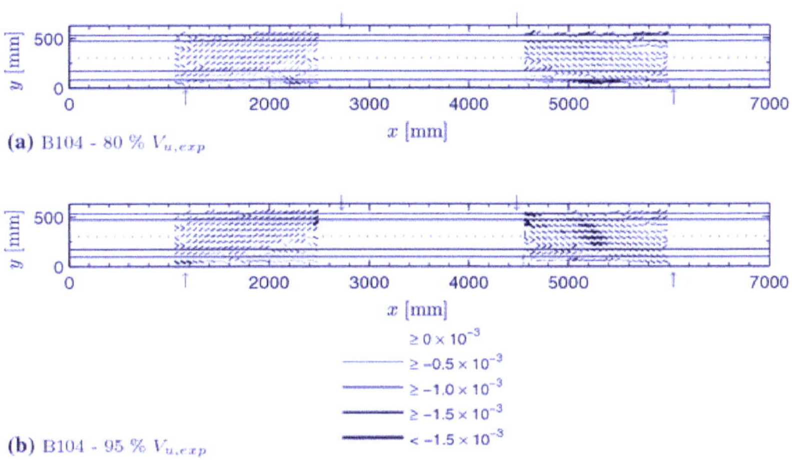

Figure 12: Experimentally determined direction and magnitude of the principal compressive strain field ϵ_2 of specimen B104 as a function of the surface coordinates and the applied load level (*Note* line where the angle $|\theta_{\epsilon 2}|$ was determined in Fig. 11b indicated with a *dotted line*).

Based on the direction and magnitude of the principal compressive strain field ϵ_2 presented in Figs. 12a and 12b, it can be observed that a direct compression strut is developed between the support point and the load application point which carries a significant amount of the applied load. The remaining part of the applied load fans out towards the bottom of the specimen. A similar observation can be made for the support point load. The force in both aforementioned fan regions is equilibrated by the vertical force in the shear reinforcement elements. A schematic representation of the identified structural behaviour as presented in Figs. 12a and12b is depicted in Fig. 13.

The identified structural behaviour as described above corresponds with the observed parabolic course of the direction of the principal compressive strain $|\theta_{\epsilon_2}|$. In view of the mechanical behaviour of the presented test specimens presented in Figs. 12a, 12b and 13, the experimentally observed failure mode should be interpreted as a splitting failure mode due to the forces in the fan regions.

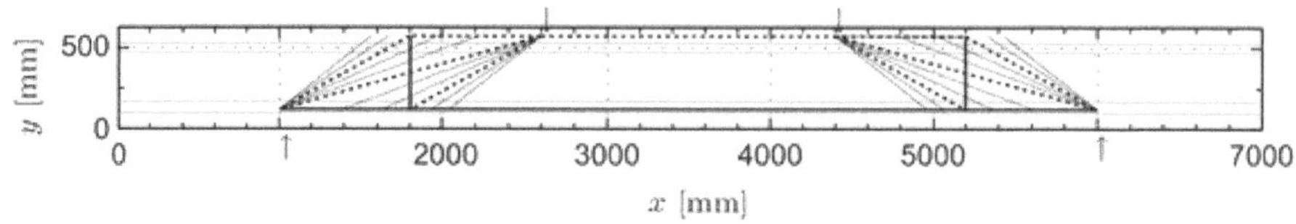

Figure 13: Schematic representation of the identified mechanical behaviour of the presented test beams (Members in compression indicated with a *dashed line*, member in tension indicated with a *solid line*).

Direct strut action is generally considered to be an important bearing mechanism for beams with a shear span-to-effective depth ratio less than or equal to 2.5 (Ramirez et al. 1998). However, the presented test beams were characterized by a shear span-to-effective depth ratio varying between 2.91 and 3.91, refer to Table 3. The identified structural behaviour, as presented in Figs. 12a, 12b and 13, provides a plausible explanation for the following experimentally observed phenomena.

- The possibility of carrying the applied shear force by means of direct strut action significantly increases the shear carrying capacity in comparison to the shear capacity obtained using the variable angle truss model as proposed by Eurocode 2. This provides a possible explanation why the current sectional shear design provisions found in EC2 performed poorly in predicting the shear capacity of the presented prestressed concrete beams.
- Figure 14 shows the typically observed profile of the horizontal strain ϵ_x at the top flange of the presented test beams. Figure 14 clearly shows that the horizontal strain at the top of the presented beams rapidly decreases to relatively low strains, and thus relatively low stresses, away from the loading point. Due to the inclined strut action, it is indeed expected that low strain values occur at the top of the specimen near the support point.

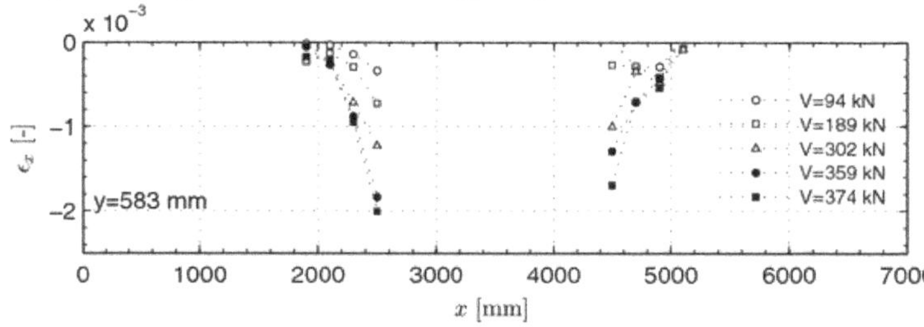

Figure 14: Typical experimentally observed horizontal strain profile at the flange of specimen B101 as a function of the applied load and location along the longitudinal x-axis.

CONCLUSIONS

This paper aims to investigate on the use of advanced optical(-numerical) measurement methods for the mechanical analysis of shear-critical prestressed concrete beams. Therefore, an experimental program consisting of nine full-scale prestressed I-shaped beams was drafted. The main investigated parameters were the amount of prestressing, the amount of longitudinal reinforcement and shear reinforcement and the shear span-to-effective depth ratio respectively. All specimens were subjected to a load-controlled four-point bending test until failure. During the experimental research, the use of two advanced optical(-numerical) measurement methods, i.e. *3D coordinate measurement machines* (CMM) and *stereo-vision digital image correlation*(3D-DIC), was explored. Firstly, the experimental setup was elaborated in detail. Specifically in the case of the DIC technique, a novel technique to apply numerically synthesised strain sensor patterns, i.e. speckle patterns, in a controlled way was presented. The presented technique allows for the application of tailor-made strain sensor patterns to virtually any given object's surface. A reference measurement was performed in unloaded state to asses the measurement precision. Both techniques were found to be comparable in terms of displacement and strain resolution. The maximum standard deviation of the in-plane displacements was equal to approximately 20×10^{-3} mm for the CMMs whereas a value of approximately 30×10^{-3} mm was found for the DIC technique. Moreover, the expected value of the strains occurring during the experiments well exceeded the observed noise levels on the in-plane horizontal and vertical strains. It can thus be concluded that both techniques are well suited for assessing the structural behaviour of the reported test specimens. However, due to the brittle and highly energy releasing failure modes observed during the tests on specimens failing in shear, the DIC technique is preferred over the CMM technique since the latter technique requires relatively expensive IR LED sensors to be glued onto the concrete side surface which can sustain damage at the moment of failure.

All specimens were designed to fail in shear. However, seven specimens failed in a brittle manner due to shear (diagonal tension failure mode) whereas two specimens failed due to bending in a ductile manner. The experimental results were compared to analytical calculations according to the current design procedures found in Eurocode 2 (EC 2). Based on the work presented in this paper, it can be concluded that EC 2, adopting the variable angle truss model, in general significantly underestimates the experimentally determined failure load. Omitting all partial safety factors and using average material strength properties rather than characteristic values resulted in an average experimental-to-predicted failure load ratio equal to 1.61 (coefficient of variation equal to 55.4 %%). Based on the extensive amount of experimental displacement and deformation data, it was found that the applied load was primarily carried by means of a direct compression strut in combination with fan regions contrary to the model adopted by EC 2 which assumes that a compression field with constant inclination along the member's axis is developed. The identified structural behaviour for the reported test specimens can be used to optimize current shear design provisions as proposed by codes of practice.

REFERENCES

1. American Concrete Institute. (2011). Aci 318–11 building code requirements for structural concrete and commentary.
2. Balázs, G. L. (2010). A historical review of shear. In V. Sigrist, F. Minelli, G. Plizzari, & S. Foster (Eds.), *Bulletin 57: Shear and punching shear in RC and FRC elements* (pp. 1–13). Lausanne, Switzerland: Fédération Internationale du Béton (fib).
3. Bossuyt, S. (2012). Optimized patterns for digital image correlation. In Proceedings of the SEM International Conference and Exposition on Experimental and Applied Mechanics, volume 3: Imaging Methods for Novel Materials and Challenging Applications, Costa Mesa, CA, USA, Society for Experimental Mechanics.
4. Bureau for Standardisation NBN. (2009a). Nbn en 12390–3.
5. Bureau for Standardisation NBN. (2009b). Nbn en 12390–5.
6. Bureau for Standardisation NBN. (2010). Nbn en 1992-1-1 anb.
7. Bureau for Standardisation NBN. (2014). Nbn en 12390–13.
8. Burg, J., Peeters, K., Natsakis, T., Dereymaeker, G., Vander Sloten, J., & Jonkers, I. (2013). In vitro analysis of muscle activity illustrates mediolateral decoupling of hind and mid foot bone motion. *Gait and Posture*, *38*(1), 56–61.
9. Canadian Standards Association. (2004). Csa a23.3 design of concrete structures (csa a23.3-04), 2004.
10. Collins, M. P. (2010). Improving analytical models for shear design and evaluation of reinforced concrete structures. In V. Sigrist, F. Minelli, G. Plizzari, & S. Foster (Eds.), *Bulletin 57: Shear and punching shear in RC and FRC elements* (pp. 77–92). Lausanne, Switzerland: Fédération Internationale du Béton (fib).
11. Cooreman, S., Debruyne, D., Coppieters, S., Lecompte, D., & Sol, H. (2008). *Identification of the mechanical material parameters through inverse modelling* (pp. 337–342). Non-Destructive Testing: Emerging Technologies in.
12. Cooreman, S., Lecompte, D., Sol, H., Vantomme, J., & Debruyne, D. (2007). Elasto-plastic material parameter identification by inverse methods: Calculation of the sensitivity matrix.*International Journal of Solids and Structures*, *44*(13), 4329–4341.
13. De Roeck, G., Wens, L., & Jacobs, S. (2004). 3d static and dynamic displacement measurements by a system of three linear ccd-units. In P. Sas & M. De Munck (Eds.), *International Conference on Noise and Vibration Engineering (ISMA)* (pp. 2289–2300), Leuven, Belgium: KU Leuven.
14. Deconinck, F., Desmet, W., & Sas, P. (2004). Increasing the accuracy of mdof road reproduction experiments: Calibration, tuning and a modified twr approach. In P. Sas & M. De Munck (Eds.),*International Conference on Noise and Vibration Engineering (ISMA)* (pp. 709–722), Leuven, Belgium: KU Leuven.
15. Destrycker, M. (2012). Experimental validation of residual stress simulations in welded steel tubes with digital image correlation. PhD thesis, KU Leuven,

Department of Civil Engineering.

16. European Committee for Standardization (CEN). (2004). Eurocode 2: Design of concrete structures—part 1–1: General rules and rules for buildings.

17. Fédération Internationale du Béton (fib). (2010). Shear and punching shear in RC and FRC elements—Workshop 15–16 October 2010, Salò (Italy), volume 57 of Bulletin, Lausanne (Switzerland).

18. Ivanov, D., Ivanov, S., Lomov, S., & Verpoest, I. (2009). Strain mapping analysis of textile composites. *Optics and Lasers in Engineering*, *47*(3–4), 360–370.

19. Jeong, J.-P., & Kim, W. (2014). Shear resistant mechanism into base components: Beam action and arch action in shear-critical rc members. *International Journal of Concrete Structures and Materials*, *8*(1), 1–14.

20. Knauss, W. G., Chasiotis, I., & Huang, Y. (2003). Mechanical measurements at the micron and nanometer scales. *Mechanics of Materials*, *35*(3–6), 217–231.

21. Lantsoght, E., van der Veen, C., Walraven, J., & de Boer, A. (2013). Shear assessment of reinforced concrete slab bridges. *Structural Engineering International*, *4*, 418–426.

22. Lava, P., Cooreman, S., Coppieters, S., De Strycker, M., & Debruyne, D. (2009). Assessment of measuring errors in dic using deformation fields generated by plastic fea. *Optics and Lasers in Engineering*, *47*(7–8), 747–753.

23. Lava, P., Cooreman, S., & Debruyne, D. (2010). Study of systematic errors in strain fields obtained via dic using heterogeneous deformation generated by plastic fea. *Optics and Lasers in Engineering*, *48*(4), 457–468.

24. Lava, P., Coppieters, S., Wang, Y., Van Houtte, P., & Debruyne, D. (2011). Error estimation in measuring strain fields with dic on planar sheet metal specimens with a non-perpendicular camera alignment. *Optics and Lasers in Engineering*, *49*(1), 57–65.

25. Lecompte, D. (2007). Elasto-Plastic material parameter identification by inverse modeling of static tests using digital image correlation. PhD thesis, Vrije Universiteit Brussel, Department of Mechanics of Materials and Constructions.

26. Müller, H., Breiner, R., Anders, I., Mechterine, V., Curbach, M., Speck, K., et al. (2013). *Code-type models for concrete bahviour* (Vol. 70)., Bulletin Lausanne, Switzerland: Fédération Internationale du Béton.

27. Nakamura, E., Avendano, A. R., & Bayrak, O. (2013). Shear database for prestressed concrete members. *ACI Structural Journal*, *110*(6), 909–918.

28. Ramirez, J. A., et al. (1998). Recent approaches to shear design of structural concrete. *Journal of Structural Engineering-Asce*, *124*(12), 1375–1417.

29. Sas, G., Blanksvard, T., Enochsson, O., Taljsten, B., & Elfgren, L. (2012). Photographic strain monitoring during full-scale failure testing of ornskoldsvik bridge. *Structural Health Monitoring-an International Journal*, *11*(4), 489–498.

30. Srikar, G., Anand, G., & Suriya Prakash, S. (2016). A study on residual compression behavior of structural fiber reinforced concrete exposed to moderate temperature using digital image correlation. *International Journal of Concrete*

Structures and Materials, pp. 1–11.

31. Sun, S. (2007). Shear behavior and capacity of large-scale prestressed high-strength concrete bulb-tee girders. PhD thesis, University of Illinois at Urbana-Champaign, Department of Civil and Environmental Engineering.
32. Sutton, M. A., Orteu, J. J., & Schreier, H. W. (2009). *Image correlation for shape, motion and deformation measurements: basic concepts, theory and applications.* New York, NY: Springer.
33. Valerio, P., Ibell, T., & Darby, A. P. (2011). Shear assessment of prestressed concrete bridges.*Proceedings of the Institution of Civil Engineers-Bridge Engineering, 164*(4), 195–210.
34. Van Paepegem, W., Shulev, A. A., Roussev, I. R., De Pauw, S., Degrieck, J., & Sainov, V. C. (2009). Study of the deformation characteristics of window security film by digital image correlation techniques. *Optics and Lasers in Engineering, 47*(3–4), 390–397.
35. Wang, Y., Lava, P., Debruyne, D., & Van Houtte, P. (2011). Error estimation of dic for heterogeneous strain states. *Advances in Experimental Mechanics VIII, 70,* 177–182.

13
CHAPTER

LONG-TERM MONITORING OF REINFORCED EARTH STRUCTURES USING DISTRIBUTED FIBER OPTIC SENSING

Florian Moser, Werner Lienhart, Helmut Woschitz, Hartmut Schuller

Keywords: Brillouin sensing, BOFDA, Field application, Calibration, Long-term monitoring

INTRODUCTION

The Semmering Base Tunnel, which is currently under construction, is one of the core infrastructure projects of Austria. Moreover, it is part of the European TEN-T Core Network Corridor that connects the Baltic and the Adriatic Sea. The Semmering Base Tunnel with a total length of 27.3 km will realize a high-speed rail connection according to the latest requirements for international rail traffic [2] in 2025. Prior to tunnel excavation, several preparatory works were necessary, including a disposal site for 4.25 million m^3 of excavated material. This disposal site is located in the so-called "Longsgraben", an uninhabited narrow valley close to "Fröschnitzgraben", one of the main access points for tunnel construction. Before the tunnel material fills up the narrow valley, the flowing stream had to be relocated from the valley floor to an up to 50 m higher position. For this purpose, reinforced earth structures were built over a length of more than 1.3 km and heights of up to 25 m on the steep slope of the valley.

Because material from 12 tunnel construction access points will be deposited in this one disposal site, any malfunction of this construction will result in a delay of the whole tunneling project. Thus, an appropriate monitoring program was needed.

In this paper, we present the developed monitoring solution to assess the operating grade of the used reinforcement. We focus on the calibration results of the fiber

Citation: Florian Moser, Werner Lienhart, Helmut Woschitz, Hartmut Schuller, Long-term monitoring of reinforced earth structures using distributed fiber optic sensing, DOI 10.1007/s13349-016-0172-9.

optic-based monitoring system and the strain results of more than one year of field measurements.

REINFORCED EARTH STRUCTURES AT THE DISPOSAL SITE

The excavation material from tunneling will fill up the valley at the disposal site Longsgraben up to more than 50 m. After filling, the new terrain will be afforested and the in advance relocated stream will mark the new valley bottom, see Fig. 1. To realize the relocation, reinforced earth structures are used over a length of about 1.3 km at the steep valley side. The reinforced earth structures feature thereby a slope angle of 60°. A special challenge is the final section of the relocation construction, where the stream is guided back to its original position. In this section, the new stream bed has a slope angle of more than 50 % (Fig. 2). A maintenance road is located at the crown of the reinforced earth structure. This road is reachable via a 25-m-high reinforced earth ramp at the final section of the relocation construction (Fig. 2). The earth structures for the relocation and access ramp use geogrids from Huesker Synthetic as reinforcement. The geogrids were dimensioned with the maximum value of tensile strength and strain behavior taken from both, conventional analysis and finite element simulations [8].

A specially designed monitoring system was installed to determine the structural behavior during the installation and after completion of the construction works. The monitoring program included weekly conventional geodetic measurements in various cross sections of the stream relocation. In the critical final section of the relocation construction a fiber optic strain measurement system was additionally installed to monitor the operating grade of the geogrids.

Further details on design and construction of the reinforced earth structures and on the geotechnical monitoring can be found in [6] or in more detail in [9].

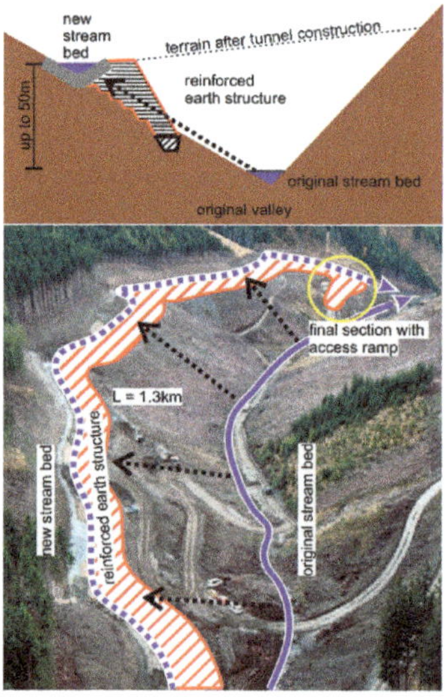

Figure 1: Reinforced earth structure to relocate the natural stream bed at the Longsgraben valley.

MEASUREMENT SYSTEM

Since the aim of the project was to determine the operating grade of the used geogrids, a measurement system which is integrable into the earth structure and mountable to the geogrids had to be used. As the exact type of geogrids was unknown until the beginning of the construction works, none of the existing monitoring systems could be used. Based on these requirements a monitoring solution using fiber optic measurements was developed by the Institute of Engineering Geodesy and Measurement Systems (IGMS) of Graz University of Technology. The uncertainty of the fiber optic strain measurement should be less than 0.1 %. A detailed description of the used components can be found in [6].

The monitoring program focuses on four cross sections (CS). Two cross sections (CS 3A and CS 3B) are located in the final section of the relocation construction, see Fig. 2. The other two cross sections (CS 1 and CS 2) are located in the reinforced earth ramp of the maintenance road. In the cross sections, up to five levels are equipped with sensing cables. Each level is thereby subdivided by adapters into segments [2–7 level segments (ls) per level]. At the end of the construction works we have installed almost 100 measurement segments.

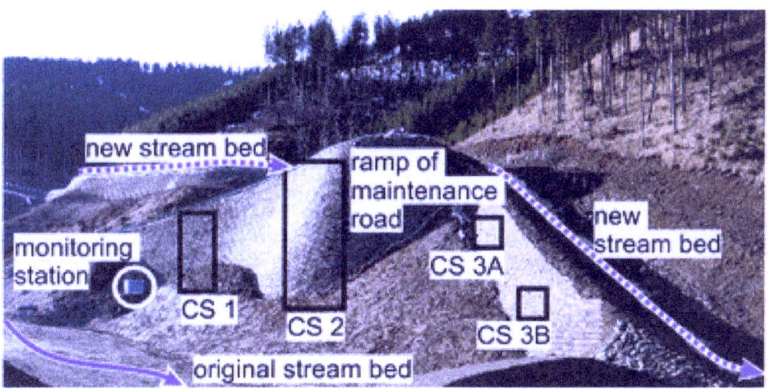

Figure 2: Monitoring cross sections (CS) at the final section of the stream relocation.

Due to the large size of the monitoring object and the high number of measurement segments it was decided to use a distributed Brillouin backscattering system. Brillouin scattering occurs when light traveling in a single-mode optical fiber is reflected by the refractive index modulations produced by acoustic waves (e.g., [3] or [7]). In our application, the scattered light is recorded and the frequency is analyzed with a spatial resolution of 0.5 m by a Brillouin optical frequency-domain analysis (BOFDA) interrogation unit (Fibris Terre fTB2505). Through determination of frequency changes between two measurement epochs it is in our setup possible to calculate strain and temperature variations.

Figure 3 shows exemplarily the Brillouin frequency spectrum for a 60-m-long fiber. Clearly visible is the center frequency of about 10.55 GHz of the unstrained fiber. In this example, strain was applied to different sections (25, 36 and 42 m) of the fiber, which results in significant shifts of the Brillouin frequency.

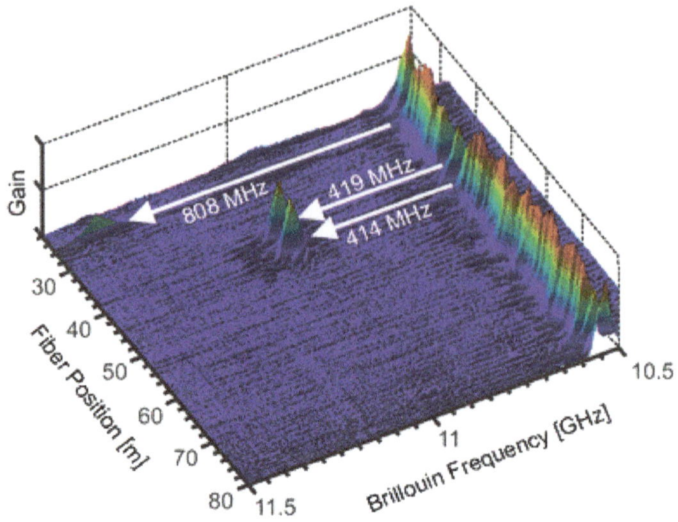

Figure 3: Brillouin frequency spectrum along an optical sensing fiber.

Since the BOFDA measurement principle requires a closed loop, each measurement level of the reinforced earth structure consists of a forward and a return path, see Fig. 4 left. The forward path was pre-strained during installation and is consequently sensitive to both temperature and strain. The stress-free installed return path is only sensitive to temperature changes. By combining both sides a temperature compensation is realized. However, the measurements cannot be performed if the fiber loop is interrupted at any location. Therefore, installations in harsh environments require a robust sensing cable to be embedded into the earth structure and a well-trained installation crew. In our application, a total of 2 km of the sensing cable BRU-Strain V4 from Brugg Cables were installed. Using specially designed adapters an appropriate connection to the geogrids could be realized. Extensive laboratory investigations confirmed a reliable strain transfer from the geogrid to the fiber core [6].

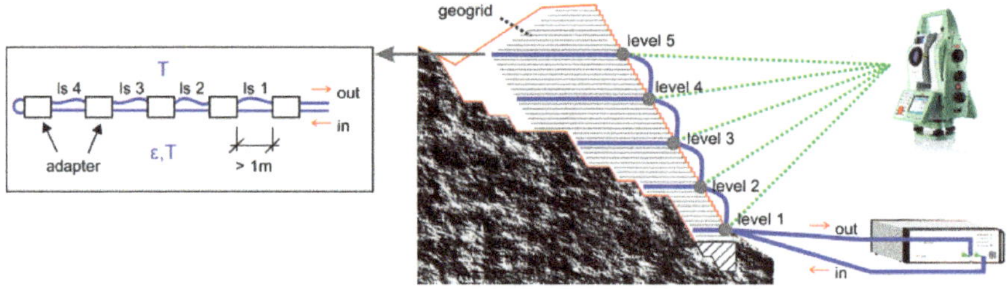

Figure 4: Monitoring levels in CS 2 with installed fiber optic sensing cable (*right*); level segments (ls) in level 5 (*left*).

SYSTEM CALIBRATION

The detected Brillouin frequency linearly depends on longitudinal strain [3] and temperature [4]. Therefore, the change in the detected Brillouin frequency $\Delta\nu_B$ between two measurements is a function of strain Δ_ϵ and temperature Δ_T variations:

$$\Delta\nu_B = \left.\frac{\partial\nu_B}{\partial\epsilon}\right|_{T=const.} \cdot \Delta\epsilon + \left.\frac{\partial\nu_B}{\partial T}\right|_{\epsilon=const.} \cdot \Delta T \tag{1}$$

We introduce the linear strain and temperature coefficients k_ϵ and k_T for the partial derivations

$$k_\epsilon = \left.\frac{\partial\nu_B}{\partial\epsilon}\right|_{T=const.} \tag{2}$$

$$k_T = \left.\frac{\partial\nu_B}{\partial T}\right|_{\epsilon=const.} \tag{3}$$

Accurate values for the strain k_ϵ and temperature k_T sensitivities are essential for the accurate determination of longitudinal strain and temperature.

Strain Calibration

The sensor testing process was performed as a sample testing procedure. Since the whole 2 km sensor cable came from the same production series, the calibration value of a section of the cable was considered valid for the whole sensing cable. The testing process was carried out under laboratory conditions in June 2013 prior to the field installation. The sensor cable was installed between two specially designed adapters in the test facility (Fig. 5) with a distance of 1 m between the adapters. The adapter design ensures that the applied strain is axial to the sensor. The strain of the sensor cable was increased in five steps from 0 to 3 %. The reference strain was measured by displacement transducers with a precision of about 10 µm. Since the finite element simulations of the reinforced earth structures showed a high extension of the geogrids, the testing process was performed up to a strain of about 3 %. Figure 6 shows the result of the calibration process with two runs, which took about 3 h. The results of both runs coincide well, with a maximum deviation of 15 MHz. Using all data, a linear strain dependence of 479 MHz/% was determined by these measurements with a precision of 5 MHz/%. This value is in agreement with the approximate 500 MHz/% stated by the manufacturer [1]. Using the manufacturer coefficient, however, would result in an error of 4 percent of the actual value.

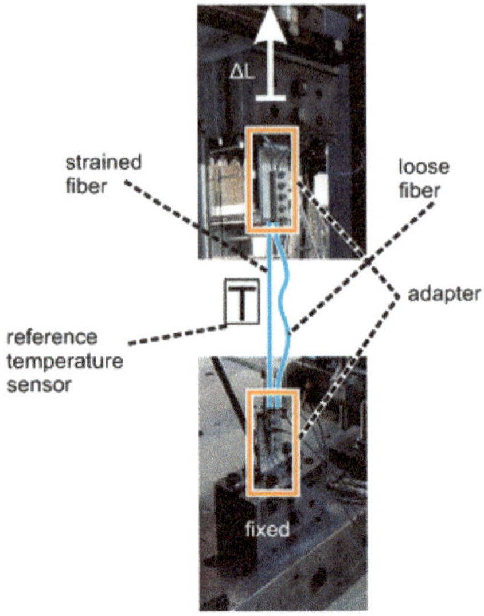

Figure 5: Strain coefficient determination: sensor testing facility.

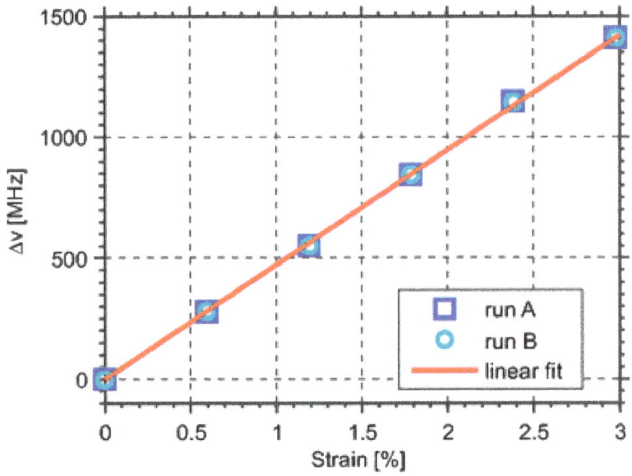

Figure 6: Strain coefficient determination: calibration results.

Temperature Calibration

According to [5] the temperature dependency of the Brillouin frequency shift k_T in jacketed single-mode silica fibers consists of the temperature dependency of the Brillouin frequency shift of the bare fiber k_T^* and the frequency shift induced by the thermal strain of the jacketed fiber

$$k_T = \left(k_T^* + k_\epsilon \cdot \frac{\partial \epsilon}{\partial T}\right) \tag{4}$$

The thermal strain in a strain-free fiber is thereby caused by the difference in the thermal expansion coefficients of the bare fiber and the coating material [5]. Temperature changes Δ_T can be calculated from the measured frequency shifts of the strain-free fiber if the combined effect k_T is known

$$\Delta T = \frac{\Delta \nu_B}{k_T} \tag{5}$$

The temperature changes are needed for the subsequent compensation of the measured strain values. To determine k_T, a sensor sample of 10 m length was carefully installed strain free in a climate chamber. In this chamber the temperature was changed in seven steps from 0 to +60 °C, which was the expected range in our application. The reference temperature was measured with two calibrated PT100 temperature sensors. Figure 7 shows the climate chamber and Fig. 8 the results of seven measured temperature cycles. The calculated temperature coefficient of the jacketed fiber k_T is 2.41 MHz/K with a precision of 0.01 MHz/K. This value is approximately twice as high as the typical value of the bare fiber ($k_T^* \sim$ 1.2 MHz/K, see e.g., [5]).

Figure 7: Temperature coefficient determination: climate chamber with sensor cable and reference temperature sensors (PT100).

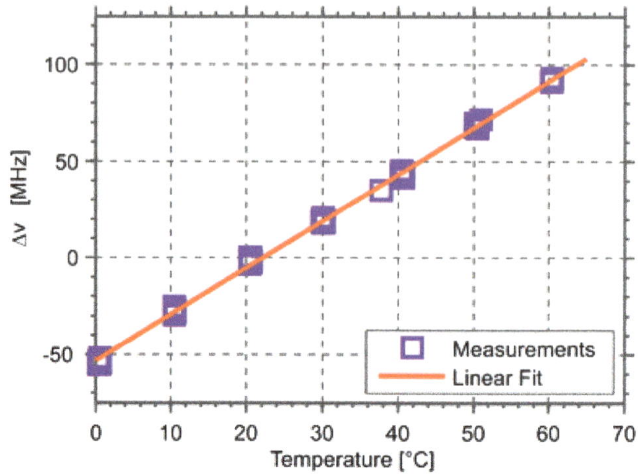

Figure 8: Temperature coefficient determination: measurement results.

FIELD MEASUREMENTS

At the disposal site all measurement segments (ls) in the different levels inside the cross sections (CS) can be measured simultaneously from a central monitoring station. The Brillouin frequencies of the entire 2 km of sensing cable are scanned by a single measurement. The software developed at our institute automatically allocates the measured frequencies to the level segments of the cross sections and calculates the temperature compensated frequency changes. Finally, the determined calibration functions are used to convert this frequency changes into strain values.

Spatial Allocation

As expected we observed a discrepancy of the spatial allocation of the level segments in measurements of different epochs. Figure 9 shows the detected frequencies of the segments of level 2 of cross section CS 2 for the measurements in February and June 2014. It can be seen that the two signals are shifted by ΔL of about 0.5 m. Such an effect could be caused by various factors. Since the determination of the position along the fiber uses the propagation velocity and the runtime of the signal, it is dependent on the used refractive index *n* of the sensor fiber. The refractive index, however, is varying with temperature. As a consequence temperature changes between two measurement epochs cause a runtime change. Another possible factor is a length change in prior level sections caused by a deformation of the reinforced earth structure. A further factor can be a physical length change of the coating material with temperature in the fiber sections prior to the strained segments.

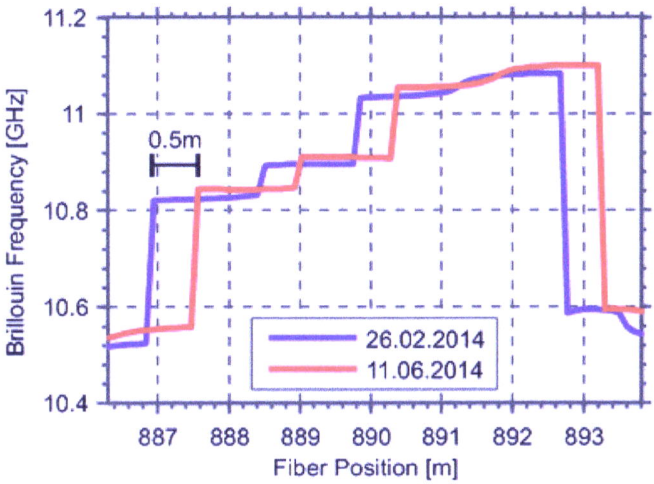

Figure 9: Spatial allocation discrepancy in the strained section of the 2nd level in CS 2 between the measurements in February and June 2014.

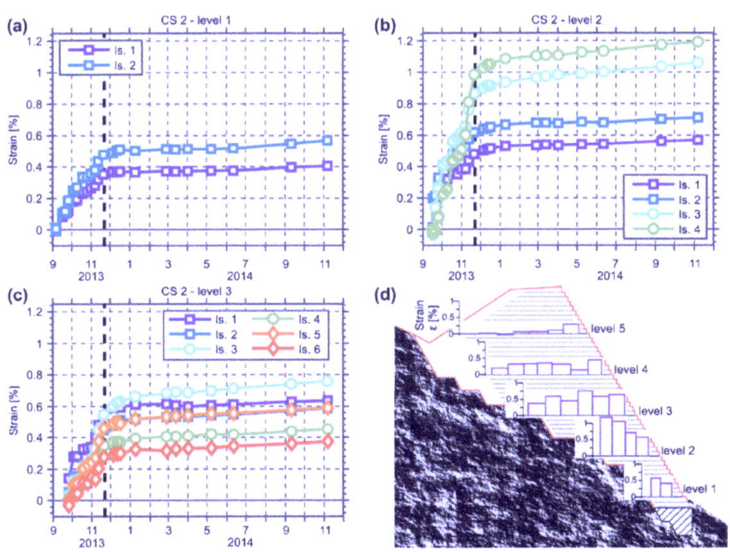

Figure 10: Strain development of the segments (ls) in the levels 1–3 in cross section CS 2; the*vertical bold dashed line* marks the end of the construction works; the sub-figure in the *bottom right* (**d**) shows the accumulated strain distribution inside the reinforced earth structure in November 2014.

In our application, different sections of the sensing cable experience different temperature changes. Connecting cable sections on the surface of the structure are exposed to direct solar irradiation and consequently large temperature differences (Δ_T approx. 30 K) over the year. Cable sections inside the reinforced earth structure experience low temperature differences (Δ_T approx. 0–10 K) over the year. A correction with a single adjusted refractive index or a correction with a single value for the expansion of the coating material is, therefore, not possible.

For these reasons we corrected for our application the fiber position in post-processing for each level between two measurement epochs by cross correlating the corresponding signals of the level.

Strain Evolution in Cross Section CS 2

Figure 10a–c shows some temperature and spatially corrected strain results of cross section CS 2 and the strain distribution (Fig. 10d) inside the earth body for a period of about 1 year. In this cross section five geogrid levels were equipped with the fiber optic sensing cable. Displayed is the strain evolution of the first three levels and their position inside the reinforced earth body. The measurements were carried out once a week during the construction phase (until end of November 2013) and sporadically afterwards. Obviously, construction works in the upper levels result in an increase of the geogrid strain in the lower levels of the reinforced earth structure. During the construction phase the strain increases thereby strongly in all segments. After the completion of the reinforced earth body the strain still continues to increase, but at a much lower rate.

Due to this continuing increase of the strain values and to gather experiences about the performance of the system over several years, the monitoring program will be continued until the disposal site is completely filled with the tunnel excavation material.

Figure 10d shows the accumulated strain inside the reinforced earth structure for the different segments in November 2014. The obtained results show an inhomogeneous strain distribution inside the reinforced earth structure. The highest observed strain can be seen in level 2 of cross section CS 2 with about 1.2 %. Nevertheless, this is much lower than the strain predicted by the finite element analysis, which shows the importance of field measurements for model validation. A comparison between the forces predicted by the finite element analysis and the measured fiber optic values can be found in [6] or in more detail in [8].

CONCLUSION

In this paper, we presented a robust monitoring system to measure the inhomogeneous strain distribution inside reinforced earth structures. The developed solution is, especially because of the modular adapters, flexible and can be used with many types of geogrids. It also survived the harsh installation process and is 1 year after installation, with strains up to 1 % and higher, still fully operational.

We also demonstrated the importance of calibration for reliable and accurate distributed fiber optic strain and temperature measurements. Both coefficients for temperature and strain sensitivity have been determined with high precisions ($\sigma_{k_T} = 0.01$ MHz/K and $\sigma_{k_\varepsilon} = 5$ MHz/%). Finally it has to be noted that the correct spatial allocation is crucial in long-term monitoring applications.

ACKNOWLEDGMENTS

We want to thank the Austrian Federal Railways (ÖBB) for the funding of this innovative monitoring project, especially the departments of ÖBB-Research and Development as well as ÖBB-Infrastruktur AG: Tunneling, Surveying and Data Management. We also want to thank Ferdinand Klug, Dietmar Denkmaier, Stefan Lackner, Gian-Philipp Patri, Gerhard Kleemaier, Rudolf Lummerstorfer, Matthias Ehrhart and Christoph Monsberger for their valuable support during the field installation.

REFERENCES

1. Brugg Cables AG (2012) BRUSENS STRAIN V4. VERS. 2012/06/19 REV. 01: Data Sheet
2. Gobiet G (2013) The new Semmering base tunnel—project overview. Geomech Tunnel 6(6):680–687. doi:10.1002/geot.201310022
3. Horiguchi T, Kurashima T, Tateda M (1989) Tensile strain dependence of Brillouin frequency shift in silica optical fibers. IEEE Photon Technol Lett 1(5):107–108
4. Kurashima T, Horiguchi T, Tateda M (1990) Thermal effects of Brillouin gain spectra in single-mode fibers. IEEE Photon Technol Lett 2(10):718–720
5. Kurashima T, Horiguchi T, Tateda M (1990) Thermal effects on the Brillouin frequency shift in jacketed optical silica fibers. Appl Opt 29(15):2219–2222
6. Lienhart W, Moser F, Schuller H, Schachinger T (2014) Reinforced earth structures at Semmering base tunnel—construction and monitoring using fiber optic strain measurements. In: 10th International Conference on Geosynthetics (10ICG)
7. Nöther N (2010) Distributed fiber sensors in river embankments: advancing and implementing the Brillouin optical frequency domain analysis, vol 64. BAM Dissertationsreihe
8. Schuller H, Schwingshackl I, Moser F, Lienhart W (2014) Semmering Basistunnel neu—geotechnisches Monitoring mit faseroptischen Messsystemen beim Bau von Bewehrte-Erde-Stützkonstruktionen in der Deponie Longsgraben. 29. Christian Veder Kolloquium: Stützmassnahmen in der Geotechnik—Bemessung, Ausführung, Langzeitverhalten 51:107–122
9. Schuller H, Schwingshackl I, Schachinger T, Moser F, Lienhart W (2014) Semmering Basistunnel neu - faseroptische Messsysteme zur Erfassung der Dehnungen beim Bau von Bewehrte-Erde-Stützkonstruktionen in der Deponie Longsgraben. Messen in der Geotechnik 2014(98):291–311

14
CHAPTER

RAILWAY TRACK STIFFNESS MEASUREMENT

Ping Wang, Li Wang, Rong Chen, Jinhui Xu, Jinmang Xu, Mingyuan Gao

Keywords: Railway, Track structure, Track stiffness, Measurement, Maintenance, Vibration

INTRODUCTION

Track structure, a load bearing and transmitting structure of railway system, has stiffness to provide necessary elasticity. Track structure comprises several layers and each layer has its own stiffness, namely the track component stiffness. The overall track stiffness is usually adopted to characterize the total stiffness of all layers and reflect the stiffness characteristics of the whole track structure [1, 2]. The stiffness measurement of a track, generally, refers to measuring the overall stiffness. In laboratory tests, the stiffness of certain track component can be measured, such as fastener stiffness and ballast stiffness [3, 4].

Discrete rail support model and continuous rail support model are two classical models for the traditional mechanical analysis of railway track [1, 2]. In the discrete rail support model, the stiffness of fastener supporting track is a spring constant, while in the continuous rail support model, it is denoted by track modulus, which is calculated with the spring constant divided by the spacing between centers of discrete supports. The stiffness measurement method based on the Winkler elastic foundation model uses track modulus to characterize the overall track stiffness.

The track stiffness can also be classified into dynamic stiffness and static stiffness [2, 5, 6]. The resistance of track structure to deformation under a static load is called as static stiffness, which, usually, is determined by the deformation degree of track

Citation: Ping Wang, Li Wang, Rong Chen, Jinhui Xu, Jinmang Xu, Mingyuan Gao, Railway Track Stiffness Measurement, DOI: 10.1007/s40534-016-0104-8.

structure under the static load. The dynamic stiffness refers to the resistance of track structure to deformation under certain dynamic excitation, which is determined by the natural frequencies of track structure. The track dynamic stiffness reflects the supporting performance of the vibrating track structure and therefore is an important factor influencing the wheel–rail interaction and train running performance. However, the static stiffness is always adopted instead for the traditional mechanical analysis of the wheel–rail system. The rapid development of high-speed railway passenger transportation, heavy-loaded train transportation, and railway engineering technologies makes the theoretical analysis and practical measurement of dynamic stiffness significant.

Suppose that under load P the rail beam will deform. If the deformation amount is y, the track static stiffness K will be

$$K = \frac{P}{y}. \quad (1)$$

Track structure is characterized by material nonlinearity and geometrical nonlinearity [7]. For example, the rubber pad in the fastening system is of material nonlinearity, and voids are distributed between the sleeper and ballast. Thus, the track static stiffness is also nonlinear, as shown in Fig. 1.

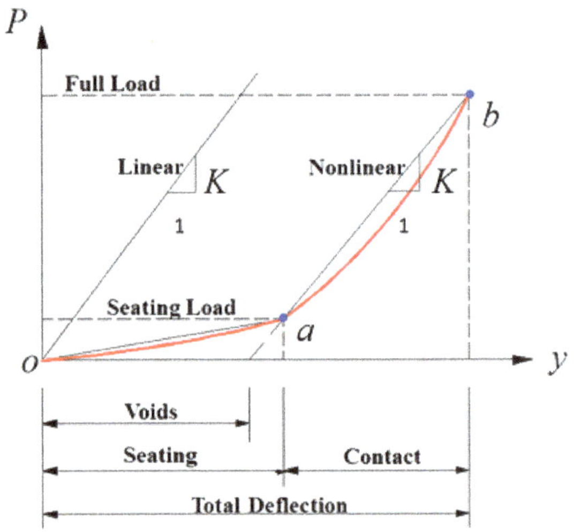

Figure 1: Load–deflection diagram illustrating nonlinearities.

In order to take into account the nonlinearities of track structure, the track static stiffness is usually calculated by the slope of the linear part of the loading history curve [2, 8]. In this sense, the tangent stiffness and secant stiffness on the loading curve or unloading curve can be defined:

$$K_{a-b} = \frac{P_b - P_a}{y_b - y_a}, \quad (2)$$

where a and b can be defined according to the specific requirement. For example, define b as the final loading state and a as the exact release state of the fastener clamping force.

The dynamic stiffness is influenced by many factors, such as the applied load and the frequency of excitation [9–11]. In order to conduct a quantitative study on the influence of the factors like loading frequency and amplitude on the track dynamic stiffness, a vertical harmonic force $P = P_0 e^{j\omega t}$ is exerted on the rail. If the dynamic displacement at the force acting point is $Z = Z_0 e^{j(\omega t+\varphi)}$, the track dynamic stiffness can be defined as the ratio of the acting force amplitude to the dynamic displacement amplitude:

$$k_d = \frac{|P|}{|Z|}. \quad (3)$$

Under this definition, the track dynamic stiffness can be adopted to reflect the stiffness characteristics in steady forced vibration at a certain frequency.

In mechanical engineering, the dynamic stiffness is always studied in frequency domain. Similarly, the dynamic force and dynamic displacement can be processed through Fourier transform or to use the transfer functions by assuming that the stiffness is linear to some extent. Thus, the track dynamic stiffness can be defined as the ratio of dynamic load to dynamic displacement in frequency domain:

$$k(f) = \frac{F(f)}{z(f)}. \quad (4)$$

Now, the track dynamic stiffness is a complex value, with the amplitude and phase [8, 12]. The track dynamic stiffness can also be measured and analyzed through its reciprocal, namely the track dynamic flexibility.

SIGNIFICANCE OF TRACK STIFFNESS MEASUREMENT

Track stiffness is an important factor influencing the safety and stability of train operation and the vibration and deformation of track structure as well as the dynamic response of substructures like subgrades and bridges [13–22].

Researches in relation to track stiffness mainly focus on three aspects:

- the analysis of the influence of track stiffness on the vehicle-track-substructure coupled dynamics;
- the nonlinear characteristics of track component stiffness, such as the nonlinearities of the rubber pad; and
- the evaluation methods of track stiffness and the reasonable stiffness for track structure and its components.

Stiffness Design of New Railway Lines

Stiffness design of new railway lines is to determine overall track stiffness and track component stiffness to meet the operation conditions of new railways. The measurement of track stiffness can help evaluate the stiffness design of new railways and present the relevant modification opinions.

Daily Maintenance of Railway Lines

Track stiffness is also an important parameter for maintenance of railway lines in addition to the track geometric irregularity [23]. For most maintenance methods, the track geometric irregularity of railway lines is always adopted as the diagnostic parameter. The track geometry can be adjusted by ballast tamping to keep it within the limited range. However, the performance of ballasts, mostly, is dependent on the performance of bottom subgrade or other substructures. Ballast tamping cannot help improve the performance of substructures, such that track geometric irregularity repeatedly occurs.

The measurement of track stiffness for maintenance purpose covers four aspects:

- *Low track stiffness* The railways built in soft soil areas and bad soil zones tend to experience serious rail displacement and fast cracking of track components after a period of operation. These diseases are mainly caused by poor stiffness of soil subgrade and therefore it should be strengthened. It is very important for checking the railway lines in time to keep safe train operation.
- (*Variable track stiffness* The stiffness always changes abruptly at the railway–bridge, railway–tunnel, and bridge–tunnel sections, and therefore transition zones shall be set [24]. Besides, when track geometric irregularity is deteriorated, support for the track will be uneven.
- *Virtual track stiffness* Diseases, such as dirty ballast, hanging sleepers, and loose fasteners, can cause deformation and holes, leading to virtual stiffness for the rail support. The virtual stiffness represents potential risk to safe train operation.
- *Assortative stiffness* The unmatched supporting stiffness of the right and left rails will result in uneven rail displacement, which can cause center bending of sleepers as well as the other unreasonable stress and cracking problems of track components.

The measurement of track stiffness can help direct the routine maintenance of railway lines and provide a reliable basis for making optimal maintenance strategies.

TRACK STIFFNESS MEASUREMENT METHODS

Standstill Measurement

The standstill measurement of overall track stiffness means that a measurement point is decided in advance, and then the overall track stiffness can be calculated by measuring displacement and vertical force exerted on this point.

The overall track stiffness, generally, can be measured based on four methods: (1) traditional hydraulic jack-loading method [1, 2], (2) impact hammer method [25, 26], (3) FWD method [9, 27], and (4) TLV method [2, 5, 28–31]. In the traditional hydraulic jack-loading method, a certain force is exerted on the rail and then rail deflection is measured with a displacement meter or a dial indicator, and thus the force–displacement curve can be obtained (see Fig. 2). The overall track stiffness can then be calculated according to stiffness definitions like secant stiffness or tangent stiffness.

Figure 2: Jack-loading diagram and vertical rail deflection measurement.

The traditional hydraulic jack-loading method has been adopted since the beginning of the twentieth century. Figure 3 shows a flatbed trailer loaded with I-typed bars used by the Talbot Committee as a reaction wall to measure the track stiffness in 1918 [1].

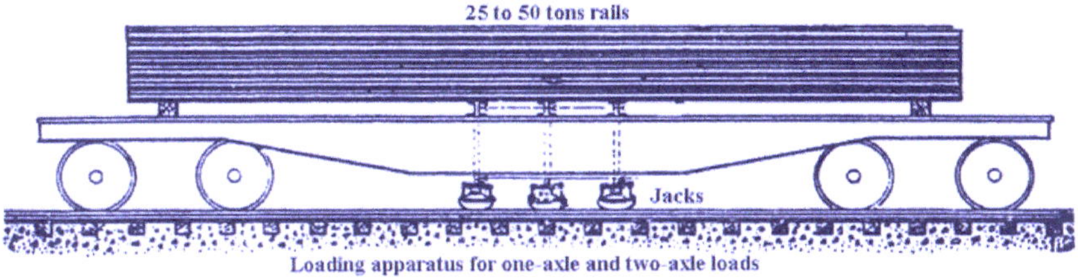

Figure 3: Track stiffness measured by the Talbot committee of the USA (1918).

In the impact hammer method, the track vibration is measured with acceleration transducers installed on rail or sleepers (track plates) after an impulse load is exerted on the track with an impact hammer. The hammer head is equipped with a force transducer to measure the impulse and thus the transfer function of the track can be obtained. At last, the track component stiffness and the overall track stiffness can be calculated through parameter identification (see Fig. 4). Typically, the impact hammer method can cover a frequency interval of 50–1,500 Hz, which depends on the material of the hammer head [32]. A soft-rubber hammer head is suitable for lower frequencies than a hard-metal one. The impact hammer method is not reliable for the frequencies less than 50 Hz.

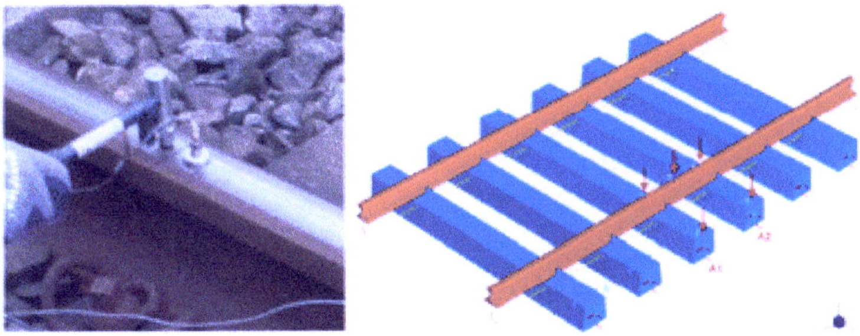

Figure 4: Track stiffness measurement using the impact hammer method.

In the FWD (falling weight deflectometer) method, a mass impacts a track and the vibration response is measured at the same time. Afterwards, the track stiffness characteristics can be explored through transfer function calculation or other methods. The FWD method, to a certain extent, can reflect the impact effect on a track when a train is running at a high speed. The standard FWD method uses a 125 kN free-falling mass to impact the track. The vibration response of track is generally measured with velocity transducers or geophones.

The principle for the track loading vehicle (TLV) method is the same as that of the traditional hydraulic jack-loading method. However, this method is easier and can provide a larger vertical force (see Fig. 5). Organizations that have TLVs include Transportation Technology Center, Inc. (TTCI) and DECAROTOR of the USA, the South Africa BSSM, Delft of Holland, the railway department of Sweden, etc.

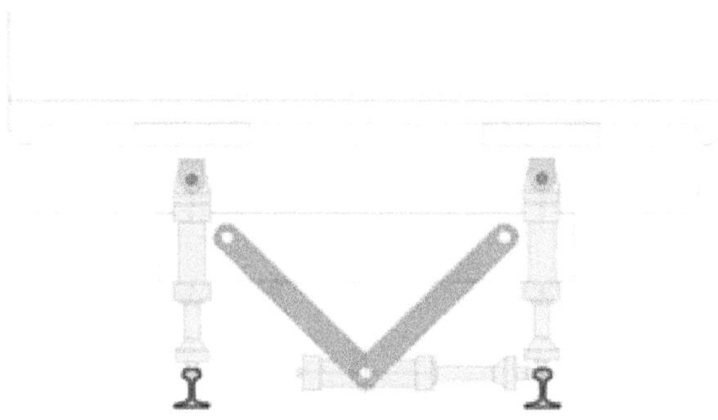

Figure 5: Track loading vehicle.

Queensland University of Technology in Australia improved a movable track stiffness measurement vehicle based on common TLV. This vehicle is composed of three cars. The front one is a six-axle towing locomotive (weight of 90 tons) to tow the vehicle to the specified measurement point. The middle one is a buffer car for buffering the influence of the front car on the rear one during the measurement. The rear car is equipped with measurement instruments and is the key part of the vehicle. The rear car weighs 57 tons totally (each part of 14.3 tons) and is equipped with hydraulic jacks and displacement transducers to exert vertical force and measure track deflection basin. When the vehicle reaches the specified measurement point, the hydraulic jacks extend from both sides of the rear car and act on the track separately. The deflection basin is recorded by 22 transducers to calculate the track modulus.

Continuous Measurements

By the time of November 2015, there are 61 high-speed railway lines in China, and the total mileage has reached 12,000 km. The rapid development of high-speed railway lines has made the accurate and continuous measurement of track stiffness extremely important.

The above methods for standstill measurement of overall track stiffness are arduous, time consuming, and not suitable for a long-distance and multi-point measurement.

The maintenance of railway lines requires continuous track stiffness measurement equipment. Therefore, many organizations have started to develop vehicles for continuous track stiffness measurement.

As early as in 1997, China Academy of Railway Sciences (CARS) put forward the conception of a vehicle for track elasticity measurement [33]. This vehicle comprises a heavy car in the front and a light car in the rear. The axle load of the heavy car can be varied by adjusting the number of concrete blocks in the range of 150–250 kN. Thus, the influence of different train axle loads on the measured results can be obtained. The light car weighs 40 kN and is used to eliminate the clearances between the rail and sleeper as well as between the sleeper and ballast bed. The measurement equipment on both heavy and light cars is the same except for the axle load.

This vehicle measures track geometric irregularities, which is similar to the chord measurement method. The main purpose of the vehicle is to measure the elastic deformation y_K of tracks, as shown in Fig. 6.

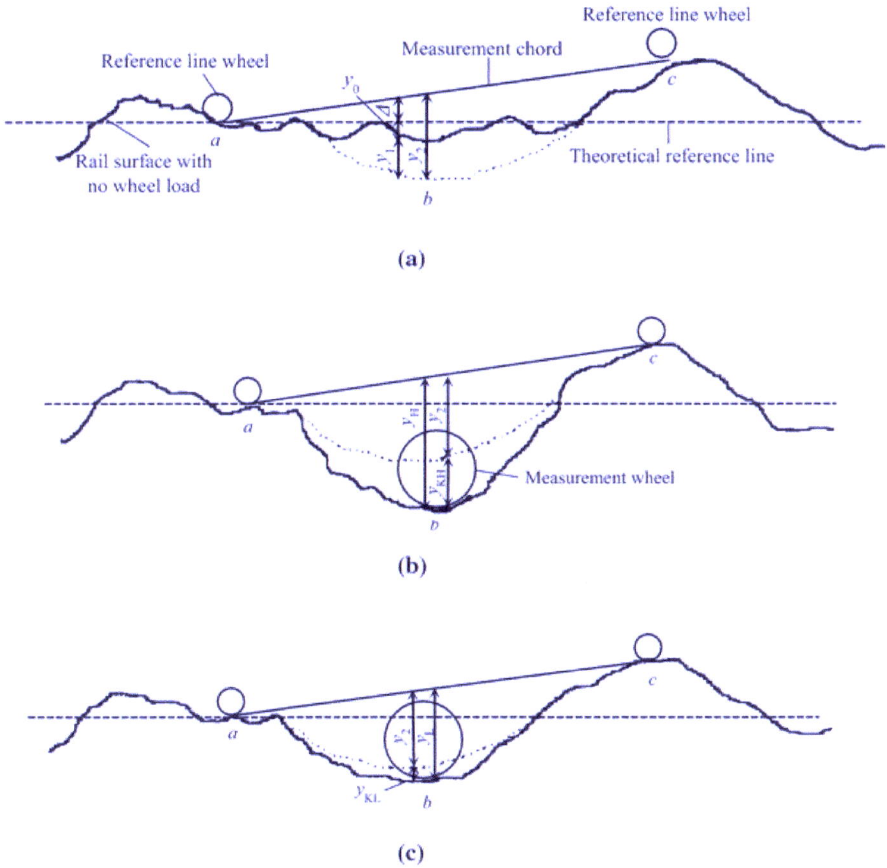

Figure 6: Principle for track elasticity measurement of CARS-stiffness equipment. **a** Measurement with no wheel load on track. **b** Measurement with the wheel load of heavy car *a* on

track. **c** Measurement with the wheel load of light car *b* on track. Δ is reference line error. y_0 rail surface irregularities when there is no load on track, y_1 hidden gaps between rail, sleeper, and ballast, y_2 sum of *triangle*, y_0, and y_H chord measured value under the wheel load of heavy car *a*, y_L chord measured value under the wheel load of light car *b*, y_{KH} track elastic settlement under wheel load of heavy car *a*, y_{KL} track elastic settlement under wheel load of light car *b*.

As shown in Fig. 6, the track stiffness can be expressed as

$$K = \frac{\Delta P}{\Delta y} = \frac{P_A - P_B}{y_{KH} - y_{KL}} = \frac{P_A - P_B}{y_H - y_L}. \quad (5)$$

Transportation Technology Center, Inc. in Pueblo, Colorado (USA), also developed a stiffness measurement vehicle, with similar measurement principle to CARS [34, 35]. As shown in Fig. 7, this vehicle comprises a heavy car, a light car, and a towing locomotive. The loading range of the heavy car is 1–55 kips (4–267 kN). The stiffness is measured under 10 kips (44 kN) or 40 kips (178 kN). The load for the light car is less than 3 kips, and the track geometric irregularity is measured under 2 kips (8.9 kN). This vehicle operates twice, under the static loads of 178 and 44 kN, respectively, and in two cases the static track irregularity must be excluded. In the dynamic measurement, it finds out areas with high or low stiffness (especially low-stiffness areas to identify ballast or subgrade diseases) and then sprays yellow coatings for performing standstill stiffness measurement. Under the load of 178 kN, the displacement measurement involves the whole track and subgrade, while under the load of 44 kN, it includes the rail, sleeper, and ballast. The measurement speed can reach 16 km/h.

Figure 7: Track stiffness measurement vehicle of TTCI.

In addition, the measurement principle for the stiffness measurement vehicle developed by Swiss Federal Railways (SBB 2007) is also similar to the above-mentioned ones [8]. This vehicle comprises a light car and a heavy car (see Fig. 8). The weight of the light car can be neglected. The weight of the heavy car is 20 tons.

The vehicle speed is 10–15 km/h. Heidenhain LS 220 transducers are used for low-pass (cut-off wavelength of 10–20 m). The accuracy of displacement measurement can reach 0.2 mm.

Figure 8: Track stiffness measurement vehicle of SBB.

At the beginning of the twenty-first century, the University of Nebraska (USA) commenced the research on track modulus measurement system [8, 36, 37]. This measurement system can also be used to explore the track modulus through measuring the vertical rail displacement according to the laser measurement method. The measurement vehicle is equipped with two laser sources, the vertical rail displacement can be obtained through measuring the distance between two laser lines, and then the track modulus could be calculated. The measurement system is shown in Figs. 9 and 10.

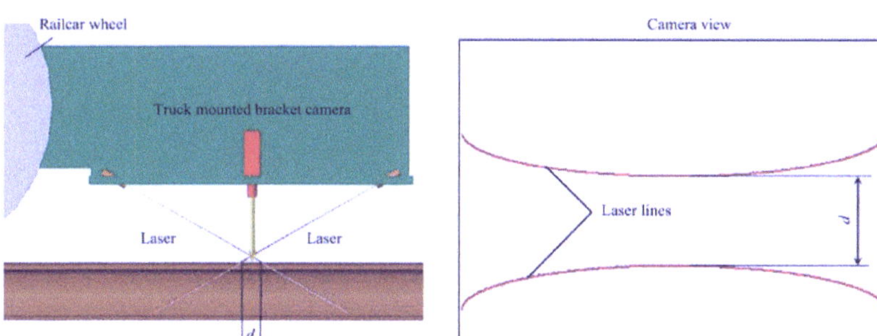

Figure 9: Measurement approach of UNL-stiffness equipment.

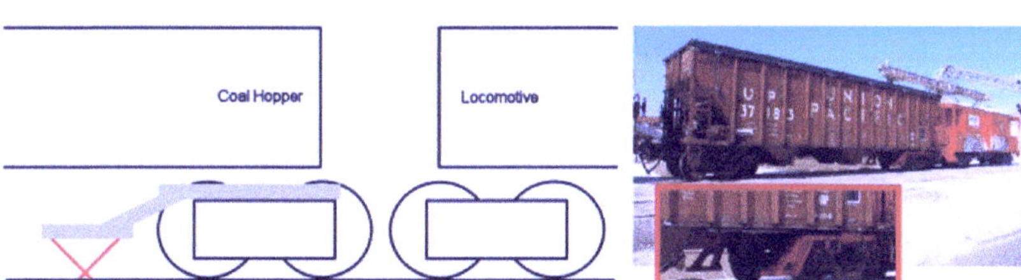

Figure 10: Mounting position of the measurement system of UNL.

The measurement principle is shown in Figs. 11 and 12, where d is measured according to the camera view analysis and L_1, L_2, θ_1, θ_2, and H are all known values. After y_r is obtained, an equation is established based on the Winkler foundation model. Solving this equation, one can obtain the track modulus.

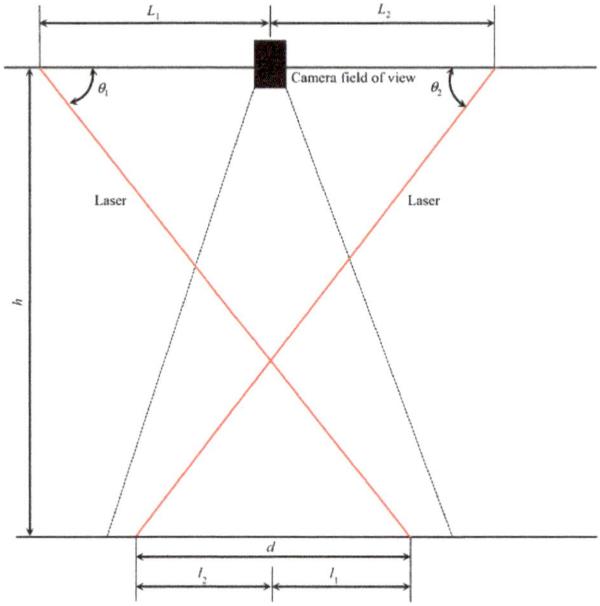

Figure 11: Sensor geometry of UNL-stiffness equipment.

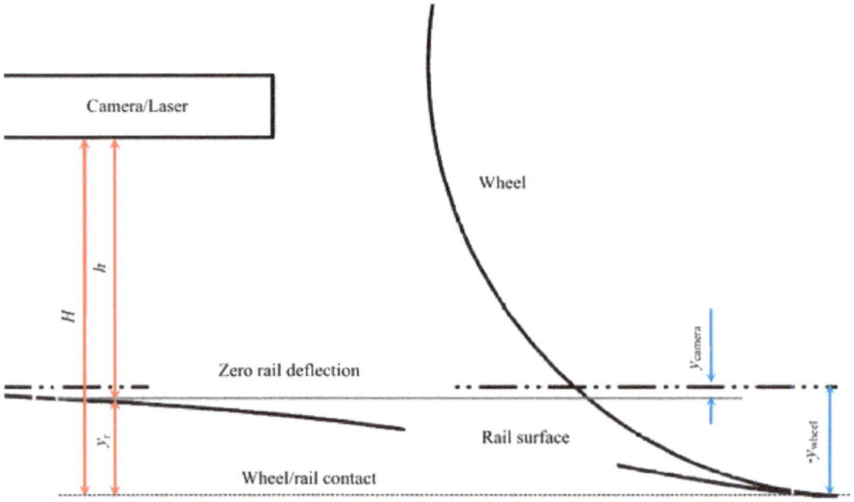

Figure 12: Rail deflection and sensor measurement of UNL-stiffness equipment.

The ZOYON Technology Co., Ltd of Wuhan University (China), is now investigating a deformation rate-based track stiffness measurement method to be applied in the movable stiffness measurement vehicles [38]. This method was first introduced by scholars of Delft University of Technology in Holland; however, no prototype car has been made [2, 39–41].

According to this method, during load moving, the deflection basins at load-acting points are similar to each other. The ratio of the vertical track deflection rate at one point of a deflection basin to the load moving rate along the railway line is the slope of track deflection line at this point (see Fig. 13). Namely,

$$w'(x) = \text{Slope} = \frac{V_{\text{deflection}}}{V_{\text{ds}}}, \quad (6)$$

where $w'(x)$ refers to the vertical track deflection rate obtained based on the Winkler foundation beam model.

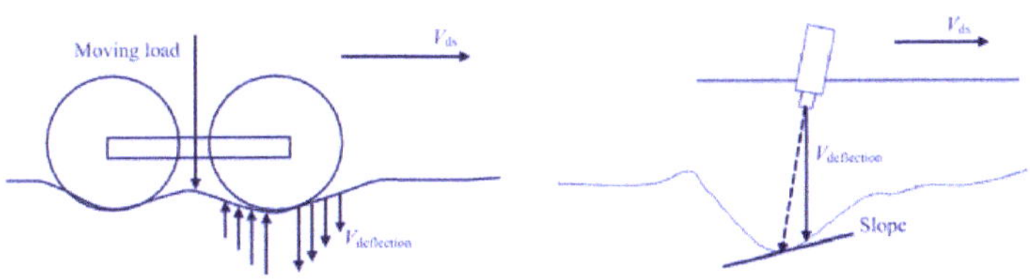

Figure 13: Principle for movable track stiffness measurement of ZOYON.

To measure the track deflection rate, the high-speed deflectograph system adopts laser Doppler sensors attached to a moving railway vehicle traveling at speeds of up to 130 km/h. The sensor layout and laser projection positions are shown in Fig. 14.

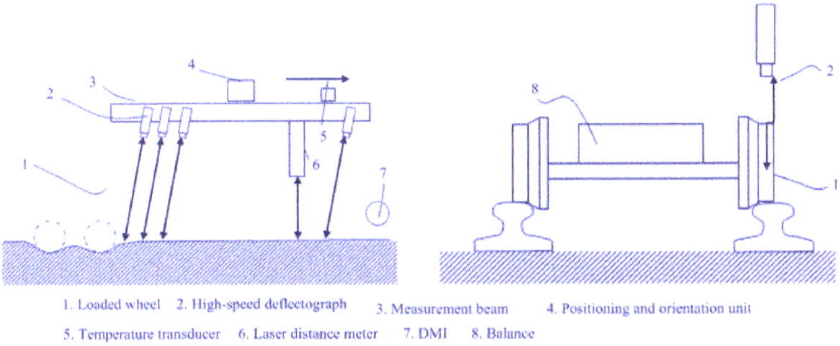

Figure 14: Sensor layout and laser projection position of ZOYON.

The European Research for an Optimised BALlasted Track (EUROBALT II) project indicates that track stiffness is another import parameter for optimal long-term maintenance strategies in addition to track geometric irregularity (Meissonnier 2000).

This project promotes Banverket (a railway department of Sweden) to develop a trolley for continuous measurement of vertical track stiffness, the measurement principle of which is shown in Fig. 15 [8, 36, 42–49].

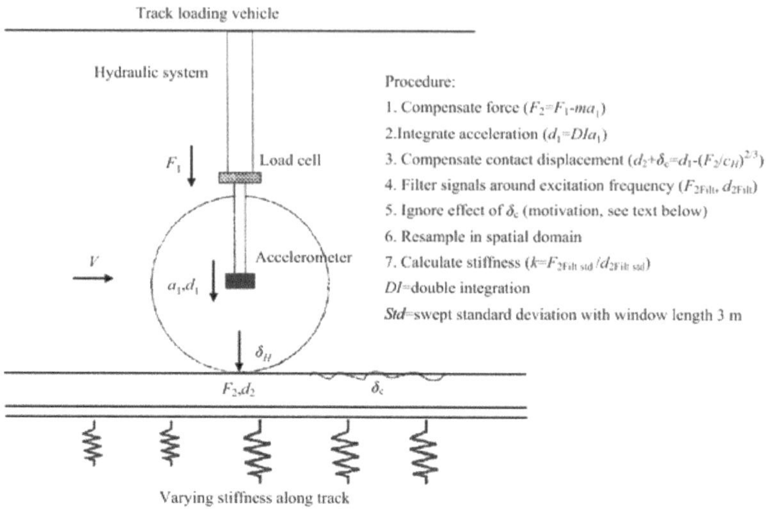

Figure 15: Principle of Banverket's continuous track stiffness measurement method.

The static load, dynamic load, and maximum speed of this trolley are 60 kN, 20 kN, and 30 km/h, respectively. This trolley can only run on straight sections and not on narrow curves or switches. This trolley can be excited with different frequencies but only one frequency for each run. This trolley has been used for many on-site measurements, which shows good repeatability and reproducibility.

Figure 16: Measurement equipment of RSMV developed by KTH.

Royal Institute of Technology (KTH) in Sweden built a new vehicle, called rolling stiffness measurement vehicle (RSMV), a rebuilt two-axle freight wagon. The RSMV, much more advanced than the prototype trolley, has one battery plate, one hydraulic system, and two oscillating mass bodies (see Fig. 16). The parameters of RSMV are listed: weight of each body of 4,000 kg, measurable axle weight of 180 kN or higher, max. oscillating amplitude for dynamic load of 60 kN, and measurement frequency of 50 Hz. The measurement speed can reach up to 50 km/h. Figure 17 shows the measurement principle of the RSMV, and its two sides are symmetrical. The measurement principle is similar to that of the trolley.

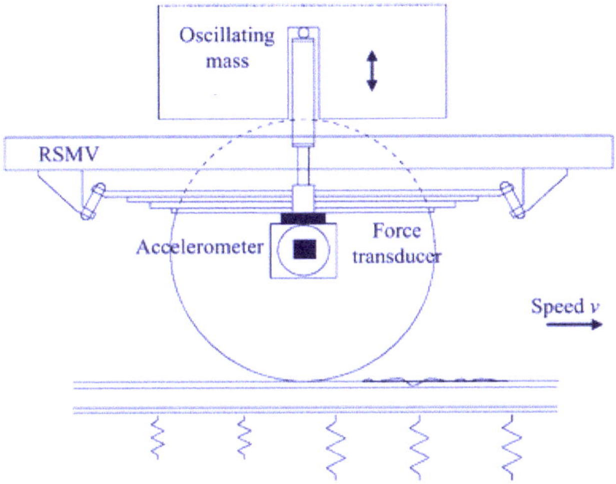

Figure 17: Measurement principle of RSMV of KTH (single side, symmetrical two sides).

In order to compare the measurement results of the RSMV with that of the prototype trolley, a comparison test was performed, showing that, for those railway sections with the overall track stiffness less than 150 kN/mm, the results were almost the same, while for those hard sections, the results are quite different.

Sponsored by the Innotrack project (D2.1.9 INNOTRACK 2009), Portancemetre for measuring the overall track stiffness [8, 50, 51] was developed by the Centre d'Experimentation et de Rechrche (CETE–NC, Grand Quevilly, France) and Engineering Department (SNCF, Paris, France). This measurement vehicle comprises two parts, one is the core measurement system of demonstrator and the other is the technical carriage system. The technical carriage system carries the energy supply equipment, hydraulic system, and electronic devices. The core measurement system is installed with Type 417 single-axle wheel pairs. The technical carriage system is installed with Y25C bogie wheel pairs. The main transducers of the Portancemetre include unsprung mass accelerometer, chassis accelerometer, phase sensor (synchronous signal), and incremental distance encoder. Figure 18 shows the on-site Portancemetre measurement in Rouen, France.

Figure 18: On-site measurement with Portancemetre in Rouen, France.

A camera with a linear CCD sensor is mounted under the technical carriage to record and monitor the measured track surface. All measurements are monitored by the portable computer placed in the locomotive.

Figure 19 shows the measurement principle for Portancemetre.

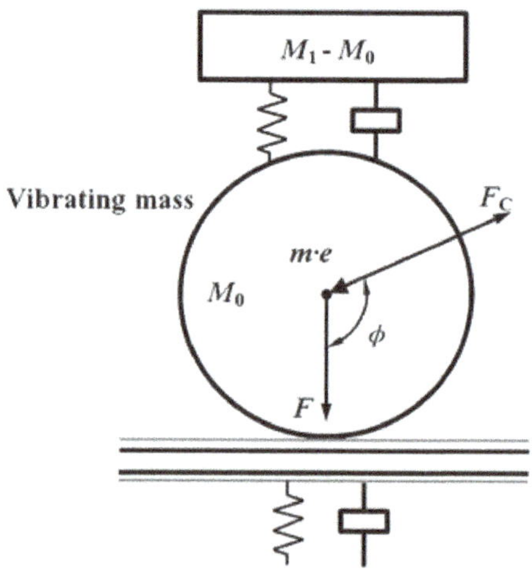

Figure 19: Schematic of the mechanical vibrating wheel of Portancemetre.

The force of the measurement system exerted on a rail can be calculated by

$$F = M_1 g = M_0 \Gamma_b + (M_1 - M_0)\Gamma_c + me\omega^2 \cos\phi, \tag{7}$$

where M_1 refers to the total mass, M_0 the unsprung mass, Γ_b the vertical acceleration of the vibrating wheel, Γ_c the vertical acceleration of the suspended mass, $m{\cdot}e$ the eccentric moment of the unbalanced system, ω the angular velocity of rotation, and φ the angle of rotation.

The vertical displacement z can be obtained by calculating the double integration of the vertical acceleration of the unsprung mass:

$$z(t) = \iint \Gamma_b(t)\mathrm{d}t\mathrm{d}t\,. \tag{8}$$

The overall track stiffness, over an average time period, can be obtained according to wheel–rail force and vertical displacement.

Figure 20 shows the typical force–displacement hysteresis curve. The static load and dynamic load per rail are about 50 and 20 kN, respectively.

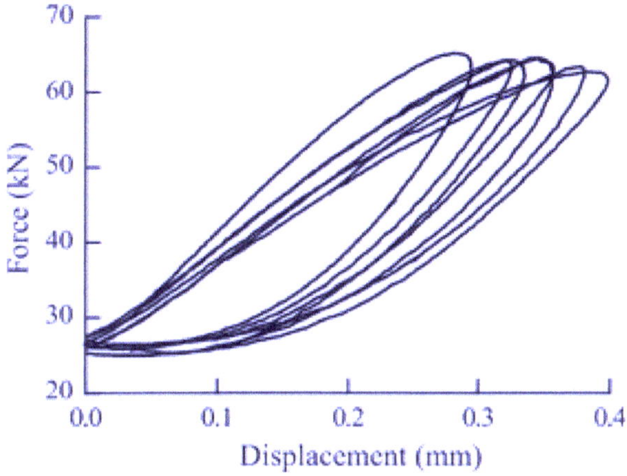

Figure 20: Vertical force–displacement curve of "Des Jardins" track by Portancemetre.

ANALYSIS AND DISCUSSION OF MEASUREMENT METHODS

The standstill measurement is suitable for monitoring track stiffness and long-term stiffness of special sections; however, the development of continuous track stiffness measurement vehicles is still the main way for realizing long-distance stiffness measurement of whole railway lines. Table 1 shows the performance parameter comparison of the above track stiffness measurement vehicles.

Table 1: Comparison of track stiffness measurement vehicles

No.	Measurement concept	Measurement accuracy	Measurement speed	Research progress
1	Standstill measurement, static stiffness	High accuracy for vertical measurement; unable to know the stiffness change along the line	Very slow	Mainly applied for heavy haul railway lines
2	Continuous measurement, static stiffness under dynamic load	Improved measurement accuracy due to enhancing displacement measurement	Adjustable, below 60 km/h usually	Mature

3	Continuous measurement, static stiffness under dynamic load, high-speed deflectograph	Unsure	130 km/h, even higher	Successfully applied to the road modulus measurement and the track modulus measurement is in progress
4	Continuous measurement, dynamic stiffness, eccentric wheel excitation, hysteresis curve measurement	High accuracy because of hysteresis curve fitting, stiffness and damping parameter identification	Slow, approximate 6 km/h	Data analysis and vehicle optimization
5	Continuous measurement, dynamic stiffness, eccentric wheel excitation	High accuracy and able to identify many railway line stiffness problems	50 km/h	Multi-line measurements and database construction

From Table 1, the existing continuous measurements of track stiffness have four problems.

Inconsistent Measurement Concepts

At present, the measurement methods target at different stiffness: static stiffness, dynamic stiffness, or the track component stiffness, which makes it rather difficult to compare the measurement results of these measurement methods.

For instance, the exciting frequency is unnecessary for static stiffness measurement, while for dynamic stiffness measurement, it is requisite. Likewise, when the deformation rate (TU Delft and ZONYON) is to be measured, the vehicle shall run as fast as possible; when the steady dynamic stiffness is to be measured (CETE-NC and SNCF), the vehicle speed cannot be too high. Hence, the measurement of track stiffness, firstly, requires a unanimous stiffness notion that can reflect the track stiffness characteristics completely.

Relatively speaking, the track stiffness notions of the RSMV (KTH, Sweden) and Portancemetre (CETE-NC and SNCF, France) are more advanced. The notions indicate that the track stiffness relates to the exciting frequency, which agrees with the conclusions of many literatures [52–54]. Other notions are based on the Winkler support model, which are not suitable for exploring deep information about track stiffness though they may solve some practical engineering problems.

Incapability for High-Accuracy Measurement

All the above-mentioned measurement methods are not suitable for high-accuracy measurement. The accuracy of the laboratory test results are acceptable; however, the accuracy of field test results is not satisfactory. The measurement accuracy mainly depends on the accuracy of force and displacement sensors or acceleration measurements. The force measurement includes strain measurement and pressure intensity measurement of hydraulic system. The strain measurement is more sensitive and accurate. Laser method is more suitable for displacement measurement, which provides high accuracy. Although acceleration transducers have high accuracy, in continuous track stiffness measurement, noise (especially low-frequency noise) can greatly affect the measurement. In addition, error cannot be eliminated when the displacement is calculated through the double integration. Hence, for the acceleration measurement method, appropriate data processing and signal analysis technologies are required to ensure high displacement accuracy. In fact, the continuous track stiffness measurement can be graded according to the desired accuracy (10, 1, 0.1 kN/mm and so on) to save cost and reasonably use the measurement sources.

Low Measurement Speed

Measurement speed is always very important for continuous track stiffness measurement. Theoretically, high measurement is likely to produce more accurate results. The track deformation rate measurement method (TU Delft and ZONYON) has the quickest measurement speed. However, it has two defects: the static stiffness notion takes no account of the influence of railway stiffness irregularity and geometric irregularity, and it is hard to deal with the dilemma between the measurement speed, transducer accuracy, and equipment operation reliability. These problems limit the application of this measurement method.

The measurement speeds of the methods designed by the University of Nebraska (USA) and CARS are moderate, which are limited by the accuracy of laser displacement transducers. The RSMV designed by the KTH (Sweden) has rather low measurement speed, which is limited by the reliability of the hydraulic system as well as the accuracy of transducers.

The measurement speed of Portancemetre (CETE-NC and SNCF, France) cannot be too high because the force–displacement hysteresis curve is to be measured.

Insufficient Data Analysis

From the perspective of the excitation-response measurement, the system transfer characteristics (the stiffness information) can be obtained regardless of the excitation type as long as there is sufficient response information. Hence, the track stiffness information, to a great extent, depends on the analysis technology of data.

KTH (Sweden) takes the leading position in the analysis technology of data, which has investigated the relation between track stiffness and time frequency/spatial frequency, the phase information of track stiffness, and amplitude information.

In fact, the laser measurement method (University of Nebraska, USA, and CARS, China) can also collect more data because the real-time track vibration can be reflected through the laser displacement data.

Up to now, no quantitative research on the relation between track stiffness and exciting frequency, exciting intensity as well as the running speed has emerged, which is the main reason for inadequate analysis of the measurement data of continuous track stiffness.

CONCLUSIONS AND SUGGESTIONS

This paper summarizes the significance of track stiffness measurement, takes a wide view of track stiffness measurement methods around the world, including the standstill measurement and the continuous track stiffness measurement, and performs a comparison analysis of all the measurement methods.

- Track stiffness is an important factor influencing the safety and reliability of train operation, the vibration and deformation of track structures, and the dynamic response of substructures like subgrades and bridges. The track stiffness measurement is of great theoretical and practical significance to the design of new railway lines and especially the railway maintenance works. The measurement of track stiffness for maintenance relates to four stiffness problems: low stiffness, variable stiffness, virtual stiffness, and assortative stiffness problems.
- The standstill measurement of overall track stiffness is time consuming and hard to realize a long-distance and multi-point measurement for whole railway lines. The existing methods for continuous track stiffness measurement have four problems: a. inconsistent measurement concepts. A unified stiffness notion has not been proposed to reflect the track stiffness characteristics completely; b. poor ability for high-accuracy measurement. The measurement accuracy for force, displacement, or acceleration is hard to ensure, especially for high-speed measurement; c. low measurement speed. It is difficult to solve the dilemma between the measurement speed, transducer accuracy, and equipment operation reliability; and d. insufficient data analysis. Much more stiffness information is expected to explore using the advanced analysis technology of the measurement data.

The following suggestions are proposed considering the imperfection of existing stiffness measurements:

- A definite stiffness notion should be established to completely reflect the track stiffness characteristics.
- A theoretical and systematic research should be performed on the quantitative relation between track stiffness and the factors like exciting frequency, running speed, and vehicle axle load, as well as the influences of track stiffness and track geometric irregularities on the vibration response of the vehicle-track-substructure coupled dynamics, respectively.
- In view of the accuracy requirement for stiffness measurement, the track stiffness measurement can be categorized based on three grades, namely low accuracy (10 kN/mm, substructure disease identification, for heavy haul railways mainly), medium accuracy (1 kN/mm, potential safety hazard checking), and high accuracy (0.1 kN/mm, evaluation of track regularity, for high-speed railways mainly). For example, for continuous measurement vehicle with low measurement accuracy, single-axle vehicle can be used. The axle load should be more than 150 kN to ensure the vertical measurement for subgrade, and one span length (sleeper pitch) along the railway line can be the measurement unit. For the continuous

measurement with medium measurement accuracy, it is advised to use double-axle vehicle with axle loads of 20 and 100 kN, the vertical measurement range is limited to the track structure, and the measurement unit ranges from 1 to 4 span lengths. For continuous measurement vehicle with high measurement accuracy, it is advised to use double-axle vehicle with axle loads of 10 and 50 kN, the vertical measurement range is limited to the rail and fasteners, and 4–10 span lengths can be adopted as the measurement unit.

For transition zones (railway–bridge, railway–tunnel, and bridge–tunnel areas), turnouts, and small-radius curve sections, the track stiffness characteristics are more complicated than those of straight lines [2, 49]. Thus, apart from the measurements, the stiffness of special track sections can be monitored.

In recent years, the features of wave propagation in track structures have attracted a wide attention, and the wave theory is applied to the high-frequency vibration of rails [55]. Rail support conditions affect the transfer laws of track vibration waves, and thus continuous track stiffness measurement equipment can be developed based on the wave transfer mechanism.

In addition, the vibrations of track components and vehicle components can be collected for power generation [56–58]. Different track stiffness values cause different vibration intensities, affecting the collection of electric energy. When the stiffness measurement accuracy is undesirable because of poor noise reduction in vibration displacement and vibration acceleration signals, the vibration energy storage ability or real-time electrical signals (voltage, power, etc.) might be a good index for continuous track stiffness measurement or long-term stiffness monitoring of special sections.

ACKNOWLEDGMENTS

This research was supported by the project (51425804) of the National Science Fund for Distinguished Young Scholars of China and the National Natural Science Foundation of China (NSFC) under grants U1234201, U1334203, and 51378439.

REFERENCES

1. Kerr AD (2000) On the determination of the rail support modulus k. Int J Solids Struct 37:4335–4351
2. Esveld C (2001) Modern railway track (second edition). MRT-Productions, Delft, pp 71–90
3. Zhang GM (2002) Research on right level of track structure stiffness and track-part stiffness. China Railw Sci 23(1):51–57
4. Liu L, Wang W, Liu W (2000) Experimental study of the static stiffness of ballast. J Shanghai Tiedao Univ 21(4):1–6
5. Liu W, Wang W (2002) Study on dynamic stiffness and damp measurement of railway crushed stone ballast. J China Railw Soc 24(6):99–104
6. Ooi LE, Ripin ZM (2011) Dynamic stiffness and loss factor measurement of engine rubber mount by impact test. Mater Des 32:1880–1887

7. Sussmann TR, Ebersöhn W, Selig ET (2001) Fundamental nonlinear track load-deflection behavior for condition evaluation. Transp Res Rec 1742:01–2916
8. Berggren E (2009) Railway track stiffness—Dynamic measurements and evaluation for efficient maintenance. Ph.D. Thesis, Royal Institute of Technology (KTH), Stockholm
9. Wu TX, Thompson DJ (1999) The effects of local preload on the foundation stiffness and vertical vibration of railway track. J Sound Vib 215(5):881–904
10. Chen XP, Wang P (2006) Distribution regularity and homogenization of track rigidity for ballastless turnout. J Southwest Jiaotong Univ 41(4):447–451
11. Chen XP (2008) Study on theory and application of track stiffness in high-speed turnouts. Ph.D. Thesis, Southwest Jiaotong University, Chengdu
12. 1Liu XY (2004) Effect analysis of track stiffness on dynamic characteristics of wheel–rail system and its dynamic optimization. J Southwest Jiaotong Univ 39(1):1–5
13. Lopez Pita A, Teixeira PF, Robusté F (2004) High speed and track deterioration: the role of vertical stiffness of the track. J Rail Rapid Transit 218:31–40
14. Cai CB, Xu P (2010) Dynamic analysis of key design parameters for ballastless track of high-speed railway. J Southwest Jiaotong Univ 45(4):493–497
15. Li D, Selig ET (1995) Evaluation of railway subgrade problems, Transportation Research Record 1489, TRB. National Research Council, Washington DC
16. Selig ET, Li D (1994) Creating track modulus: its meaning and factors influencing it. In: Transportation Research Record: Journal of the Transportation Research Board, No. 1470, Transportation Research Board of the National Academies, Washington, DC, pp 47–54
17. Frohling RD, Scheffel H, Ebersohn W (1996) The vertical dynamic response of a rail vehicle caused by track stiffness variations along the track. Veh Syst Dyn 25:17–31
18. Zhai JQ (2007) Test and analysis of vibration propagation caused by the subway train. Dissertation, Tongji University
19. Wu JZ (2012) The impact of vibration on the building structure caused by metro and isolation research. Dissertation, South China University of Technology
20. Wang P, Xu JH, Wang L (2014) Effect of track stiffness on frequency response of vehicle-track coupling system. J Railw Eng Soc 9:46–52
21. Robinet A, Hosseingholian M, Quibel A, Froumentin M (2008) Track stiffness assessment. In: 8th World Congress on Railway Research, Seoul, Korea, pp 1–8
22. Zhai WM, Cai CB, Kai KY (2000) Effect of track stiffness on train running behavior. J China Railw Soc 22(6):80–83
23. Ebersöhn W (1995) Substructure influence on track maintenance requirements. Ph.D. Dissertation, Department of Civil and Environmental Engineering, University of Massachusetts at Amherst
24. Konstantinos G, Spyridon T (2012) Transition zone between ballastless and

ballasted track: influence of changing stiffness on acting forces. Proc Soc Behav Sci 48:3548–3557

25. Kaewunruen S, Remennikov A (2005) Application of experimental modal testing for estimating dynamic properties of structural components. In: Proceedings of Australian structural engineering conference 2005, Newcastle
26. Kaewunruen S, Remennikov A (2005) Monitoring structural degradation of rail bearing pads in laboratory using impact excitation technique. In: Proceedings of the 1st international conference on structural condition assessment, monitoring, and improvement, Perth, pp 389–394
27. Burrow MPN, Chan AHC, Shein A (2007) Falling weight deflectometer based inverse analysis of ballasted railway tracks. In: Geotechnical engineering, proceedings of the Institution of Civil Engineers, Issue GE3, pp 169–177
28. Kerr AD (1983) A method for determining the track modulus using a locomotive or car on multi-axle trucks. In: Proceedings AREA, vol 84
29. Kerr AD (1987) On the vertical modulus in the standard railway track analysis. Rail International, November Issue
30. Kerr AD On the reduced area method for calculating the vertical track modulus. In: Proceedings AREA, vol 86
31. Esveld C (1980) Track stiffness measurements using an adapted tamping machine. Rail International
32. D2.1.9 INNOTRACK (2009) Adapted "Portancemetre" for track structure stiffness measurement on existing tracks, Paris
33. Wangqing W, Geming Z, Kaiming Z, Lin L (1997) Development of inspection car for measuring railway track elasticity. In: Proceedings from 6th international heavy haul conference, Cape Town
34. Li D, Thompson R, Kalay S (2002) Development of continuous lateral and vertical track stiffness measurement techniques. In: Proceedings from railway engineering, London
35. Thompson R, Marquez D, Li D (2001) Track strength testing using TTCI's track loading vehicle. Railw Track Struct 97:15–17
36. McVey B, Norman C, Wood N, Farritor S, Arnold R, Fateh M, El-Sibaie M (2005) Track modulus measurement from a moving railcar. In: Proceedings of the AREMA annual conference, Chicago, September 25–27, 2005
37. Norman C, Farritor S, Arnold R, Elias SEG, Fateh M, Sibaie ME (2004) Design of a system to measure track modulus from a moving railcar. In: Proceedings from railway engineering, London
38. 3Information from http://old.zoyon.com.cn
39. Rasmussen S, de Man AP (2000) The high speed deflectograph and the hammer excitation test: two measurement techniques for determining track performance. Rail Eng Int 2000(1):13–16
40. Rasmussen S, Man AP (1999) Measurement techniques for track performance.

In: Proceedings of conference on innovations in the design and assessment of railway track, Technical University of Delft, Delft

41. Rasmussen S, Krarup JA, Hildebrand G, Non-contact deflection measurement at high speed. In: 6th International conference on the bearing capacity of roads, railways and airfields, 24–26 June, 2002, Lisbon
42. Berggren E, Jahlénius Å, Bengtsson B-E (2002) Continuous track stiffness measurement—an effective method to investigate the structural conditions of the track. In: Proceedings of the conference railway engineering—2002, London, 3–4 July, 2002. ISBN: 0-947644-49-0
43. Berggren E, Jahlénius Å, Bengtsson B-E, Berg M (2005) Simulation, development and field testing of a track stiffness measurement vehicle. In: Proceedings of 8th international heavy haul conference, Rio de Janeiro, 13–16 June, 2005. ISBN: 0-646-33463-8
44. Berggren E, Kaynia A, Dehlbom B (2010) Identification of substructure properties of railway tracks by dynamic stiffness measurements and simulations. J Sound Vib 329:3999–4016
45. Berggren E (2006) Measurements of track stiffness and track irregularities to detect short waved support conditions. In: Proceedings of international conference on railway track foundations, Birmingham, 11–13 September, 2006. ISBN: 0-704426-00-5
46. 4Berggren E, Li M, Spännar J (2006) A new approach to the analysis and presentation of vertical track geometry quality and rail roughness. In: Proceedings of 7th international conference on contact mechanics and wear of rail/wheel systems (CM2006), Brisbane, 24–27 September, 2006. Also in journal of WEAR, vol 265, pp 1488–1496
47. 4Berggren E (2009) Efficient track maintenance—methodology for combined analysis of condition data, submitted to the conference IHHA 2009. Candidate paper to Special Issue of J Rail Rapid Transit
48. Li M, Berggren E (2009) A study of the effect of global track stiffness and its variations on track performance: simulation and measurement, submitted to the conference IHHA 2009. Candidate paper to Special Issue of J Rail Rapid Transit
49. Smekal A, Berggren E, Silvast M (2006) Monitoring and substructure condition assessment of existing railway lines for upgrading to higher axle loads and speeds. In: Proceedings of 7th World Congress on Railway Research, Montreal, 5–7 June, 2006
50. Alexander R, Sakdirat K Laboratory measurements of dynamic properties of rail pads under incremental preload. In: 19th Australasian conference on the mechanics of structures and materials, Christchurch, pp 319–324
51. Hosseingholian M, Froumentin M, Robinet A (2006) Feasibility of a continuous method to measure track stiffness. In: Proceedings from railway foundations conference, Birmingham
52. Fenander A (1997) Frequency dependent stiffness and damping of railpads. Proc

Inst Mech Eng Part F 211:51–62

53. Fenander A (1998) A fractional derivative railpad model included in a railway track model. J Sound Vib 212(5):889–903
54. Frohling RD (1998) Low frequency dynamic vehicle/track interaction: modeling and simulation. Veh Syst Dyn 28:30–46
55. Ryue J, Thompson DJ, White PR, Thompson DR (2008) Investigation of propagating wave types in railway tracks at high frequencies. J Sound Vib 315:157–175
56. Pourghodrat A, Nelson CA, Hansen SE, Kamarajugadda V, Platt SP (2014) Power harvesting systems design for railroad safety. Proc IMechE Part F 228:504–521
57. Wang JJ, Penamalli GP, Zuo L (2012) Electromagnetic energy harvesting from train induced railway track vibrations. In: Proceedings of IEEE/ASME international conference on mechatronics and embedded systems and applications (MESA), pp 29–34
58. Gao MY (2015) Design and verification of a rail-borne electromagnetic transducer for railway energy harvesting. Research report, Key Laboratory of High-speed Railway Engineering, Ministry of Education, Southwest Jiaotong University, Chengdu

15

CHAPTER

OPTIMIZATION OF WHEEL–RAIL CONTACT FORCE MEASUREMENT SYSTEMS

S. Papini, L. Pugi, A. Rindi, E. Meli

Keywords: Wheel–rail interaction, Contact force, Strain gauge

INTRODUCTION

In order to evaluate the ride quality of a railway vehicle, the vertical and lateral contact forces have to be measured. In the reference frame as shown in Fig. 1, the three components of the measured contact force are indicated: the longitudinal force X, directed along the x-axle in the longitudinal direction of rail, the lateral force Y, directed along the y-axle, and the vertical force denoted by Q.

A dynamic behavior analysis in the norm UIC518 [1, 2] prescribes the experimental measurement of Y-force and Q-force with a minimum bandwidth of 20 Hz. The X-force is also scientifically interesting to the identification and modeling of wheel–rail adhesion phenomena in the testing and homologation of safety relevant subsystems like the odometry for on-board wheel-slide protection system (WSP). In this article, we propose a benchmark test bogie for the three components of contact force, designed to be equipped with sensor and control systems. In order to reduce the negative influence of braking forces, the bogie layout as shown in Fig. 2 is designed, which has a standard H-shaped steel frame, inspired by a widely diffused design adopted also on coaches of ETR500 High Speed Train. To diminish the disturbances on measurements caused by braking, the disks are flanged over an auxiliary shaft connected through a suspended gearbox to the axle. This mechanical solution is usually adopted on some well-diffused locomotive like Siemens E190, typically running with a maximum service speed of

Citation: S. Papini, L. Pugi, A. Rindi, E. Meli, Optimization of Wheel–Rail Contact Force Measurement Systems, DOI: 10.1007/s40534-013-0013-z.

about 200 km/h and a 22.5 t of axle weight. Hence, this layout is considered as reliable and feasible even in the cases of augmented bogie with unsuspended masses/inertia.

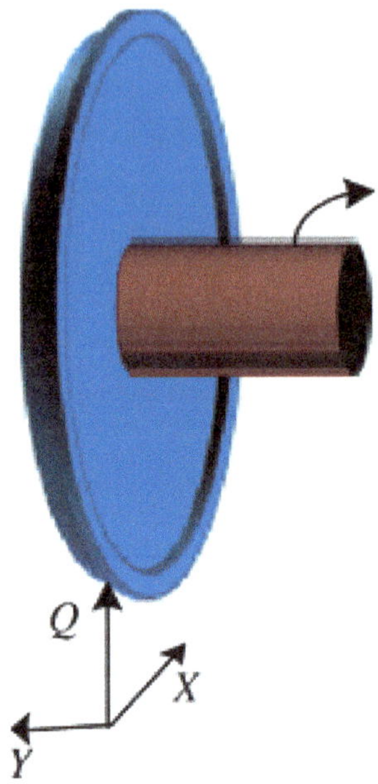

Figure 1: Wheel–rail reference system.

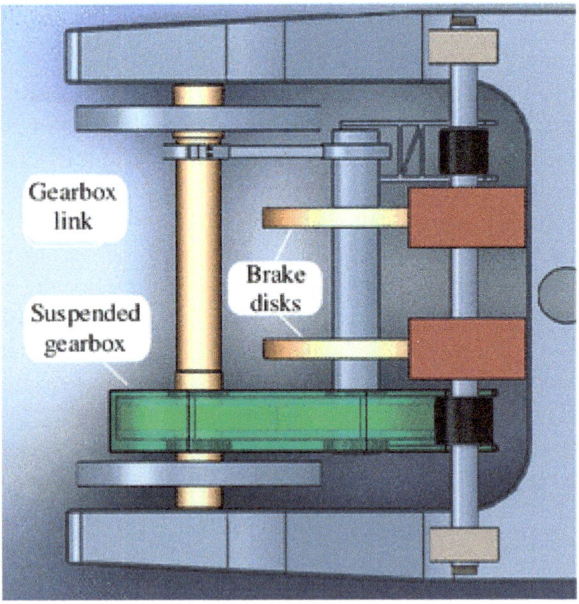

Figure 2: Bogie layout.

This article is organized as follows: In Sect. 2, the layout for the contact force measurement is introduced; Sect. 3 describes the FEM model and the relative calculation; in Sect. 4, the error sensitivity analysis is conducted: the longitudinal and vertical forces with a mathematical model, and the lateral force with a FEM model. Moreover, in Sect. 5, a study of dynamic bogie behavior is carried out in terms of frequency response functions.

CONTACT FORCE MEASUREMENT

The solution proposed in Fig. 2 insures enough space on the axle to place sensor and other telemetry devices on the shaft. Here, a classical layout in which the three contact force components are acquired independently on its own sensor system is supposed:

- Longitudinal forces X are reconstructed in terms of the torque exchanged along the axle [3];
- Lateral force Y measurements are performed by instrumenting the lateral deformation of the wheel using the methods in Ref. [4];
- Vertical forces Q are measured by the estimation of the shear stress in different sections of the axle; the shear stress is evaluated by comparing bending stresses on adjacent instrumented sections [4].

Longitudinal Force

Longitudinal forces X are estimated from torque measurements on two instrumented sections on the axle. Both sections are located between wheels, and the torque load is applied by braking or traction system (see Fig. 3). Torque is measured using the Wheatstone bridge [4, 5]. The strain gauge layout assures the rejection of disturbances such as spurious load due to axial forces, bending, and thermal-induced deformations. The longitudinal force X_i of the i-th wheel is calculated by

$$X_i = -\frac{M^i_{\text{tor}}}{r_{\text{w}}}, \tag{1}$$

where r_{w} is the rolling radius, adopted as a constant, and M^i_{tor} is the torque load applied by braking or traction system on the i-th section. The torque load is calculated with traditional expression for a hollow shaft:

$$M_{\text{tor}} = \frac{\pi G \varepsilon_{45^\circ} \left(d^4_{\text{est}} - d^4_{\text{int}}\right)}{8 d_{\text{est}}}, \tag{2}$$

where ε_{45° is the strain gauge deformation taken on a 45° helix and its polar moment of inertia is $J_p = \pi\left(d^4_{\text{est}} - d^4_{\text{int}}\right)/32$ d_{est} and d_{int} are the external and internal diameters of the axle, respectively, and G is shear modulus.

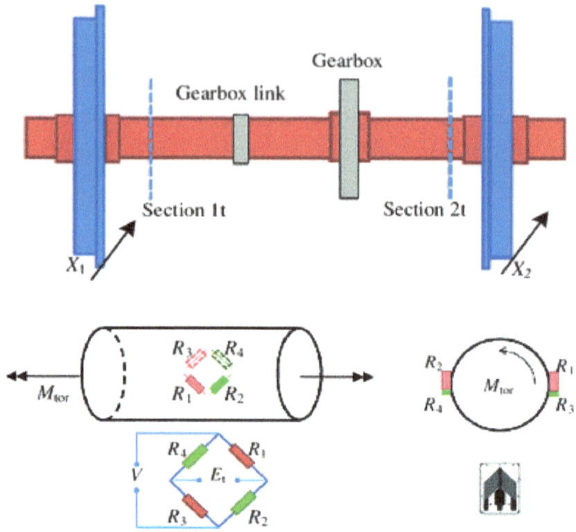

Figure 3: Strain gauge layout for the measurement of *X*.

Vertical Force

The vertical contact force on each wheel was obtained by measuring the axle bending torque with strain gauges through the compression of primary suspension. The vertical component Q_i of the contact force on the *i*-th wheel is evaluated by imposing the corresponding equilibrium relation calculated according to the simplified scheme of Figs. 4 and 5:

$$\left.\begin{aligned} Q_1 &= T_3 - V_1 + m_3 g, \\ Q_2 &= T_6 - V_2 + m_6 g, \end{aligned}\right\} \tag{3}$$

where m_3g and m_6g are the weights of the corresponding axle and bogie parts which are delimited by sections 3 and 6, respectively; T_3 and T_6 are the shear loads, respectively, applied in sections 3 and 6. Shear is defined as the derivative of the bending effort M along the axle; hence, it can be measured according to (4) as the ratio between the measured increment of the bending ΔM_{fi} along the axle to a known length Δx_i:

$$T_i = \frac{\mathrm{d}M}{\mathrm{d}x} = \frac{\Delta M_{fi}}{\Delta x_i}; \tag{4}$$

v_1 and v_2 are the vertical forces transmitted by primary suspension as shown in Fig. 4, which can be also measured by a load cell.

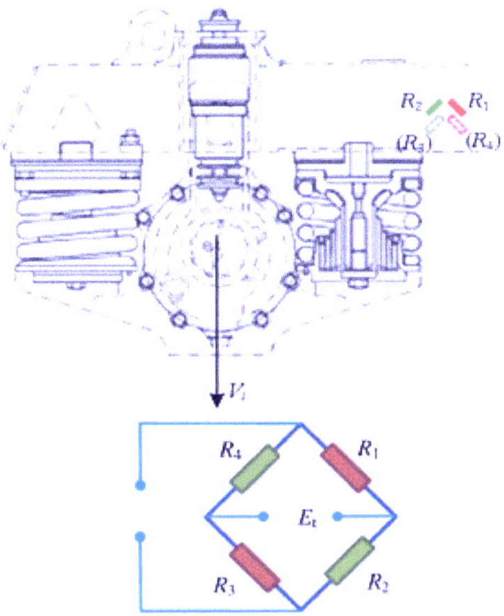

Figure 4: Strain gauge positioning on suspension (simplified scheme).

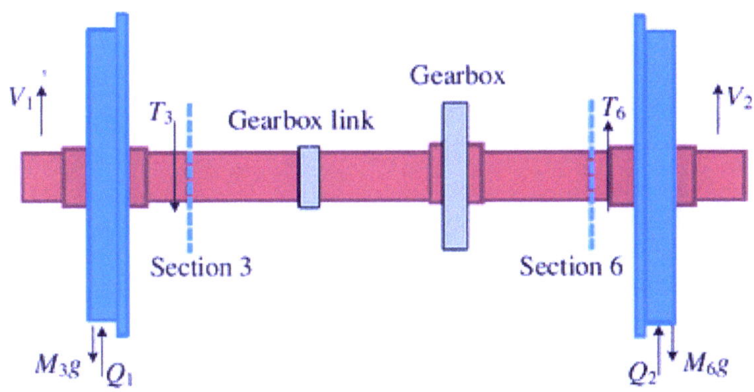

Figure 5: Measurament sections for Q-force.

This solution should be preferred especially to reduce encumbrances of the measurement system. In this case, toroidal load cells can be inserted under the springs of the primary suspension system. The bending torque is expressed with the function of longitudinal deformation ε_f:

$$M_f = \frac{\pi E \varepsilon_f (d_{\text{est}}^4 - d_{\text{int}}^4)}{32 d_{\text{est}}}, \qquad (5)$$

where E is Young's modulus.

Lateral Force

Lateral forces are estimated trough the bending moment by an array of strain-gauges on two different circular arrays as shown in Fig. 6. The lateral force Y_1 on a wheel is calculated by

$$Y_1 = \frac{M_I - M_{II}}{r_{11} - r_{12}}, \tag{6}$$

where M_I and M_{II} are the bending moments on two measurement radius. The radii of the two circumferences on which strain gauges are placed have to be optimized to increase the sensitivity of the sensors to the lateral forces and to eliminate cross-sensitivity effects against spurious forces:

$$\begin{aligned} M_I &= Y_1 \cdot r_{11} \pm Q_1 \cdot \Delta b_1, \\ M_{II} &= Y_1 \cdot r_{12} \pm Q_1 \cdot \Delta b_1. \end{aligned} \tag{7}$$

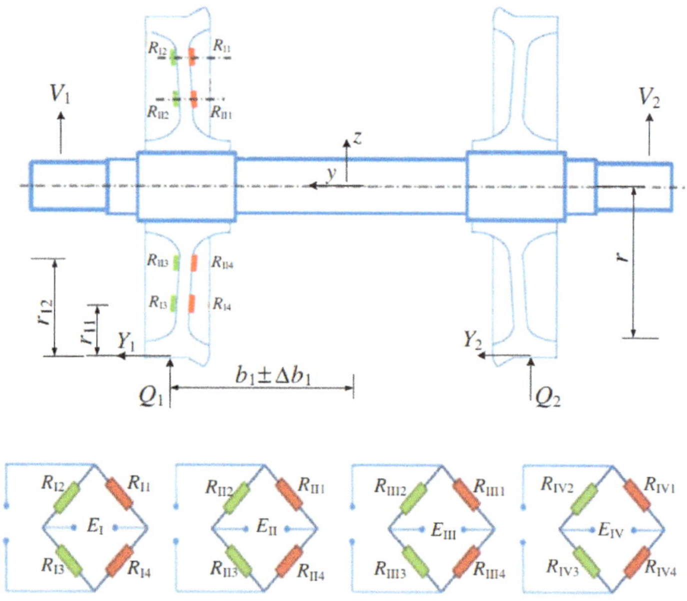

Figure 6: Strain gauge layout for *Y*-force.

CONTACT POINT POSITION

With the complete sensor layout used to calculate vertical and lateral components of the wheel rail contact forces (Q and Y), it is also possible to estimate the contact point position by imposing static equilibrium of the forces applied on the axle.

FEM MODEL

In order to evaluate the influence of strain gauge position on measurement, a complete FEM model of the axle and bogie is developed using MSC Nastran–Patran™. Preliminary simulations for a complete wheelset with two braking disks are performed to evaluate the stress–strain distribution, and consequently, to find an optimal strain gauge layout for the measurement of lateral forces *Y*. The strain gauges have to be placed on two concentric circumferences with radiir_1 and r_2 as shown in Fig. 7.

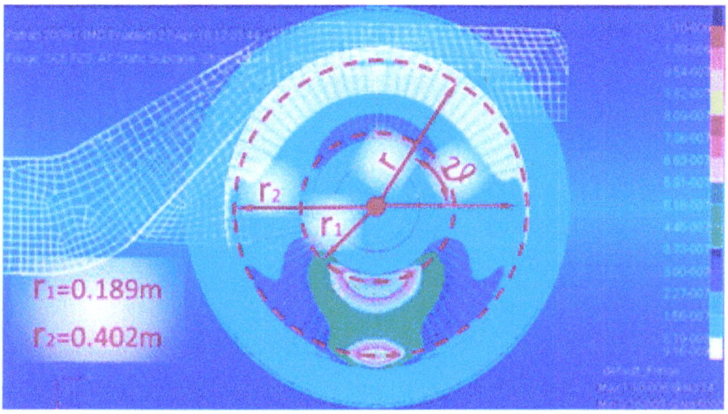

Figure 7: Radial strain on wheels (referred to the loading condition of N_2).

Values of r_1 and r_2 are optimized to improve the sensitivity of the measurement system to *Y* and to minimize the influences of other forces applied to the wheels such as vertical and lateral forces.

In particular, the optimization is performed with three different loading conditions, N_1, N_2, andN_3, and three different positions *A*, *B*, and *C* for a single contact point (see Fig. 8). As a consequence, in the optimization of r_1 and $r_{2,}$ nine FEM simulations have to be performed, in which both the applied forces and the position of a single contact point are changed. Values of applied forces *X*, *Y*, and *Q* in the three loading conditions N_1, N_2, and N_3 are chosen according to realistic operating conditions [6–8].Tables 1, 2 shows the values of the relative percentage error of *Y* measurements e_{ij}, which is defined as

$$e_{ij} = 100\frac{\Delta Y_{ij}}{Y_{ij}}, \tag{8}$$

where Y_{ij} denotes the value of the *Y* force considering the *i*-th contact point and the *j*-th loading condition; for instance, Y_{B2} represents the nominal condition in which the measurement system is calibrated: contact point position corresponding to case *B* and loading conditions corresponding to case 2. ΔY_{ij} is the absolute estimation error between measured and real values of Y_{ij}.

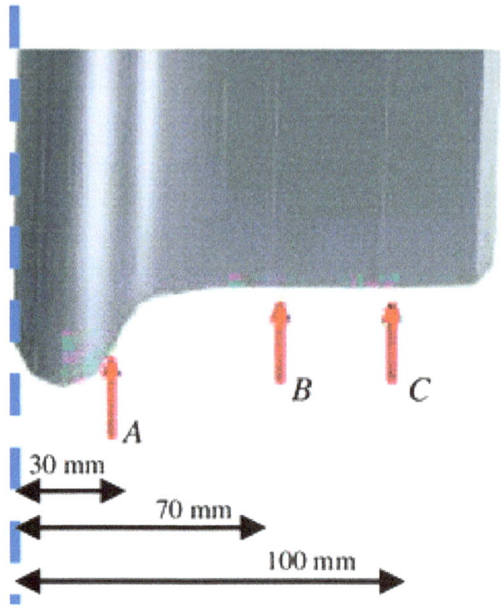

Figure 8: Contact *point* positions corresponding to the three differrent loading conditions N_1, N_2, and N_3.

Table 1: Vertical (Q), lateral (Y), and longitudinal (X) forces corresponding to the three loading conditions

Loading conditions (N)			
Forces	N_1	N_2	N_3
Y	30,000	30,000	30,000
Q	5,500	75,000	55,000
X	22,500	22,500	100

Table 2: Evaluation of indexes e_{ij} under different applied loadings and with different positions of the contact point

Contact point positions	**Loading conditions**		
	N_1 **(%)**	N_2 **(%)**	N_3 **(%)**
A	11	13	11
B	2	0	2
C	−7.9	−14	−7.9

In Table 2, values of e_{ij} corresponding to the optimal strain gauge layout (r_1 = 0.189 andr_2 = 0.402 mm) are shown. It is interesting to notice that the values of the relative error e_{ij} of Yforce estimation are slightly disturbed by a change of the loading conditions (N_1, N_2, and N_3). On the other hand, the proposed measure layout is more influenced by the equivalent contact point position which might cause relative estimation errors more than 10 %.

UNCERTAINTY ANALYSIS

Uncertainty analysis was performed according to the standard UNI CEI ENV 1300 [9] and the nomenclature definitions in UNI CEI 70099 [10].

In particular, the X-force and Q-force are evaluated with explicit functions in subsect. 2.1 and2.2. As it is not possible to evaluate an explicit relationship between the applied Y values and the corresponding measurement, the sensitivity analysis is performed by a numerical approach, by means of which the results of FEM model simulations are used to evaluate how disturbances and parametric uncertainties of the system affect the reliability and the precision of the measurement.

Uncertainty Analysis of Longitudinal Force Measurement

The X-force and Q-force as measurands are defined by the explicit relationships with a known set of independent quantities x_i. We define X and Q by considering the axle as a Bernoulli beam:

$$X = f_X(x_1, x_2, \ldots .x_N),$$
$$Q = f_Q(x_1, x_2, \ldots .x_N). \qquad (9)$$

Each quantity x_i is subjected to a standard uncertainty deriving from measurement errors or by natural tolerances when assuming system parameters as constants. The combined standard measurement uncertainty u of a generic quantity y is defined as

$$u = \sqrt{\sum_{i=1}^{N} \left[\left(\frac{\partial y}{\partial x_i} \right) \cdot u_i \right]^2}, \qquad (10)$$

where u_i is uncertainty of the i-th parameter. Supposing that independent variables are affected by a Gaussian distribution of uncertainties with a coverage factor equal to 2, the following relation is applied to calculate the expanded measurement uncertainty:

$$U = k \cdot u = \sqrt{\sum_{i=1}^{N} \left[\left(\frac{\partial y}{\partial x_i} \right) k u_i \right]^2}$$
$$= \sqrt{\sum_{i=1}^{N} \left[\left(\frac{\partial y}{\partial x_i} \right) k_i u_i \right]^2}, \qquad (11)$$

$$U = \sqrt{\sum_{i=1}^{N} \left[\left(\frac{\partial y}{\partial x_i} \right) U_i \right]^2}. \qquad (12)$$

Table 3 shows the results for the X-force during a braking maneuver with a deceleration of about 1–1.2 m/s^2 and a tangential force on each axle of about 10–15 kN.

The measurements of longitudinal forces X are affected by errors which are strongly influenced by wheel–rail adhesion factor μ:

$$\mu = \frac{X}{Q}. \qquad (13)$$

From a physical point of view, the vehicle adhesion coefficient should be the minimum wheel–rail friction factor that assures the transmission of the tangential force X if the contribution of rotating inertias is neglected. As Fig. 9 shows, the relative precisions of the X measurements rapidly decrease in the degraded adhesion conditions or when small longitudinal forces X between wheel and rails are exchanged.

Table 3: Estimated uncertainty in the X-force measurement

Parameters	Values	Uncertainty	**Uncertainty in (%)**
Shaft external diameter D_e (mm)	165	0.5	0.3
Elastic modulus E (MPa)	206,000	6,000	2.9
Shear modulus G (MPa)	79,8435	23,256	2.9
Strain-gauge deformation (m/m)	566	2	3.5
Wheel radius r (mm)	520	0.75	0.1

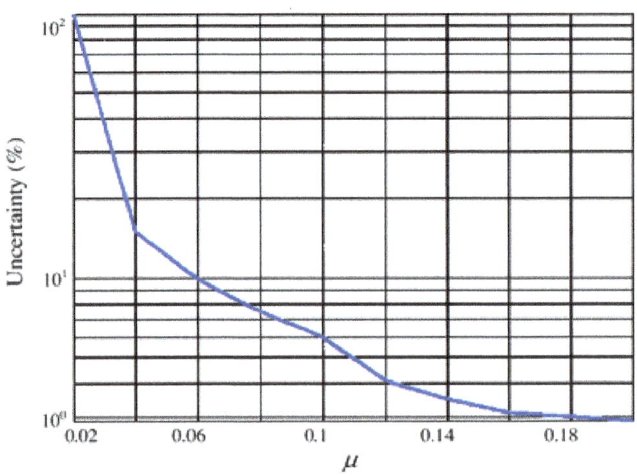

Figure 9: Relative uncertainty on longitudinal force with respect to adhesion behavior.

The reason for the unacceptable precision performances for low values of the wheel–rail friction factor lies in the high torsional stiffness of the axle compared with the applied torques and X. For low μ values, longitudinal forces are quite low, and consequently, axle deformations are quite negligible and affected by heavy errors.

In order to measure small longitudinal forces in degraded adhesion tests, brakes or the auxiliary shaft in Fig. 2 is equipped with sensors to measure applied braking torques. In this way, it is possible to accurately estimate the total force of X exchanged between both wheels and rails. For low adhesion tests that are usually performed on straight lines, these kinds of results/measurements of the total X force are valuable. In particular, degraded adhesion tests are often performed to verify performances of Wheel Slide Protection (WSP) systems installed on passenger coaches [11].

Uncertainty Analysis of Vertical Force Measurement

Also for the measurement of vertical forces Q, a sensitivity analysis is performed. The sensitivity analysis is conducted by introducing a symmetric vertical load discharge factor ΔQ_r, which is defined as a ratio between the absolute vertical load variation and the vertical load on wheel:

$$\Delta Q_r = \Delta Q_{r_1} = \Delta Q_1 / Q_i = -\Delta Q_{r_2} = \Delta Q_2 / Q_i. \tag{14}$$

In Fig. 10, the graphical behavior of the estimated maximum uncertainty between the two wheels, as a function of ΔQ_r, is shown. Note that with more load transfer between wheels, the uncertainty increases. The sensitivity analysis results of the vertical force Q are shown in Table 4, where i and j correspond to the two measurements sections (numbered as section numbers 3 and 6) according the scheme of Fig. 5. The analysis is performed with some known uncertainties such as errors on strain measurements, geometric tolerances, and partially known material properties.

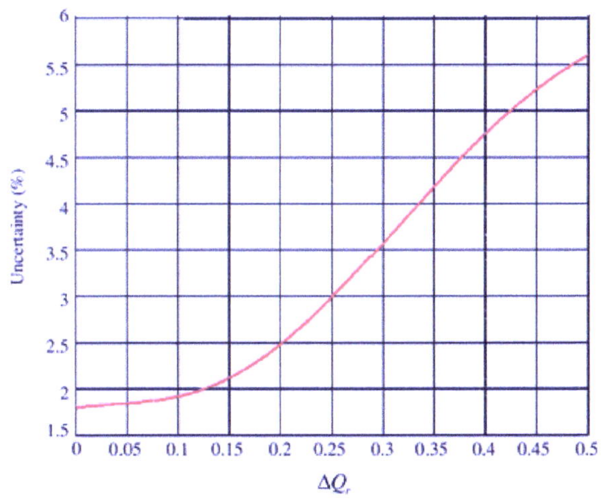

Figure 10: Relative uncertainty in the function of ΔQ_r.

Table 4: Estimated uncertainty of Q-force measurements

	Value	Uncertainty	Uncertainty (%)
Strain gauge deformation i (m/m)	−119.2	2	1.7
Strain gauge deformation j (m/m)	−117.2	2	1.7

Shaft external diameter D_e (mm)	165	0.5	0.3
Elastic modulus E (MPa)	206,000	6,000	2.9
Mass i–j shaft M (kg)	675	10	1.5

Uncertainty Analysis of Lateral Force

For the measurement of Y, it is not possible to apply an analytic expression. Thus, the uncertainty analysis is performed using the FEM model. In particular, the analysis is performed considering linear and angular errors of strain gauge positioning as shown in Fig. 11, where $\Delta r = \pm 1$ mm, $\Delta\theta = \pm 1°$, $\Delta\alpha = \pm 2°$.

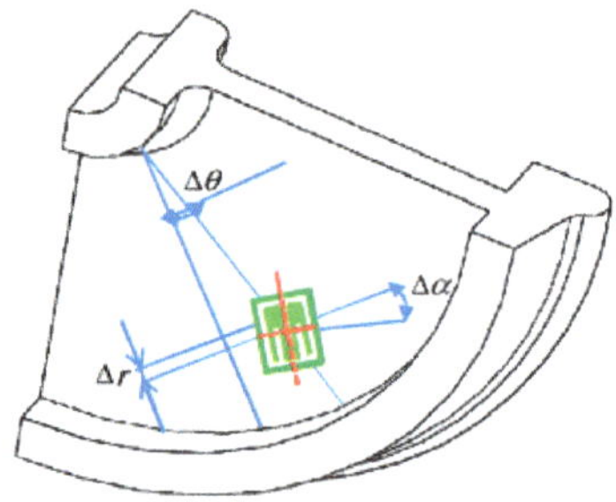

Figure 11: Placement errors of strain gauge.

FEM model produces strain results which are defined over a discrete population of nodes. In order to perform a sensitivity analysis, continuous derivatives of strain with respect to strain gauge›s positioning have to be performed. As a consequence, techniques to obtain a smart interpolation of calculated stress and strain along the wheel surface have to be applied. With a polar reference system centered on the rotation axis of the wheel, strain results have to be interpolated over a grid of 936 nodes corresponding to 26 radial distances r and 36 angular positions.

Two different interpolation techniques are used:

- Standard triangular interpolation: the generic properties y_p for a point p is calculated as the weighted sum of the calculated y_{p1}, y_{p2}, and y_{p0} on the three nearest nodes p_0, p_1, and p_2. The simplified scheme is shown in Fig. 12.

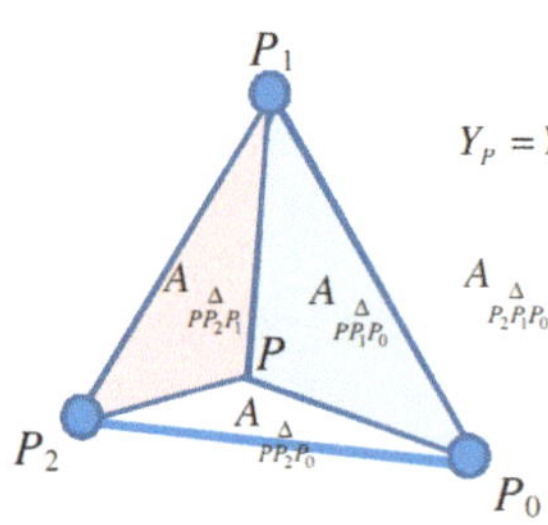

$$Y_P = Y_{P_2}\left(\frac{A_{\Delta PP_1P_0}}{A_{\Delta P_2P_1P_0}}\right) + Y_{P_1}\left(\frac{A_{\Delta PP_2P_0}}{A_{\Delta P_2P_1P_0}}\right) + Y_{P_0}\left(\frac{A_{\Delta PP_2P_1}}{A_{\Delta P_2P_1P_0}}\right)$$

$$A_{\Delta P_2P_1P_0} = A_{\Delta PP_1P_0} + A_{\Delta PP_2P_0} + A_{\Delta PP_2P_1}$$

Figure 12: Standard triangular interpolation.

- Inverse distance weighted interpolation (IDWI) [12]: the interpolation is performed on a subset of the complete population of nodes, with a weighting function which is inversely proportional to the squared distance between the interpolation point p and the corresponding node p_i. The subset population is chosen among the n nearest nodes with respect to p where $n$is the size of the chosen subset population. Figure 13 demonstrates how the IDWI interpolation gradually converges to a very high precision with the increase of size n.

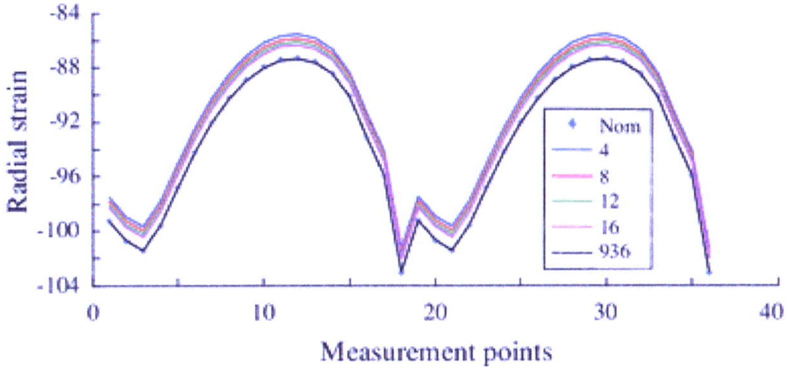

Figuure 13: IDWI interpolation of Radial microstrain FEM results.

After several tests, IDWI is chosen since it produces desirable results on the polar grid used for the FEM model of the bogie. As clearly shown in Fig. 13, the precision of the IDWI interpolation gradually improve as the number of nodes n used for interpolation increases.

Using the interpolated results of the FEM model, it is possible to calculate how predicted tolerances on strain gauge position affect the precision of Y measurements as shown in Table 5.

Table 5: Relative uncertainty on Y measurements due to placement tolerances of strain gauges

Placement error	**Uncertainty %**		
	Max	**Min**	**Mean**
Radial error Δr	1.02	0.28	0.57
Tangential error $\Delta \varepsilon$	0.88	0.12	0.42
Inclination angle $\Delta \alpha$	0.60	−1.45	0.86
Total	2.54	−1.05	0.2

FREQUENCY RESPONSE ESTIMATION

Using the FEM model of the bogie, it is possible to approximately evaluate the bandwidth of the proposed measurement system in terms of transfer functions. In Figs. 14 and 15, the calculated transfer functions $P_Y(\omega)$ and $P_Q(\omega)$ between Y and Q forces and the corresponding measurements performed by the proposed system are shown. Notice that for both Y and Q measurements, the bandwidth is limited to about 20 Hz. This bandwidth limitation is mainly caused by a structural mode-eigenfrequency of the bogie located at about 22.5 Hz. The modal shape of the bogie vibrating at 22.5 Hz is shown in Fig. 16.

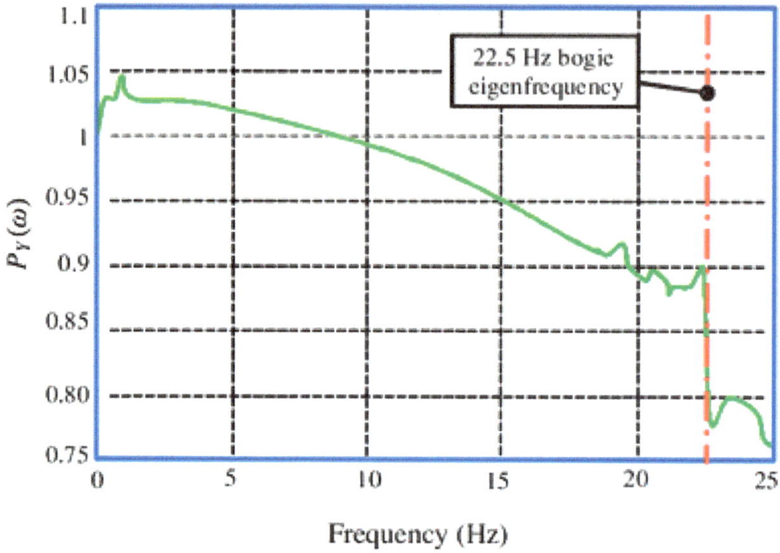

Figure 14: The transfer function between Y force and corresponding strain measurement.

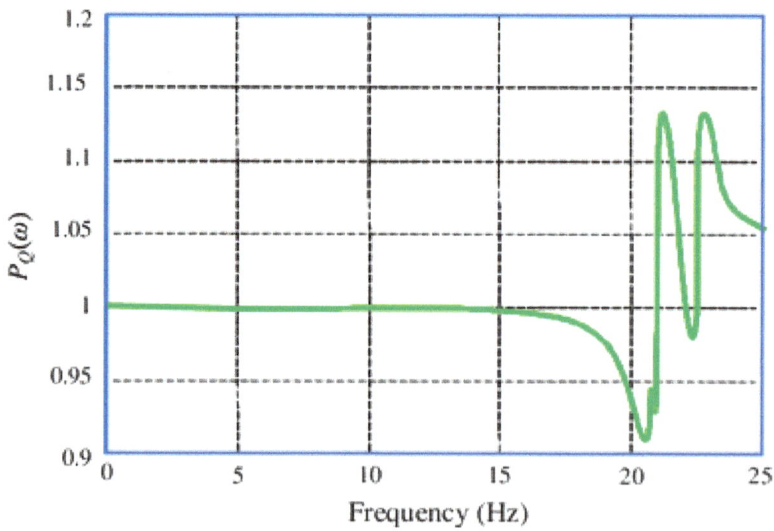

Figure 15: The transfer function between Q force and corresponding strain measurement.

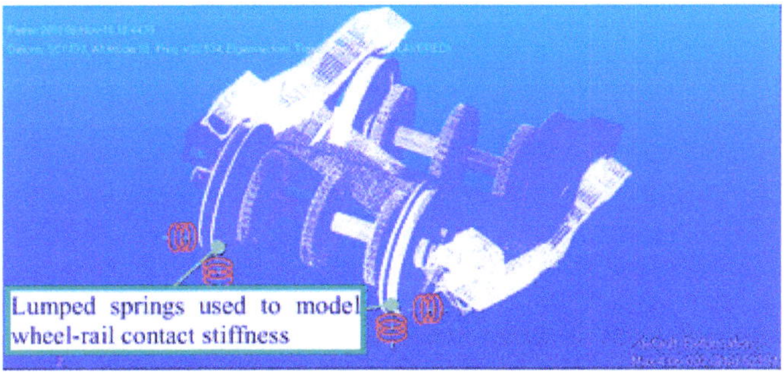

Figure 16: Modal response of the bogie at 22.5 Hz.

CONCLUSIONS AND FUTURE DEVELOPMENTS

In this study, a railway bogie model was proposed for the contact force measurements, and its performance has been evaluated.

The proposed model has the following features:

- Disturbances introduced by braking (both thermally and mechanically induced deformations) are not considered.
- The proposed approach is based on a static FEM model of the bogie, and the most accurate results are obtained by introducing mixed flexible-multibody models of vehicles and railway line.
- The influence of strain gauge position on the measurement of contact force was evaluated based on a synergy between FEM model bogie and sensitivity analysis of contact force measurements against several uncertainty factors. Among the various factors, the position of the contact point is one of the most important factors. Calculation of contact point positions is directly estimated in the case of a single wheel–rail contact point. For the identification of multiple contact points, an undirect approach should be applied. We are also considering different approaches such as the inverse identification approach [13] and investigating the feasibility of using an estimator based on a modal approach [14].
- From the calculation of the frequency response of the system, a strong coupling between the flexible behaviors of rails and vehicle is clearly recognized. As a consequence, the real bandwidth of the contact force measurement system has to be verified taking into account the dynamic response of both systems.

Research activities will be further extended to introducing a more accurate model of the flexible three-dimensional contact between rail and wheels using FEM. In particular, attention needs to be directed to the possibility of identifying some typical disturbance pattern associated with known singularities or failure that should be identified such as sleeper voids [15] or wheel flats [16]. Moreover, neural networks, fuzzy, or other approaches such as wavelet analysis should be applied to identify separately different kinds of defects [17–19]. Also further studies need to be performed to identify in a fast and smart manner the noise introduced in the acquisition of measurements by the process of analog-to-digital conversion, to accord with the approach proposed by Balestrieri [20].

REFERENCES

1. UIC 518, Testing and approval of railway vehicles from the point of view of their dynamic behaviour—safety—track fatigue—running behaviour, 4th edn, September 2009
2. UNI EN 14363 Railway applications-testing for the acceptance of running characteristics of railway vehicles—testing of running behaviour and stationary tests, 2005
3. Benigni F, Braghin S, Cervello et al. Wheel–rail contact force measurements from axle strain measurements, Railway Engineering (official journal of CIFI the association of Italian Railway Engineers), N. 12, December 2002, p 1059–1075
4. Allotta B, Pugi L. Mechatronic, electric and hydraulic actuators, esculapio, bologna ISBN: 8874884958, ISBN-13: 9788874884957
5. Berg H, Goessling G, Zueck H (1996) Axle shaft and wheel disk: the right combination for measuring the wheel/rail contact forces. ZEV Glasers Ann 120(2):40–47
6. Meli E, Malvezzi M, Papini S et al (2008) A railway vehicle multibody model for real-time applications. Vehicle Syst Dyn 46(12):1083–1105
7. Falomi S, Papini S, Rindi A, et al. (2011) A FEM model to compare measurement layouts to evaluate the wheel-rail contact forces, In: World Congress on Railway Research, Lille, France, pp 1–11
8. Malvezzi M, Pugi L, Papini S et al. (2012) Identification of a wheel–rail adhesion coefficient from experimental data during braking tests, In: Proceedings of the Institution of Mechanical Engineers, Part F: Journal of Rail and Rapid Transit
9. NI CEI ENV 13005, Guide to the expression of uncertainty in measurement, 31-07-2000
10. UNI CEI 70099 (2008) International vocabulary of metrology basic and general concepts and associated terms (VIM)
11. Pugi L, Malvezzi M, Tarasconi A et al (2006) HIL simulation of WSP systems on MI-6 test rig. Vehicle Syst Dyn 44(2006(Sup.)):843–852
12. Franke R, Nielson G (2005) Smooth interpolation of large sets of scattered data. Int J Numer Meth Eng 15(11):1691–1704
13. Uhl T (2007) The inverse identification problem and its technical application. Arch Appl Mech 77(5):325–337
14. Ronasi H, Johansson H, Larsson F (2012) Identification of wheel–rail contact forces based on strain measurement and finite element model of the rolling wheel, topics in modal analysis II. In: Conference Proceedings of the Society for Experimental Mechanics Series, Vol 31, 2012: Jacksonville, Jan 30 Feb 2, 2012 p 169–177

15. Bezin Y, Iwnicki SD, Cavalletti M et al. (2009) An investigation of sleeper voids using a flexible track model integrated with railway multi-body dynamics. In: Proceedings of the Institution of Mechanical Engineers, Part F: J Rail and Rapid Transit, 223(276): 597–607

16. Zhao X, Li Z, Liu J (2012) Wheel–rail impact and the dynamic forces at discrete supports of rails in the presence of singular rail, surface defects, In: Proceedings of the Institution of Mechanical Engineers, Part F: Journal of Rail and Rapid Transit, 226(2): 124–139

17. Chiu WK, Barke D (2005) Structural health monitoring in the railway industry: a review. Struct Health Monit 4(1):81–93

18. Bracciali A, Lionetti G, Pieralli M (2002) Effective wheel flats detection trough a simple device. Techrail Workshop 2:14–15

19. G. Yue, Fault analysis about wheel tread slid flat of freight car and its countermeasure, Railway locomotive and car, 2000(4): 46–47 (in Chinese)

20. Balestrieri E, Catelani M, Ciani L et al (2012) The Student's distribution to measure the word error rate in analog-to-digital converters. J Int Meas Confed 45(2):148–154

16
CHAPTER

IMAGE ANALYSIS AND DC CONDUCTIVITY MEASUREMENT FOR THE EVALUATION OF CARBON NANOTUBE DISTRIBUTION IN CEMENT MATRIX

I. W. Nam, H. K. Lee

Keywords: Cement composites, carbon nanotubes, distribution evaluation, image analysis, electrical conductivity, fluidity

INTRODUCTION

The electrical and mechanical properties of cement composites with conductive fibers are significantly influenced by the fiber distribution (Sorensen et al. 2014; Liu et al. 2011; Lee et al.2009). The electrical properties can be enhanced by the formation of conductive fiber networks. The mechanical properties can also be improved by the effect of fibers bridging micro-cracks, whereas they deteriorated in the base of a non-uniform distribution of fibers (Sorensen et al.2014; Liu et al. 2011). Accordingly, evaluation of the fiber distribution is important in understanding its influence on the physical properties of the composite materials and making full use of fibers (Liu et al. 2011).

Evaluation methods for the fiber distribution can be classified into two categories—image analyses and electrical property analyses. An image analysis entails acquisition of images via a microscope (optical microscope, scanning electron microscope (SEM), etc.) or a transmission X-ray and image processing and analysis. Lee et al. (2009) and Kang et al. (2011) obtained cross sectional images of composite materials via an optical microscope (Lee et al. 2009; Kang et al.2011). Lee et al. (2002) and Liu et al. (2011) obtained images via digital cameras (Lee et al.2002; Liu et al. 2011). Redon et al. (1999) used X-ray photography and Fan et al. (2000) used a SEM to capture fiber distribution

Citation: I. W. Nam, H. K. Lee, Image Analysis and DC Conductivity Measurement for the Evaluation of Carbon Nanotube Distribution in Cement Matrix, DOI: 10.1007/s40069-015-0121-8.

images (Redon et al. 1999; Fan et al. 2000). On the other hand, the fiber distribution state can also be evaluated by the DC conductivity or AC impedance. Li et al. (2007) analyzed MWNT distribution states using the DC conductivity and Ozyurt et al. (2006) evaluated the degree of carbon fiber clumping and fiber orientation by AC impedance spectroscopy (Li et al. 2007; Ozyurt et al. 2006). In the case of the electrical property analysis method, it is only applicable to conductive fiber-incorporated composite materials.

The image analysis is a more direct and convincing approach in that actual fiber distribution states in the composite materials can be evaluated. The electrical property analysis method is an indirect approach for evaluation of the fiber distribution, but sometimes it is affected by other parameters (matrix type, moisture content in matrix, etc.), possibly leading to incorrect electrical property results. Accordingly, the image analysis plays an important role in the evaluation of fiber distributions and it is a critical technique that should be carried out along with an electrical property analysis (Li et al. 2007; Ozyurt et al. 2006).

Uncovered in 1991, carbon nanotube (CNT) has gained great attention from researchers attributable to its remarkable physical properties (Li et al. 2007; Kim et al. 2014). From 2010s, CNT has been used as a nano-scale fiber in cement-based composites in an effort to enhance physical properties of the composites (Kim et al. 2014). Various approaches have been explored to evaluate CNT distributions in the literature. SEM observation has been the most common approach to study the CNT distribution state, carried out in tandem with an electrical property analysis (Konsta-Gdoutos et al. 2010). However, the SEM observations may vary depending on observation spots in the same sample since it is mostly conducted with high magnification levels (up to 3000-fold). Accordingly, the CNT distribution should be observed and evaluated at low magnification levels and the evaluation results should be compared with electrical properties of the CNT-embedded composites.

In the present work, a novel image analysis method to evaluate the multi-walled carbon nanotube (MWNT) distribution in cement-based composites is proposed. In this approach, white cement was used instead of ordinary Portland cement (OPC) with MWNT in an effort to differentiate MWNT from the cement matrix. MWNT distribution images in the cement matrix materials were acquired by using an optical microscope in conjunction with image processing tools. The MWNT distribution was quantitatively assessed in terms of the planar occupation ratio of MWNT. In preparation of specimens, a novel method was adopted for dispersion of CNT in cement. This method was proposed on the basis of experimental attempts previously conducted by the authors. It was suggested that the distribution of CNT can be enhanced by means of lowering the fluidity of CNT/cement mixture at fresh state. It is notable that the suggested method does not require sonication technique, acid treatment, etc., which were widely demonstrated in the literature. To verify correlation between the MWNT distribution and flow of the mixtures, MWNT-embedded cement composites were fabricated under different flows, incorporating a constant MWNT content (0.6 wt%). The image analysis demonstrated that the MWNT distribution was significantly enhanced in the composites fabricated under a low flow condition, and DC conductivity measurement results revealed the dramatic increase in the conductivity of the composites fabricated under the low flow condition, which supported the image analysis results. The composites were also prepared under the low flow condition (114 mm < flow < 126 mm), incorporating various MWNT contents. The image analysis and DC conductivity results demonstrated remarkable enhancement in the planar occupation ratio of MWNT and the conductivity, respectively, which indicated that fabrication of the composites under the low flow condition was an effective way to enhance the MWNT distribution.

MATERIALS

MWNT, Portland cement, nylon fiber, super-plasticizer, tap water, and silica fume (SF) were used in the present work. MWNT produced through the chemical vapor deposition (CVD) growth method, a proprietary product of Hyosung Inc. (M1111), was used (Nam et al. 2012). Their purity, diameter, and aspect ratio were 96.2 %, 12.29 ± 2.18 nm, and 930 (aspect ratio was approximate value), respectively. Type I ordinary Portland cement was used in the present work. SF, a proprietary product of Elkem Inc. (EMS-970 D), contained over 90 % silicon oxide (SiO) and 80 wt% of its primary particles have a diameter greater than 5 μm. White cement was used instead of OPC when image analysis samples were fabricated. Nylon fiber, a proprietary product of Nycon fibers Inc. (NYMAX), was used in an effort to prevent cracks that occur due to shrinkage while cement matrix materials cured. Their diameter and length were 23–36 μm and 3 mm, respectively. A poly-carboxylic acid-based super-plasticizer (SP), a proprietary product of BASF Pozzolith Ltd., (Rheobuild SP8HU), was utilized in an effort to improve the workability of the fresh cement matrix materials. The true specific gravity of MWNT, Portland cement, SP, tap water, and SF was 1.32, 3.15, 1.07, 1, and 2.1, respectively.

METHODS

Image Analysis of MWNT Distribution in MWNT-Embedded Cement Composites

Specimen Preparation and Image Acquisition for Image Analysis

A novel image analysis approach was proposed in an effort to evaluate the MWNT distribution in cement matrix materials. In order to distinguish MWNT with black color from OPC, which also shows a fairly dark color, white cement was used in place of OPC when composite specimens were fabricated for the image analysis. In addition to using white cement instead of OPC, the incorporation of nylon fiber was omitted in the preparation of the samples for the image analysis since crack control was unnecessary. The constituent materials (white cement, MWNT, water, and SP) were weighed according to mixing ratios and placed together in a steel bowl.

Table 1 shows the mix proportions of composites fabricated with different W/C values, which will be dealt in Sect. 4.1.1. The mix proportions for composites fabricated under a controlled low flow condition (from 114 to 126 mm) with various MWNT contents are presented in Table 2, which will be dealt in Sect. 4.2.1. The weighed materials were thoroughly mixed for 20 min by an electric hand mixer. The electric hand mixer consisted of two curl shaped-beaters and their rotation rate was approximately 200 rpm. The fresh mixtures were poured into plastic molds that were designed in accordance with the sample size. The sample size was 10 × 10 × 80 (mm^3) for the image analysis. After pouring the fresh mixtures in the molds, they were covered with a plastic plate (or wrapping film) in an effort to avoid moisture loss of the samples and cured for 24–48 h at room temperature (18–20 °C) under a humid condition (40–50 %), and then detached from the molds.

Table 1: Constituent materials and their mix proportions of the MWNT-embedded cement composites fabricated under different flow.

Denotations	**Mix proportions (g)**						**MWNT (vol%)**
	Water	**Cement**	**SF**	**SP**	**Nylon fiber**	**MWNT**	
W26–M06	26	100	20	1.6	0 (omitted)	0.6	0.66
W30–M06	30						0.62
W34–M06	34						0.59
W38–M06	38						0.56
W42–M06	42						0.53

Table 2: Constituent materials and their mix proportions of the composites fabricated under the low flow condition (114 mm < flow < 126 mm).

Denotations		Mix proportions (g)						Flow (mm)
Group	Type	Water	Cement	MWNT		SP	Nylon fiber	
				Wt% by cement	Vol%			
LF-M	LF-M0	25	100	0	0	0.6	0.2	160
	LF-M0.3	25		0.3	0.39	1.0		126
	LF-M0.6	25		0.6	0.76	3		125
	LF-M1.0	30		1.0	1.14	4.54		120
	LF-M1.5	36		1.5	1.56	4.4		114

In an effort to produce a fracture surface, a cutting knife was used to make a cutting guide line with depth of approximately 1 mm at the center of the sample. The sample was split into two parts applying manual force by hand to the cutting guide. Each split part was sliced by means of a hand saw at 3 mm from the fracture surface. As a result, a specimen with size of 10 × 10 × 3 mm^3 was prepared from the original sample.

A magnification to cover MWNT agglomerates whose diameter ranged from 3 to 200 μm was selected by a trial and error process and determined as 50-fold. An auto stage optical microscope (Olympus MX51) was used so that visible light could be used to distinguish MWNT (black) from the cement matrix (white). The microscope system illuminates the specimen with uniform brightness and transfers the observation image to image processing software, DotSlide. Since the optical focus at each local spot of the fractured specimen changes due to unevenness of the specimen surface, the microscope system captured observation images at every 3 μm along the out of plane direction from the deepest level to the highest level of the fractured surface. By using the image processing software, focused local spots in the captured images were collected and combined to form a single image that is in focus overall. Two images of each specimen were gained through the aforementioned procedure.

Image Processing and Analysis Procedures

In microscopic images, the black amorphous agglomerates indicated MWNTs. The contrast of white cement and MWNT was seen remarkably well by the naked eye. By image processing and analysis with the commercial software, the MWNT distribution can be quantitatively evaluated. Stepwise tasks were carried out in the following order (illustrations for each step can be found in Fig. 1) with the fractured specimens.

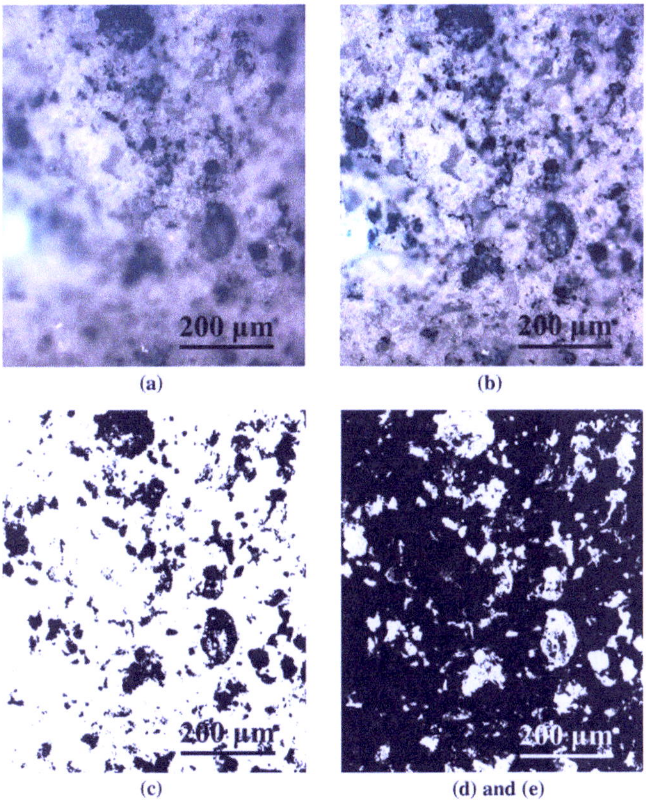

Figure 1: Image processing procedures (image size: 295 × 355 pixel): **a** Image acquisition, **b**gathering of areas that are well optically focused and assemble them together, **c**conversion of the image to a black and white image by thresholding, **d** reversal of the image, and **e** conversion of the image to a binary image.

- Image acquisition by microscope in collaboration with a digital camera
- Capture areas that are optically well focused and assemble them so a well-

focused image can be completed (conducted by DotSlide)

- Convert the image to a black and white image by thresholding
- Reverse the black and white image
- Convert the black and white image to a binary image
- Acquire data (area) of agglomerate regions

Steps (1) to (2) were conducted by the microscope system and the remaining steps were conducted using a MATLAB Image processing tool box. Automatic brightness control was applied in step (1) by the microscope system. The effect of change of brightness level on the image thresholding was considered a minor factor and hence was ignored in the present work. Otsu (1979)'s automatic thresholding method was used in the thresholding process of step (3) (Otsu 1979). The gray-level scale, which is determined between 0 and 1, was used with a 15 % reduction (gray-level scale × 0.85) since the brightness of MWNT was considered high when compared with the original image. After the reversion process (4), bright parts indicated MWNT. Accordingly, the planar occupation area of the bright parts expressed as 1 in the corresponding binary image, was summed. The planar occupation ratio of bright parts can be calculated by dividing the sum total of the bright parts by the total area of the corresponding image. The planar occupation ratio of MWNT was calculated from each of two images and an average value was determined from the ratios.

DC Conductivity of the MWNT-Embedded Cement Composites

Specimen Preparation for DC Conductivity Measurement

The preparation procedures of samples for DC conductivity measurement followed the specimen preparation procedures for the image analysis in Sect. 3.1.1. However, OPC was used as cement material in this experiment and nylon fiber was used. The constituent materials of the composites were prepared in accordance with mix proportions. In particular, the mix proportions for composites prepared with different flows and constant MWNT content are presented in Table 3, and DC conductivity of the composites will be dealt in Sect. 4.1.2. On the other hand, Tables 2 and 4 show the mix proportions of composites fabricated under the low flow condition and a high flow condition (over 250 mm), which will be dealt in Sect. 4.2.2 in terms of DC conductivity. The constituent materials were weighed according to the mix proportions and mixed by using the electric hand mixer. The resultant fresh mixture was decanted into plastic molds designed with the size 25 × 25 × 25 (mm^3). After undergoing 1 day's curing under conditions explained in Sect. 3.1.1, the samples were separated from the molds. In an effort to gain mechanical strength to prevent unexpected damages in the samples, they were submerged in water during 7 days.

Table 3: The constituent materials and their weight content ratios of the composites fabricated under different flow in an effort to understand change of the MWNT distribution by examining the DC conductivity of the composites.

Denotations	Mix proportions (g)						
	Water	**Cement**	**SF**	**SP**	**Nylon fiber**	**MWNT**	**Flow (mm)**
M06–W26	26	100	20	1.6	0.2	0.6	102
M06–W30	30						127
M06–W34	34						151
M06–W38	38						211
M06–W40	40						250
M06–W42	42						>250
M06–SP0	40			0			130
M06–SP04				0.4			119
M06–SP16				1.6			250
M06–SP32				3.2			>250
M06–SP64				6.4			>250

Table 4: The constituent materials and their mix proportions of the composites fabricated under the high flow condition (flow > 250 mm).

Denotations		Mix proportions (g)					
Group	**Type**	**Water**	**Cement**	**MWNT**		**SP**	**Nylon fiber**
				Wt% by cement	**Vol%**		
HF-M	HF-M0	40	100	0	0	1.6	0.2
	HF-M0.3			0.3	0.31		
	HF-M0.6			0.6	0.62		
	HF-M1.0			1.0	1.02		

Measurement Method

Li et al. (2007) and Xie et al. (1996) revealed that the CNT distribution in CNT-added composite materials can be assessed by examining the DC conductivity of the materials (Li et al. 2007; Xie et al. 1996). Once SEM or transmission electron microscopy (TEM) reveal that CNTs are homogeneously dispersed in the composite materials, DC conductivity evaluation can be used in an effort to identify the percolation network, which is a large cluster network spanning from one side to the opposite side without disconnection in the composite materials (Stauffer and Aharony 1994). The measurement method of DC conductivity in the present work complied with a standard method presented in SEMI MF43 (2005). As electrodes for the measurement, a pair of copper plates having dimensions of 10 × 35 × 0.5 (mm^3) was inserted in the center part of the sample with a spacing of 8 mm as shown in Fig. 1 of Nam et al. (2015). The copper electrodes were embedded when the composite mixtures were still fresh. As another pair of electrodes, both sides of the samples were coated with silver paste as shown in Fig. 1 of Nam et al. (2015). A power supply equipment (Agilent E3642A) generated DC current, and it was passed through the silver paste electrodes. Due to the supply of current in resistors, which are the MWNT-embedded cement composites, potential difference in the composites was created, and it was measured by a digital multimeter (Agilent 34410A) connected with the copper plates. The supplied current was automatically controlled by the power supplier with a limitation up to 0.2 A and the resultant voltage was produced with a limitation up to 20 V. Figure 2 illustrates the measurement method for DC conductivity.

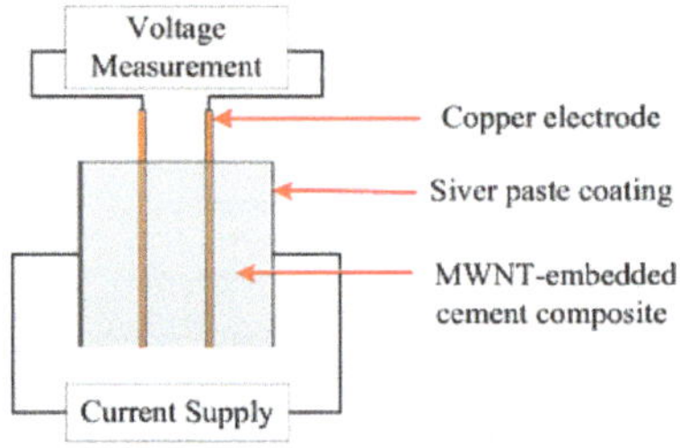

Figure 2: A schematic illustration of the DC conductivity measurement for the composites.

The resistance of the composites was calculated on the basis of Ohm's law. For the determination of DC conductivity, the calculated resistance value (*R*), cross sectional area of the electrode in contact with the composites (*A*), and the interval of the electrodes (*L*) were plugged into a following equation (Xie et al. 1996; Vance et al. 2014)

$$\sigma = \frac{1}{\rho} = \frac{L}{R \times A} = \frac{0.8\,\text{cm}}{R \times (2.5\,\text{cm} \times 1\,\text{cm})}\ (\text{S/cm}) \qquad (1)$$

Flow of the Composite Mixtures

The flow of fresh mixture of the composites was tested on the basis of ASTM C1437 (ASTM International 2013). In addition, the flow was determined from an average of mixture's diameter after dropping the flow table.

RESULTS AND DISCUSSION

MWNT Distribution in Composites Fabricated Under Different Flows

Image Analysis of MWNT Distribution State in the Composites Fabricated Under Different Flows

Through preliminary tests on MWNT-embedded cement composites, it is found that the flow of the fresh mixture may influence the MWNT distribution. Accordingly, MWNT-embedded cement composites fabricated with different W/C values were prepared and the MWNT distribution state of the composites was evaluated. The constituent materials are listed in Table 1. The mixing ratio of MWNT was 0.6 wt% in order to provide a sufficient amount of MWNT. 0.6 wt% of MWNT exceeded the percolation threshold as reported in the literature (Nam et al. 2015). Accordingly, the distribution of MWNT was expected to be pronouncedly visible. The volumetric fractions (vol%) of MWNT were also provided in the table. In determination procedures of volumetric fractions of MWNT, volume of each constituent material had to be calculated by using true specific gravity of the materials shown in the Sect. 2, then volume ratio of MWNT to total volume of the mixture was obtained. 20 % (by cement weight) SF was added in the mixtures under the consideration that SF can improve the MWNT distribution (Nam et al. 2012). The white cement was substituted for OPC and the incorporation of nylon fiber was omitted, as addressed in Sect. 3.1.1. The specimens were prepared and images of fractured surfaces of the specimens were obtained according to the method described in Sect. 3.1.2.

Figure 3 presents processed images that were obtained from the fabricated specimens. In the processed images, the total area of the MWNT agglomerates appeared to decrease with an increase of the flow (or W/C ratio). Moreover, MWNT clumps, which are excessively entangled-MWNT agglomerates, were generated as the flow increased. Based on these observations, it can be surmised that MWNT agglomerates disentangled as the flow (or W/C ratio) decreased.

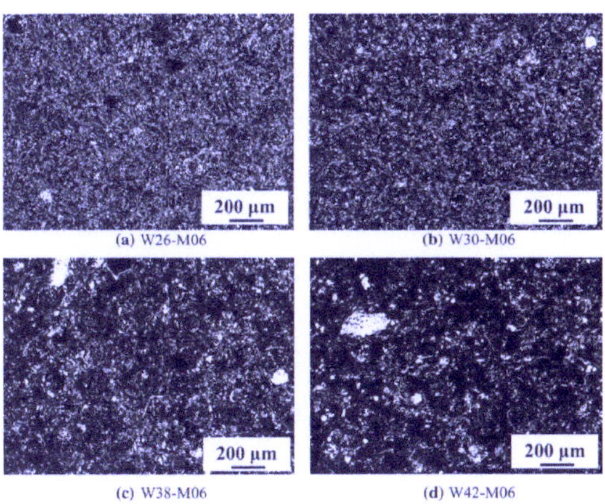

Figure 3: The fractured surface of the MWNT-embedded cement composites fabricated under different flow (W/C ratio) after having the image thresholding, reversion, and binarization processes (image size: 570 × 426 pixel): **a** W26–M06, **b** W30–M06, **c** W38–M06, and **d** W42–M06.

In an effort to express the change of the MWNT distribution states in a quantitative manner, the proportion of MWNT agglomerates, indicated by white color, to the total area of the image, was calculated by the MATLAB image processing tool box. This proportion is designated as a *q*value. The *q* value refers to the planar occupation ratio of MWNT on a fractured surface of the composite.

Figure 4 shows an increase of the q value as the W/C ratio of the cement composites decreases. The q value doubled due to a 16 % reduction of the W/C ratio. The increase of the q value was attributed to disentanglement of the MWNT agglomerates. The disentanglement of the MWNT agglomerates can be explained by variation of the microstructure of the fresh mixture. Figure 5a shows the microstructure of the fresh mixture of the MWNT-embedded cement composite prepared under a high flow condition. The distance between unhydrated cements increased as the W/C ratio (or SP/C ratio, i.e. SP versus cement ratio) increased (Daimon and Roy 1979). When the interspace among the unhydrated cements increased, highly entangled MWNTs floated in the free water. This is ascribed to hydrophobic characteristics of MWNT. As a result, the images obtained from the W38–M06 and W42–M06 samples showed the highly entangled MWNTs, as presented in the Fig. 3. On the contrary, the microstructure of the fresh mixture of the composites can be changed if the flow of the mixture is decreased. The distance between unhydrated cements in the composites prepared under a low flow condition was narrowed, as shown in Fig. 5b. If the interspace is narrowed, the MWNT agglomerates can be disentangled during the mixing process. This is attributable to the characteristic that the size of the MWNT agglomerates can be restricted by the size of the interspace among the unhydrated cements. As a result, the images obtained from the W26–M06 and W30–M06 specimens in the Fig. 3showed disentangled MWNT agglomerates.

The planar occupation ratio of MWNT, q, was increased twofold with decrease of W/C ratio from 42 to 26 % as shown in the image analysis result of Fig. 4. However, electrical conductivity of the composites is expected to increase more than hundreds of times as the W/C ratio decreases. This stems from that disconnected CNTs can be electrically connected attributable to the enhancement of MWNT distribution. Accordingly, it can be said that the twofold greater $q$value can lead to a dramatic increase in the electrical conductivity. This will be dealt in the electrical property characterization section.

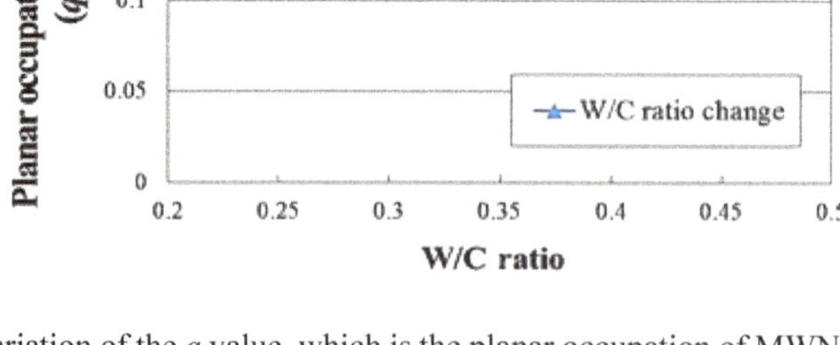

Figure 4: The variation of the q value, which is the planar occupation of MWNT on the fractured surface of the composites, as a function of W/C ratio.

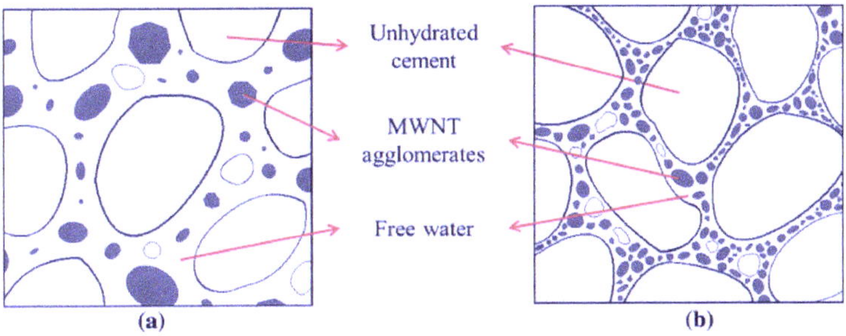

Figure 5: Micro structure of fresh mixture of the MWNT-embedded cement composites prepared under the high flow condition (**a**) and the low flow condition (**b**).

DC Conductivity of the Composites Fabricated Under Different Flows

The image analysis of the MWNT distribution state of the MWNT-embedded cement composites fabricated under different flows (or W/C) studied in Sect. 4.1.1 showed that the MWNT distribution improved as the flow of the composite decreased. In the present section, the DC conductivity of the composites was examined in an effort to understand the change of the MWNT distribution in the composites and also to verify whether it supported the image analysis results. The constituent materials and respective weight content ratios of the composites are given in Table 3. To make specimens with different flows, one group of specimens was fabricated by varying the water content, as described in Sect. 4.1.1, and an additional group of specimens was fabricated by varying the SP content in the present section. The fabrication procedures of the specimens followed descriptions in Sect. 3.2.1 and the measurement method was described in Sect. 3.2.2.

The variations in the DC conductivity of the specimens are plotted in Fig. 6a, b. The change of the DC conductivity of the specimen group prepared with different water content is shown in the Fig. 6a and the results of the specimen group prepared with different SP content are shown in Fig. 6b. An electrical percolation phenomenon can be observed in the electrical conductivity versus W/C ratio plot in the Fig. 6a. This phenomenon indicated that the water content or flow of the mixture is a crucial factor affecting the MWNT distribution of the composites, as found in the image analysis in Sect. 4.1.1. The dramatic increase of electrical conductivity with a decrease of the flow is attributed to disentanglement of the MWNT agglomerates, as shown in the image analysis. In the image analysis, the q value linearly increased with a decrease of the W/C ratio but the electrical conductivity exponentially increased with the decrease of the W/C ratio. This was due to the formation of the percolation network, which induced a remarkable increase of the DC conductivity of some orders. The Fig. 6b, which presents the electrical conductivity versus SP/C ratio plot, also shows the electrical percolation phenomenon. This indicates that the flow plays a decisive role in determining the MWNT distribution, because water content was constant in the specimen group with different SP/C ratio. It is consequently found that the flow of fresh mixture of the composites is an important factor that determines the MWNT distribution state, based on the image analysis and the electrical conductivity evaluation. The improvement of the MWNT distribution with a decrease of the flow is thought to be related to variation of the interspace among unhydrated cement, as discussed in Sect. 4.1.1.

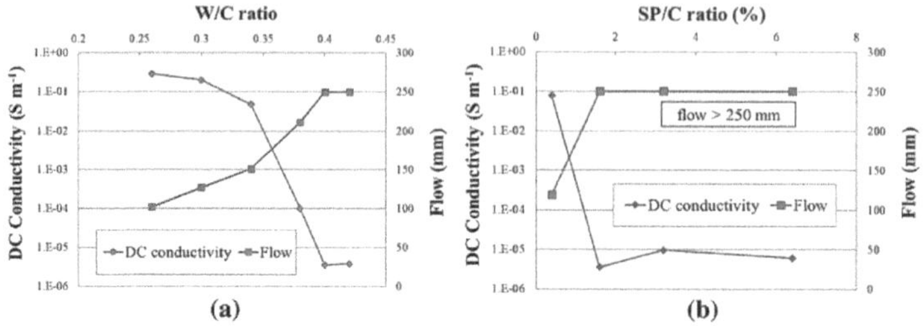

Figure 6: The DC conductivity of the MWNT-embedded cement composites fabricated as a function of W/C ratio (**a**) and SP/C ratio (**b**).

It is worth noting that the electrical conductivity of the M06–W30 was greater than that of the M06–SP04 although the flow of the M06–W30 was higher than that of the M06–SP04. This indicates that control of water content can be a more effective way of enhancing the electrical conductivity of the composites.

Therefore, two categories of specimens were prepared in the subsequent experiments. One category included MWNT-embedded cement composites fabricated under a high flow condition (flow > 250 mm) and another category included composites fabricated under a low flow condition (114 mm < flow < 126 mm) by adjusting the W/C ratio and the SP/C ratio. To understand the MWNT distribution states, the image analysis and DC conductivity measurement for the specimen groups were carried out.

Influence of the Flow on MWNT Distribution in the Composites Incorporating Various MWNT Contents

Image Analysis of the MWNT Distribution State in the Composites

It was found that the flow of fresh mixture was closely related with the MWNT distribution, as observed in Sect. 4.1.2. Accordingly, the MWNT distribution states in MWNT-embedded cement composites with various MWNT contents can be compared by setting the flow of mixtures to be consistently low (flow between 114 and 126 mm). Otherwise, the MWNT distribution is so poor that the reliability of the image analysis data obtained from the composites with various MWNT contents can diminish. To prepare composite mixtures with the low flow values, cement composites with various MWNT contents were fabricated by adjusting the content ratios of water and SP. The water to cement ratio was controlled with reference to the MWNT content in each batch. This was based on a report that MWNT tends to absorb water within its hollow structure (Striolo et al. 2005). The SP content ratio was also adjusted in consideration of the amount of cement and MWNT, but the final SP content was determined by some trials of SP addition in the mixture and flow test. The constituent materials used in the present work are listed in Table 2 but nylon fiber was not used for the image analysis specimens. The mixing ratio of MWNT ranged from 0 to 1.5 %. Figure 7 presents processed images obtained from the fabricated composites. Images observed from LF–M0 showed some particles that were thought to be impurities of the white cement. However, their planar occupation area was so small that it was negligible in a comparison study of the q value.

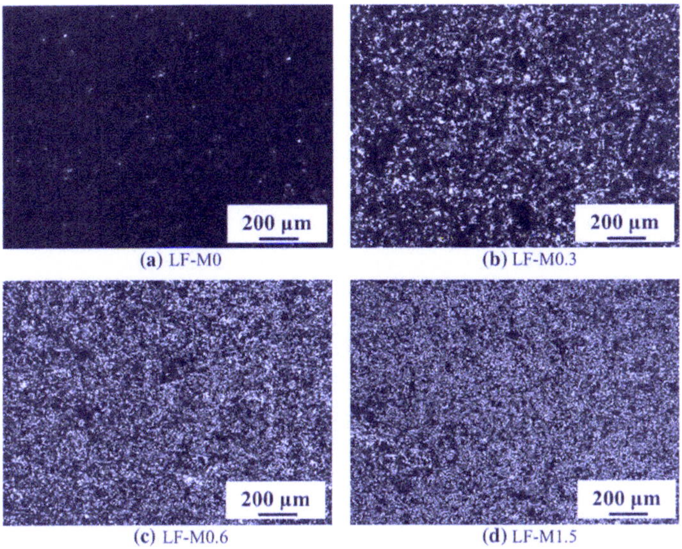

Figure 7: The fractured surface of the cement composites of the LF–M group fabricated under the low flow condition after having image thresholding, reversion, and binarization processes (Image size: 570 × 426 pixel): **a** LF–M0, **b** LF–M0.3, **c** LF–M0.6, and **d** LF–M1.5.

It is generally agreed that an increase in the MWNT content is accompanied with an increase in the total area of MWNT if it is uniformly distributed throughout composites. Such phenomena was in close agreement with the test result provided in Fig. 7 where the total area of MWNT was observed to increase with the MWNT content. In addition, the MWNT clumps were not found even when the MWNT content was increased up to 1.5 %. Observation of the images thus indicated that MWNT was satisfactorily distributed throughout the composites in the LF–M group.

In an effort to express the change of MWNT distribution states in a quantitative manner, the q value was calculated for the LF–M group, as shown in Fig. 8. The q values of the group exhibited a steady increase as a function of the MWNT content in the composites. Accordingly, a dramatic increase in the electrical conductivity is expected as the MWNT content increases.

An image analysis of the MWNT distribution state in the composites fabricated under the high flow condition was not carried out because it was not possible to present reliable q values due to the presence of the MWNT clumps, as shown in the W42–M06 specimen of the Fig. 3.

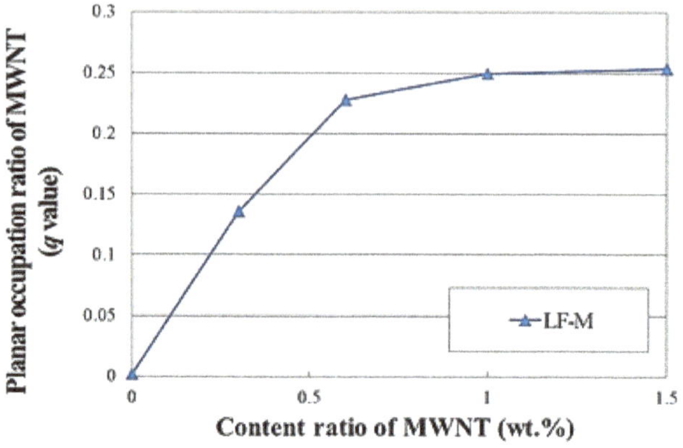

Figure 8: A comparison of the q value, which is the proportion of MWNT agglomerates to total area of the image, of the specimens in the LF–M group.

DC Conductivity of the Composites Incorporating Various MWNT Contents

The change of the DC conductivity of composites fabricated under the two different flow condition, the low and high flow conditions, with various MWNT contents was investigated here. The constituent materials and respective weight content ratios of the composites fabricated under the low flow condition are given in the Table 2. The DC conductivity of the composites in the LF–M group is shown as a function of the MWNT content in Fig. 9. DC conductivity of 1.7 S m^{-1} was attained for specimen LF–M1.5. This value was greater than the electrical conductivity of 5 mm long carbon fiber-incorporated cement composites and carbon black-filled cement composites (Wen and Chung 2007; Li et al. 2006). The high DC conductivity of the composites in the present section and the image analysis conducted in Sect. 4.2.1indicated that MWNT was well distributed in the cement matrix.

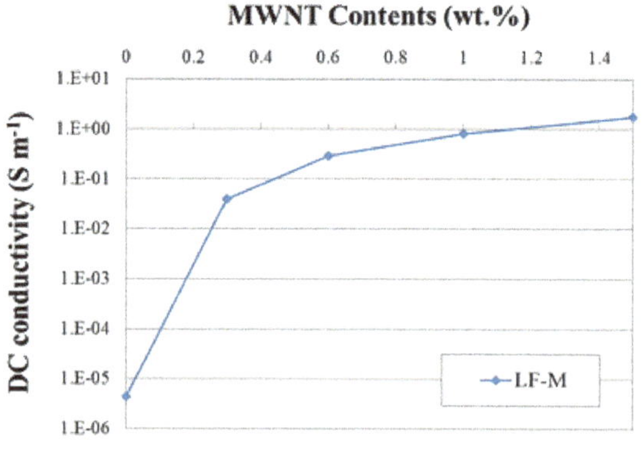

Figure 9: The experimental conductivity values of the specimens fabricated under the low flow condition (114 mm < flow < 126 mm).

The electrical percolation phenomenon of the LF–M group was found in the Fig. 9. The percolation threshold, which refers to the critical volume fraction of MWNT inducing remarkable change in the conductive phase, existed in a MWNT content range of 0–0.3 wt% for the LF–M group. Accordingly, the percolation threshold corresponded to 0.196 vol% if it is determined as the mean value of the percolation threshold range.

The percolation phenomena manifested in the MWNT-embedded cement composites fabricated under the low flow condition indicated that MWNT was well distributed throughout the composites and the MWNT distribution was consistent in the specimen group. The results supported the image analysis results obtained in Sect. 4.2.1. Based on the image analysis results and the electrical percolation phenomena, an acceptable MWNT distribution was attained by reducing the flow of the composites. Consequently, it can be concluded that maintaining the low flow of fresh mixtures of the composites is crucial to improve the MWNT distribution.

The change of the DC conductivity of the composites fabricated under the high flow condition with various MWNT contents was also investigated. The constituent materials and respective weight content ratios of the composites are given in Table 4 (Nam et al. 2012). MWNT was incorporated in each composite type at 0, 0.3, 0.6, and 1.0 wt% by weight of cement. To fabricate specimens with high flow, the W/C ratio was fixed to 0.4 in the composites and the SP/C ratio was 0.016.

Figure 10 presents the change of the DC conductivity of the composites as a function of the MWNT content. The electrical percolation phenomenon was not found in Fig. 10. The DC conductivity of all the composites did not exceed 0.001 S m^{-1}, hence all the composites appeared to be under the percolation threshold. The under percolation behavior of the DC conductivity was attributed to a poor distribution of MWNT in the composites. In addition, the DC conductivity of the composites may be dependent on many factors such as the dry condition of the specimens, void ratio, contact resistance between electrodes and the composite specimens, etc. In conclusion, it is not recommended that MWNT-embedded cement composites are fabricated under a high flow condition unless appropriate dispersion methods were carried out.

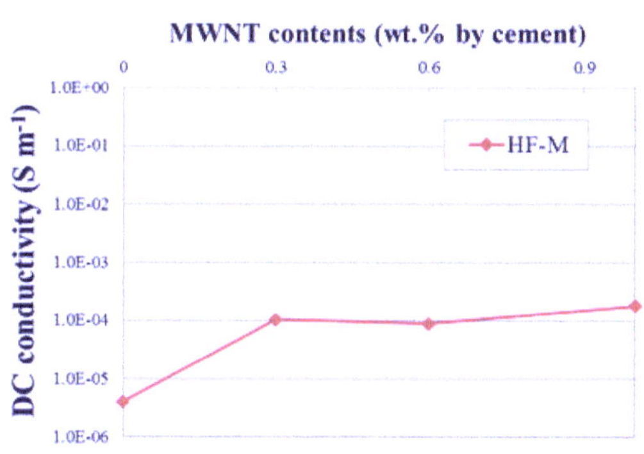

Figure 10: The variations of the DC conductivity of the specimens fabricated under the high flow condition (flow > 250 mm) are plotted as a function of the MWNT content.

CONCLUSION

In the present work, the MWNT distribution in MWNT-embedded cement composites was evaluated by a newly proposed image analysis method as well as electrical conductivity measurement. In addition, the influence of the fresh mixture's flow on the MWNT distribution in the composites was experimentally studied on the basis of the two evaluation methods. The conclusions derived from the present work can be summarized as follow.

- The MWNT distribution, which was evaluated by the planar occupation ratio of MWNT, q, was enhanced as the flow of fresh composite mixtures decreased.
- The DC conductivity of MWNT-embedded cement composites fabricated with different flows was examined and it was observed that the conductivity increased as the flow was decreased.
- The image analysis of the MWNT-embedded cement composites fabricated under the low flow condition (114 mm < flow < 126 mm) revealed that the q value linearly increased as the MWNT content increased. The linear relationship between the $q$value and the MWNT content demonstrated that fabrication of the composites under the low flow condition is an effective way to enhance the MWNT distribution.
- The DC conductivity of the MWNT-embedded cement composites fabricated under the low flow condition showed percolation phenomena, thus indicating that MWNT was well distributed in the composites, as evaluated in the image analysis.

The proposed image analysis procedures for evaluation of the MWNT distribution can be used as an evaluation method to quantify the distribution of carbon nanomaterials in the cement matrix. The conductive MWNT-embedded cement composites with percolative networks produced by controlling the flow are expected to be utilized as piezoresistive sensors, EMI shielding materials, electrostatic discharge materials, heating elements, etc. Future work will be focused on in-depth study of the origin of the flow effect on the MWNT distribution.

ACKNOWLEDGMENTS

This research was supported by the KUSTAR-KAIST Institute, KAIST, Korea and also sponsored by a research project 'Development of fabrication method for floor heating material by use of CNT-cement composites with self-heating capacity of 5W-50 °C' (Project code: 15CTAP-C098086-01) funded by Ministry of Land, Infrastructure and Transport (MOLIT) of Korea Government and Korea Agency for Infrastructure Technology Advancement (KAIA). Authors also deeply thank Prof. Hyeong-Ki Kim in Chosun University and Prof. Sung-Min Choi in KAIST for their important comments on this research and acknowledge June Lee in National NanoFab Center for cooperation in the image analysis.

REFERENCES

1. ASTM International. (2013). *ASTM C1437—standard test method for flow of hydraulic cement mortar*. West Conshohocken, PA: ASTM.
2. Daimon, M., & Roy, D. M. (1979). Rheological properties of cement mixes: II. Zeta potential and preliminary viscosity studies. *Cement and Concrete Research, 9*(1), 103–109.
3. Fan, M. Z., Bonfield, P. W., Dinwoodie, J. M., & Breese, M. C. (2000). Dimensional instability of cement bonded particleboard: SEM and image analysis. *Journal of Materials Science, 35*(24), 6213–6220.
4. Kang, S. T., Lee, B. Y., Kim, J. K., & Kim, Y. Y. (2011). The effect of fibre distribution characteristics on the flexural strength of steel fibre-reinforced ultra high strength concrete.*Construction and Building Materials, 25*(5), 2450–2457.
5. Kim, H. K., Nam, I. W., & Lee, H. K. (2014). Enhanced effect of carbon nanotube on mechanical and electrical properties of cement composites by incorporation of silica fume. *Composite Structures, 107*, 60–69.
6. Konsta-Gdoutos, M. S., Metaxa, Z. S., & Shah, S. P. (2010). Highly dispersed carbon nanotube reinforced cement based materials. *Cement and Concrete Research, 40*(7), 1052–1059.
7. Lee, B. Y., Kim, J. K., Kim, J. S., & Kim, Y. Y. (2009). Quantitative evaluation technique of polyvinyl alcohol (PVA) fiber dispersion in engineered cementitious composites. *Cement and Concrete Composites, 31*(6), 408–417.
8. Lee, Y. H., Lee, S. W., Youn, J. R., Chung, K., & Kang, T. J. (2002). Characterization of fiber orientation in short fiber reinforced composites with an image processing technique. *Materials Research Innovations, 6*, 65–72.
9. Li, J., Ma, P. C., Chow, W. S., To, C. K., Tang, B. Z., & Kim, J. K. (2007). Correlations between percolation threshold, dispersion state, and aspect ratio of carbon nanotubes. *Advanced Functional Materials, 17*(16), 3207–3215.
10. Li, H., Xiao, H., & Ou, J. (2006). Effect of compressive strain on electrical resistivity of carbon black-filled cement-based composites. *Cement & Concrete Composites, 28*(9), 824–828.
11. Liu, J., Li, C., Liu, J., Du, Z., & Cui, G. (2011). Characterization of fiber distribution in steel fiber reinforced cementitious composites with low water-binder ratio. *Indian Journal of Engineering and Materials Sciences, 18*, 449–457.
12. Nam, I. W., Kim, H. K., & Lee, H. K. (2012). Influence of silica fume additions on electromagnetic interference shielding effectiveness of multi-walled carbon nanotube/cement composites. *Construction and Building Materials, 30*, 480–487.
13. Nam, I. W., Souri, H., Lee, H. K. (2015). Percolation threshold and piezoresistive response of multi-wall carbon nanotube/cement composites. *Smart Structures and Systems* (in review).
14. Otsu, N. (1979). A threshold selection method from gray-level histograms. *IEEE Transactions on Systems, Man, and Cybernetics, 9*(1), 62–66.

15. Ozyurt, N., Woo, L. Y., Mason, T. O., & Shah, S. P. (2006). Monitoring fiber dispersion in fiber-reinforced cementitious materials: Comparison of AC-Impedance Spectroscopy and Image Analysis. *ACI Materials Journal, 103*(5), 340–347.

16. Redon, C., Chermant, L., Chermant, J. L., & Coster, M. (1999). Automatic image analysis and morphology of fibre reinforced concrete. *Cement & Concrete Composites, 21*(5–6), 403–412.

17. SEMI MF 43 (2005) Standard test methods for resistivity of semiconductor materials.

18. Sorensen, C., Berge, E., & Nikolaisen, E. B. (2014). Investigation of fiber distribution in concrete batches discharged from ready-mix truck. *International Journal of Concrete Structures and Materials, 8*(4), 279–287.

19. Stauffer, D., & Aharony, A. (1994). *Introduction to percolation theory Revised* (2nd ed.). London, UK: Taylor & Francis.

20. Striolo, A., Chialvo, A. A., Gubbins, K. E., & Cummings, P. T. (2005). Water in carbon nanotubes: Adsorption isotherms and thermodynamic properties from molecular simulation. *The Journal of Chemical Physics, 122*(23), 234712.

21. Vance, K., Aguayo, M., Dakhane, A., Ravikumar, D., Jain, J., & Neithalath, N. (2014). Microstructural, mechanical, and durability related similarities in concretes based on OPC and alkali-activated slag binders. *International Journal of Concrete Structures and Materials, 8*(4), 289–299.

22. Wen, S., & Chung, D. D. L. (2007). Double percolation in the electrical conduction in carbon fiber reinforced cement-based materials. *Carbon, 45*(2), 263–267.

23. Xie, P., Gu, P., & Beaudoin, J. J. (1996). Electrical percolation phenomena in cement composites containing conductive fibres. *Journal of Materials Science, 31*(15), 4093–4097.

17
CHAPTER

MEASUREMENT METHODS AND ANALYSIS TOOLS FOR RAIL IRREGULARITIES: A CASE STUDY FOR URBAN TRAM TRACK

Laura Chiacchiari, Giuseppe Loprencipe

Keywords: Rail corrugation, Noise disturbance, Measurement methods, Continuous measurement, Data processing algorithm, Wavebands filtering

INTRODUCTION

The interaction between rail and wheel influences safety and efficiency of railway service. The knowledge in contact phenomena, in both the surface characteristics and the contact mechanics [1, 2], may improve the accuracy in evaluating wheel–rail contact functional performance, and ensure the related prediction of local stresses. High vertical contact forces, also with the lateral and longitudinal ones, induce stresses that may produce fatigue failure of the material [3]. High contact stresses and rolling resistance induce friction heating and consequently wear [4]. Thus, traction and braking may lead to wheel sliding, which results in rail burn, wheel flat, unfavorable material phase transformation, and thermal cracks [5]. In addition to wear, fatigue, crack formation, and other permanent deformations, there is a particular type of defect which affects the rail surface called "Rail Corrugation". This is a kind of periodic wear that is often visible with the naked eyes and can be identified as the pattern on the rail has regular spacing.

In the contact area between rail and wheel, dynamic forces are exchanged: it has been noted that if the dynamic contact forces are high, then the level of corrugation will be high (though with a phase lag of around 135°) [4–9]. These dynamic phenomena may generate corrugation of the rails and the wheels and they can also cause the formation of

Citation: Laura Chiacchiari, Giuseppe Loprencipe, Measurement methods and analysis tools for rail irregularities: a case study for urban tram track, DOI 10.1007/s40534-015-0070-6.

different types of irregularities. This results in poor vehicle dynamics (when passing on a worn rail) and consequently higher contact forces, vibrations, and noises. Also in the case of perfectly smooth rail surface, this phenomenon appears as a consequence of the track discontinuities for the sleepers present (regularly spaced).

LITERATURE REVIEW

Due to the contact between the wheels and rail surfaces, the train's motion on a track creates complex dynamic phenomena in the train itself and in the track's superstructure, which generates noise and vibrations that propagate in the external environment. These phenomena cause a service life reduction in all the structural components involved, an external disturbance in the receptors, and the decay of the passengers comfort. Figure 1 shows a scheme for the systematic approach to wheel–rail interface research and development: the figure illustrates that it is essential to take into account the combination of all the following aspects: contact mechanics, materials, dynamics, as well as friction. A proper study of this scheme can guarantee profitable optimizations for the management of the wheel–rail system interface.

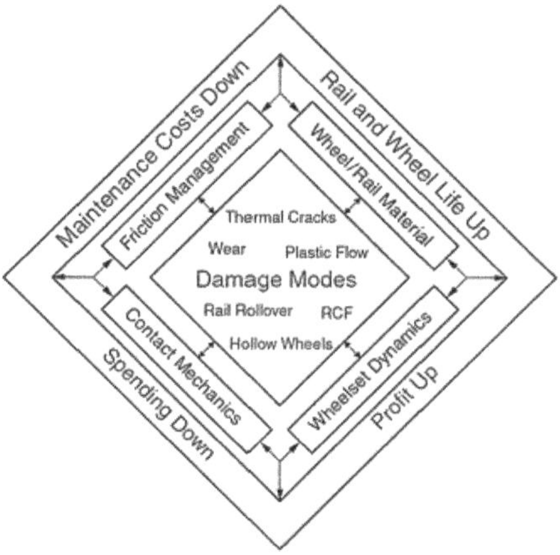

Figure 1: Systems approach to wheel–rail interface research and development [10].

Imperfections that occur at the wheel/rail contact, due to geometrical changes in the shape of rail and wheel surfaces, influence the stress distribution. In fact, the presence of these imperfections alters the pressure in the contact nominal area; this generated dynamic stress excites all the elements of the rail track system and propagates from the track through the ground, in the range of frequency excitation. Railhead irregularities occur on the surface of the rail: the profile can be worn, the railhead damaged by cracks from rolling contact fatigue phenomenon (RCF) [6], squats can take place [7], and further, short-pitch corrugation and rail welds create periodic occurrences [8, 9]. The rail corrugation on rail surface is very frequent in all railway infrastructures and is also classified as short-wave defect (long wavelength irregularities are, instead, defined as irregularities of wavelengths of 300 mm or longer, that may be either geometric irregularities in the track or on the wheel, or irregularities of the track stiffness [10]).

The importance of amplitude and wavelength dimension of rail corrugation has been demonstrated in many references. A clear correlation between the size of the railhead running surface irregularities and levels of generated noise and vibration was found by Thompson, [11] (see Fig. 2). In addition, because the correlation is related to dynamics aspects, the occurrence of defects is related to the train speed [4–8].

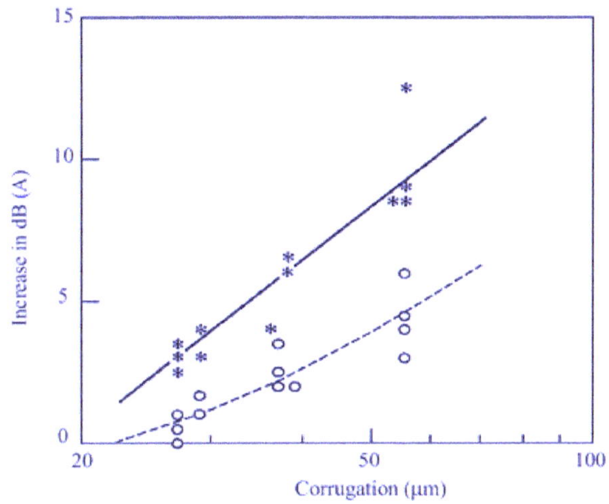

Figure 2: Correlation between corrugation height and noise [11].

Rail corrugation, also known by the term short-pitch corrugation, frequently appears in straight track, where traction or braking is particularly severe, or in small radius curves, although it may also be detected in curves of greater radius. Therefore, in urban rail transport system this problem appears frequently, especially for tram networks, which present the following critical situations:

- very accentuated wear due to exposition at both railway vehicles and other categories of traffic;
- specific geometry of the tram line that offers particular conditions for the establishment of disturbance phenomena and wear;
- location of tracks near noise-sensitive regions.

In addition, the insertion of rolling stock in this scenario contributes to developing wear: maintenance cycles lead to an uncontrolled increase in management costs that nullifies all the advantages of this transport system.

The geometry of this kind of irregularity, similar to waveform, is periodical and changes with different situations. A useful classification of the corrugations (irregularity wavelengths less than 1 m) is performed by grouping wavelength (λ) ranges:

- 10–30 mm;
- 30–100 mm;
- 100–300 mm;
- 300–1000 mm.

Rail corrugation appears to have a small wavelength when compared with the longitudinal waves of the rail.

During vehicle/track interaction, the forces are transmitted in the wheel/rail contact area: dynamic behavior is found in a wide band, ranging from very low frequencies of 0.5–1 Hz for lateral and vertical car body accelerations, to 2000 Hz, as a consequence of geometrical irregularities in rail and wheel contact. This latter portion of high-frequency excitation content typically falls in the audible frequency range, inducing the most relevant contribution to rolling noise.

Along with other contact issues, rail corrugation is a severe problem for many tramway administrations, therefore short wavelength corrugation is the major focus of this paper.

Due to the tram speed, and according to the relation between the frequencies f and the train speed V, $f = V/\lambda$, the wavelength λ on a tram rail surface is commonly around 10–50 mm.

Since the vehicle/track interaction produces a wide-ranging vibratory phenomenon, this results in different types of noise: besides the rolling noise, the ground-borne noise and the structure-borne noise are generated as reflection of ground-borne vibration [12]. In general, the effect of ground-borne vibration is related to people discomfort (high-frequency vibration affects concentration ability, low-frequency vibration may cause muscular or internal organ injury, both depending on exposure time), rather than damage to buildings. Noise limitation can avoid structural damage of the buildings, but is also important to prevent fatigue damage due to cyclical vibration [13–16].

RAIL CORRUGATION: TYPES AND CAUSES OF FORMATION

The phenomenon of rail corrugation occurs in all rail systems, with different sizes and naturally takes place even when smooth contact surfaces are involved. In this way, Grassie [17] proposed two distinct mechanisms for the generation of dynamic actions (Fig. 3). The first is called wavelength-fixing mechanism, that appears also in the case of ideal geometrical contact as a consequence of structural discontinuities in railway track configuration (e.g., as rails are mounted on discrete supports, the sleepers). For this reason, dynamic loads arise altering the wheel–rail interaction, regardless of the geometric imperfections on the rolling railhead surface. These dynamic forces propagate in all the train and track structural components, determining damage mechanism that occurs primarily on the wheel–rail contact surfaces. In consequence, permanent deformations on rolling surface are generated and gradually increases the generation of the dynamic forces in next train passages.

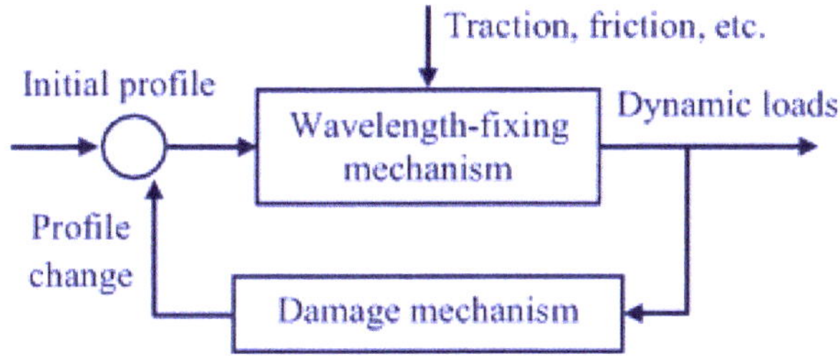

Figure 3: A general corrugation mechanism [19].

Irregularities present on wheel surface, will not be involved in this study (for which refer i.e., [18]). Both the generation and distribution of rail surface irregularities have a random nature.

Like irregularities of the road pavements, using appropriate instruments riding on the surface of the rail, measurements of profiles can be collected. These measurements show the variability of the irregularities of a longitudinal section of the rail surface compared to an ideal plane of reference. A proper analysis of the profiles allows to define the reference levels of rail degradation, and ensures to relate them to damage mechanisms, corrective solutions, levels of generation of noise and vibration, etc.

These longitudinal irregularities can be attributed to the following causes:

- pinned–pinned resonance,
- rutting due to second torsional resonance,
- P2 resonances.

The first type of wear is caused by pinned–pinned resonance of a track vibration, occurs mostly in straight tracks (indifferently on two rails), and in high rail of curves, especially in curves of greater radius, as consequence of the vibration of the rail bounded into two sleepers. This kind of corrugation may also appear in the track for light axle load traffic (less than 200 kN). Furthermore, in some circumstances, this defect is more evident in the correspondence of the sleepers [19].

The second type of corrugation occurs primarily in curves (on the inner rails) and in straight tracks often subject to traction or braking particularly severe. The name "rutting" was chosen because of the characteristic transverse grooves ("ruts") on the railhead [20]. The presence of local imperfections, as welds and joints, acts like triggering factor, determining the position of the corrugation along the rail. The damage of the inner rail in the curve is caused by a mechanism of the driven axles [19]: the phenomenon starts when the one wheel traction ratio is close to the friction limit, thus causing the slip and then the oscillation of the other wheel in correspondence with the second natural frequency of torsional resonance [21], as shown in Fig. 4.

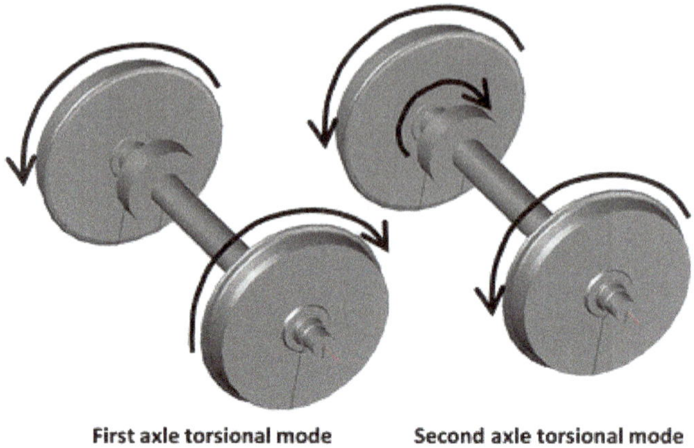

Figure 4: Torsional resonances of wheelsets [19].

The result of this phenomenon is wear of the rail with a geometric wavy development, with uniform wavelengths and appearance, and depth that can quickly reach tenths of a millimeter [19].

The last generation cause of corrugation comes from P2 resonance and interests straight tracks and outer rails in curves: this kind of wear is mainly due to the vehicle "bouncing" on track, and the continuous vertical displacements of the wheelset, excited by the different stiffness that the unsprung masses perceive passing over discontinuities [22]. It occurs often on the tramlines (Fig. 5).

Figure 5: Wear on tramway and railway tracks.

MEASUREMENT OF RAIL IRREGULARITIES

Corrugation is not a single phenomenon with a single cause and a single solution or treatment [20]. Cross studies should be effectuated to its right classification. The method to identify the causes is based on the survey of the damaged rail profile with appropriate instruments (Fig. 6).

Figure 6: Instruments for measuring the longitudinal profile of the rail (rail measurement).

In order to examine the longitudinal profiles of the rails, two different types of instrumentations are used:

- contact instruments;
- non-contact instruments.

The first are able to measure, one rail at a time, different part of a track, permitting to choose the length of each single measuring track and to select time by time the position of the sensor in the width of the railhead, guaranteeing extreme accuracy of 0.001 mm RMS (root mean square).

The other ones, because mounted on vehicles of high speeds (5–140 km/h), record profiles of both rails simultaneously and provide uninterrupted measurements but with lower accuracy than the previous (0.003 mm RMS).

In addition, these instruments measure the irregularities under load and, therefore, in the same operating conditions of trains; however, they should solve the problem of inertial reference frame being vehicle-integrated measuring systems. This may cause some problem in finding the reference plane to which measurements of rail irregularities are referred.

For these reasons, contact instruments are generally preferred over the non-contact ones, to measure the longitudinal profiles of the rails.

The University of Cambridge and the British Rail Research [17] collaboratively developed the first model of CAT (Corrugation Analysis Trolley) in 1970 and then it was used in many railway infrastructures. This instrument, falling within the category of contact instrument, has a contact sensor (tip or roller sensor) and an odometer. The CAT works hand-pushed, riding the rail, allowing to detect the longitudinal profile on the surface of the rail with vertical precision of 0.01 μm [17]. This instrument, in addition to measuring the surface of the rail in terms of longitudinal profile, is able to return the acoustic roughness, the roughness of the rail connected to the generation of the noise (expressed in dB), and being provided with a dedicated software used for the calculation [23–25].

Another type of apparatus able to read rail profile, belonging to the category "not in contact", is the ORPMS (Optical Rail Profile Measurement System) that is based on optical triangulation realized through laser rays and assisted by mini cameras. For the diagnosis of track defects during the last few years, several trains instrumented capable of continuous measurement and all the geometric parameters of interest including the profile of the rail, have been designed. For example the HSRCA system (High Speed Rail Corrugation Analyser), a not-in-contact system with Axle-box Accelerometers, is extremely reliable even at relatively high speeds (maximum recommended running speed of 120 km/h) [26].

Three different ways can be used to evaluate the condition of a rail surface starting from recorded profilometric data: a method based on the spectral analysis, a method based on the assessments of percentage of exceeding, and expeditious methods.

ANALYSIS OF PROFILOMETRIC DATA BY MEANS OF SPECTRAL ANALYSIS

The direct measurement and spectral analysis-based processing method form the basis of the shown resolution (Fig. 7). The presented algorithm is entirely in line with the UNI EN 15610:2009 one [27].

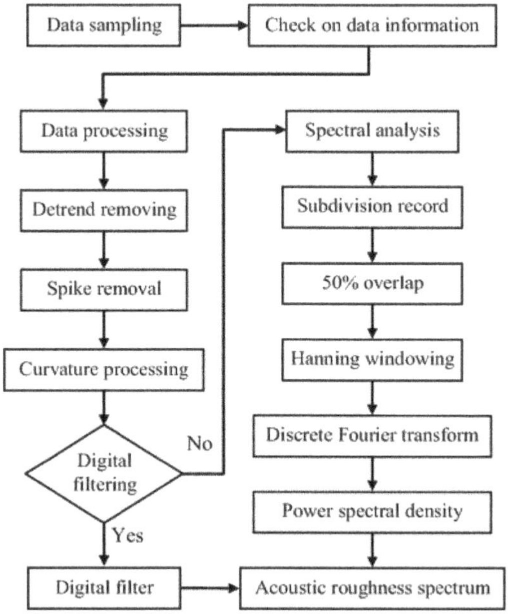

Figure 7: Algorithm of the method (UNI EN 15610:2009) [27].

Method A determines the one-third octave band acoustic roughness spectrum $\tilde{r}(\lambda)$. The data must be processed to remove some unwanted "pit and spike", to adapt the measurement (if chosen) through an appropriate curvature processing that takes into account the particular roundness of the sensor adopted in the test measures, and at last, to apply the discrete Fourier transform (DFT) to the function determining power spectral density (PSD) in the frequency domain and produce a one-third octave level roughness spectrum as a function of wavelength averaging the magnitude suitably.

A detailed scheme for method A is shown in Fig. 7.

Localized rail defects are not significant in the assessment of the acoustic roughness related to the rolling noise component, therefore they should not be included in the application.

After removing moisture and other contamination from the railhead surface, determining the length along the rail of interest, the transverse width (w_{ref}), the relative distance to the field face of the rail (d_{ref}), and setting the instrument, the measurement can start.

According to Fig. 7, these are three steps to follow:

- check data information to edit out the data relating to any rail joints, rail head defects, and welds, for not affecting the final spectrum analysis;
- remove pits and spikes in the records;
- correct the measure by the curvature processing that takes into account the form of the probe, before conducting spectral analysis.

The next step is to calculate the one-third octave band spectrum for each acoustic roughness record (amplitude of the acoustic roughness expressed as a function of the wavelength λ), and, finally, estimate the same spectrum in dB. The technique of spectral analysis adopted for road profiles to treat the rail profiles was used [28].

An example of Wavelength-RMS disp graph is shown in Fig. 8. Central frequencies in the calculus are chosen, the same as the International Standard ISO 8608:1995 [29] provides, and to create, in a numerical way, a one-third octave band spectrum, the Technical Specification ISO/TS 13473-4 [30] about Spectral analysis of surface profiles was followed. To select the alternative method (Method B), the one-third octave band values shall be obtained by applying digital one-third octave band filters directly to the roughness data.

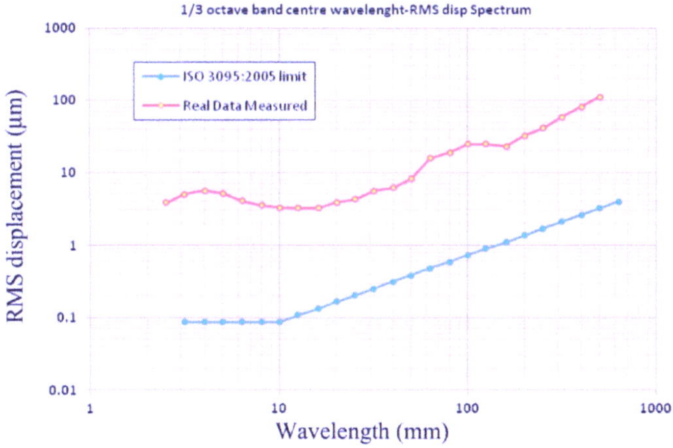

Figure 8: One-third octave band spectrum calculated through the algorithm proposed by ISO 8608 [29], on a recorded measure in a depot of rolling stock.

The obtained spectrum $\lambda-\tilde{r}(\lambda)$ thus can be compared with standard limits according to the needs: limits that are not univocally established if not for rail used during vehicle-type testing (Technical Specifications for Interoperability, Decision 2006/66/EC, [31]) or for measurement of noise emitted by rail-bound vehicles (ISO 3095:2005, [32]), and particularly for railway field [12].

Figure 9 shows both acoustic roughness spectra of a tramway rail and standard limits for railway rail.

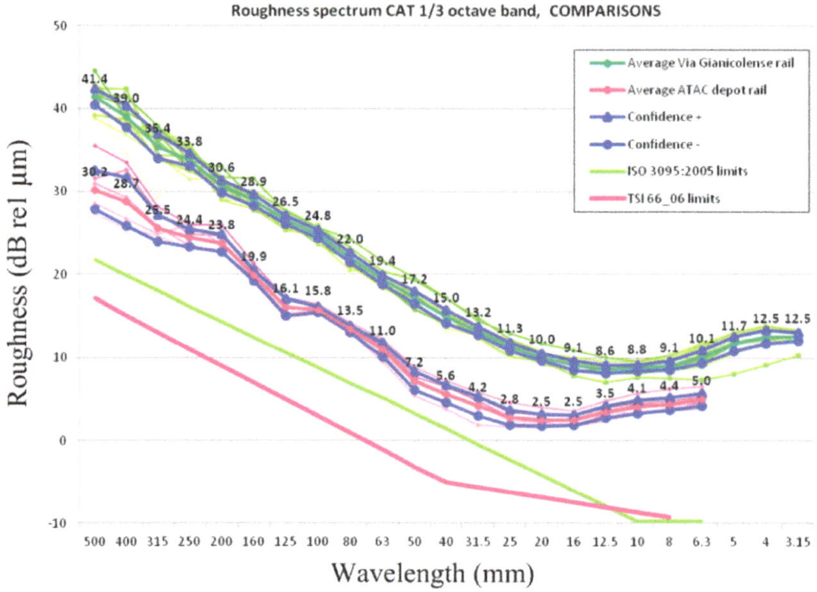

Figure 9: Example of roughness spectra (whose values are given in dB ref. 1 µm) of standard limit curves and of measures carried out with the corrugation analysis trolley.

To calculate the acoustic roughness spectrum the following steps are necessary:

- Detrend removing process: to remove the best straight-line fit from distance vector.
- Spike removal process: to remove narrow upward spikes that are regarded as being linked with the presence of small particles of foreign matter on the rail surface; identify the spikes by the criteria $d^2r/dx^2 < -10^7$ µm/m^2 and a change of sign for dr/dx, calculate the first and the second derivative of r(x), the roughness acoustic vector, identify the edges of each spike as being the extreme points (x_1 and x_2) on either side of the maxima or minima, for which abs(dr/dx) becomes less than 5×10^3 µm/m; calculate the width w of the spike with the formula w = abs(x_2–x_1), and where h > w^2/a (h = height of the spike, a = 3 m) spike shall be removed by linear interpolation between x_1 and x_2.
- Curvature process: to account for the effect of the small radius of the probe tip compared to that of the wheel, taken as the r'(x_i) obtained from the sum between

the maximum values intercepted in the difference $r(x)-C_i(x)$ (in particular the roughness function $r(x)$ and the curve $C_i(x)$, r = 0,375 m, centered at x_i above $r(x)$ function), and the $r(x_i)$, the roughness value in that specific i (Fig. 10).

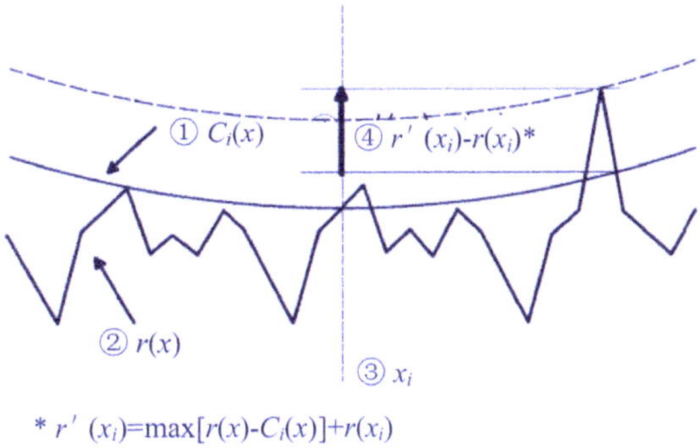

Figure 10: Criterion applied to position x_i

- Windowing: to avoid discontinuity at the edges of each discrete signal due to the DFT is based on the hypothesis that the input discrete signal repeats itself with a period equal to the signal length. The single signals may be linked only if at the edges of each discrete signal, there are the same values equal to zero. The windowing reduces the signal to zero at the edges limiting the attenuation of the signal out of the edges. In general, the equation for determining the windowed profile $Z_{i,win}$ starting from original profile Z_i is given:

$$Z_{i,\text{win}} = \frac{w_i Z_i}{\sqrt{\frac{1}{N}\sum_{i=0}^{N-1} w_i^2}} \text{ for } i = 0, 1, \ldots, N-1, \tag{1}$$

where w_i is a set of coefficients that changes the original profile minimizing the unexpected effect known as "leakage".

- Discrete Fourier Transform (DFT): for the rail profile, according to the classic utilization for a time-domain signal, DFT involves a transformation from the distance domain (in m) to the frequency domain, in particular spatial frequency domain (in cycle/m or $1/\lambda$ where λ is the wavelength). The result of the DFT is a constant bandwidth narrow band spectrum Δf with complex values, calculated by:

$$Z_k = \frac{1}{N}\sum_{i=0}^{N-1} Z_{i,\text{win}} e^{-j\left(\frac{2\pi k}{N}\right)i}, \tag{2}$$

where j imaginary unit, $j^2 = -1$ used for transform the N real numbers $Z_{i,win}$, i = 0,1,... ,N-1, inN complex numbers Z_k, k = 0,1,...,N–1. The bandwidth Δf depends on the discrete profile length $L = (N-1)\Delta x$ and is equal to:

$$\Delta_f = \frac{1}{L}. \quad (3)$$

The spatial frequency scale of Z_k starts at 0 with steps equal to Δf until $f_{max} = (N/2 - 1)$ Δf. The others Z_k complex values shall be not used because this function is symmetrical.

- Power spectral density (PSD): PSD function shows the frequency distribution of the power associated with the signal per unit frequency. PSD is obtained by dividing the squared amplitude of each narrow band by the spectral bandwidth Δf according to equation:

$$Z_k^{\mathrm{PSD}} = \frac{2|Z_k|^2}{\Delta f} \text{ for } k = 0, 1, \ldots, \left(\frac{N}{2} - 1\right), \quad (4)$$

expressed in $m^2/m^{-1} = m^3$.

- Roughness spectrum: a constant bandwidth spectrum to summarize the frequency content of the signal (that is specifically the longitudinal profile) is created. A one-third octave band spectrum shall be synthesized on the basis of the narrow band spectrum. Each one-third octave band value shall be calculated as the energy sum of the narrow band values from the DFT. The same signal examined in third-octave bands can evidence the presence of maxima or minima of the acoustic roughness levels that in octave bands would not be visible, providing more detailed information. The bands are sets of frequencies characterized by constant percentage bandwidth: calling the center frequency f_c and f_l (low) and f_h (high) frequencies below and above, for an octave band is delimitated with respect to the following equation:

$$\frac{f_h - f_l}{f_c} = \frac{1}{2^{1/2}}. \quad (5)$$

The exponent 1/6 is used to define the one-third octave band.

To filter the signal of profiles, a Butterworth filter can be used.

The Butterworth filter is a type of signal-processing filter designed to have more flat frequency response in the pass band.

The function of power spectral density (in frequency domain) is

$$A^2(\omega) = k^2/1 + (\omega/\Omega)^{2n}, \quad (6)$$

where ω is related to frequency through the relation $\omega = 2\pi f$ and the max value of $A(\omega)$ is in $\omega = 0$, where it is just equal to the constant k. The integer n is called "order" of the filter, and $\Omega$is the cutoff frequency (approximately the −3 dB frequency).

To design a numerical Butterworth filter a simple code written in MATLAB® has been used.

The order of the filter and the cutoff frequency W_n should be defined in advance in the code (here, the filter is designed to be a 2n-order filter).

Figure 11 shows the magnitude and phase responses of the 6th order Butterworth filter.

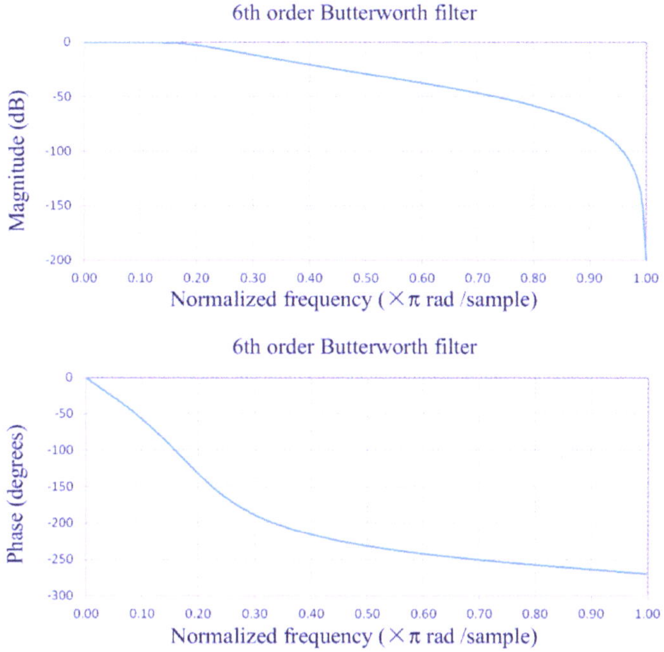

Figure 11: Magnitude and phase responses of the 6th order bandpass filter used to analyze the displacement signal with cutoff frequency of 200 cycles/m (corresponding to $\lambda_{min} = 5$ mm).

The acoustic roughness level is given by the following expression:

$$L_r = 10\log\left(r_{\mathrm{RMS}}^2 / r_0^2\right), \quad (7)$$

where

- L_r is the acoustic roughness level in dB,
- r_{RMS} is the root mean square of rail roughness in μm,
- r_0 is the reference of rail roughness ($r_0 = 1$ μm).

Using this decibel scale, a roughness value of 1×10^{-6} m = 0 dB, 3.2×10^{-6} m = 10 dB, 10×10^{-6} m = 20 dB, etc. [34].

EVALUATING PROFILOMETRIC DATA BY MEANS OF PERCENTAGE EXCEEDANCE

Another evaluation criterion used for the analysis of the profiles is the method of "percentage exceedance". For each reference interval λ, the curves of cumulative frequency are constructed compared with amplitude values expressed in microns RMS. For example, in Fig. 12, the red curve shows the corrugation range 30–100 mm. For a displacement of 20 μm (RMS) corresponds an exceedance more than 45 % on the rail and for a displacement of 40 μm (RMS) corresponds an exceedance by around 15 %. Such information may be used as a reference for management of rail grinding.

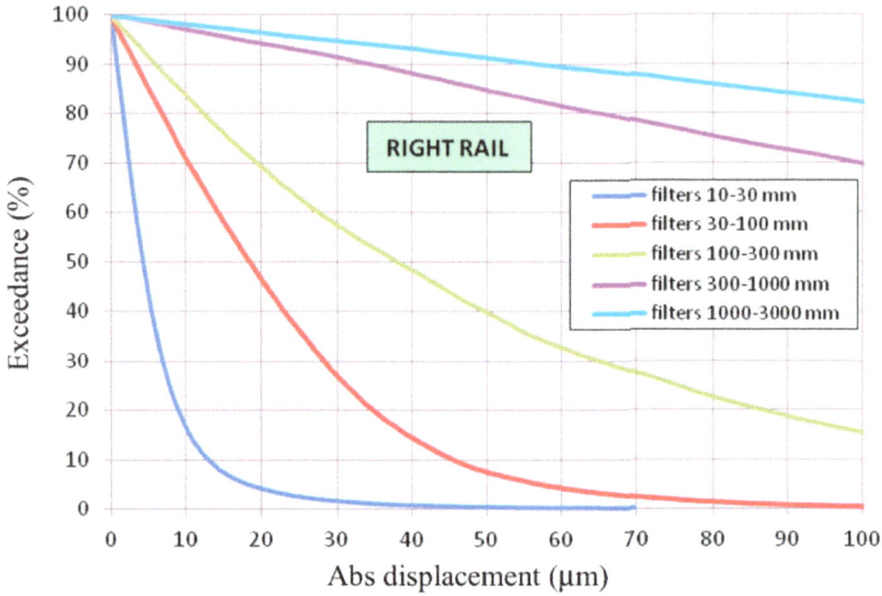

Figure 12: Graph of the percentage exceedance on a tramway section in Rome.

This procedure is given in Annex C of the standard on the acceptance of the quality of grinding work (UNI EN 13231-3:2012, [35]).

ASSESSMENT OF PROFILOMETRIC DATA BY MEANS OF EXPEDITIOUS METHODS

Some administrations of public transport, for each meter of profile recorded, decide the need for grinding the rail if the defect height exceeds the limits of 0.3–0.4 dmm with 8–10 cm of wavelength.

A tram rail section, by about 60 m, was measured by means of CAT. The profile recording, filtered in the range of wavelength λ=30–100 mm, is shown in Fig. 13a. In Fig. 13b, a 1 m long sub-section is produced, showing the trend that remains under the displacements of ±300 μm.

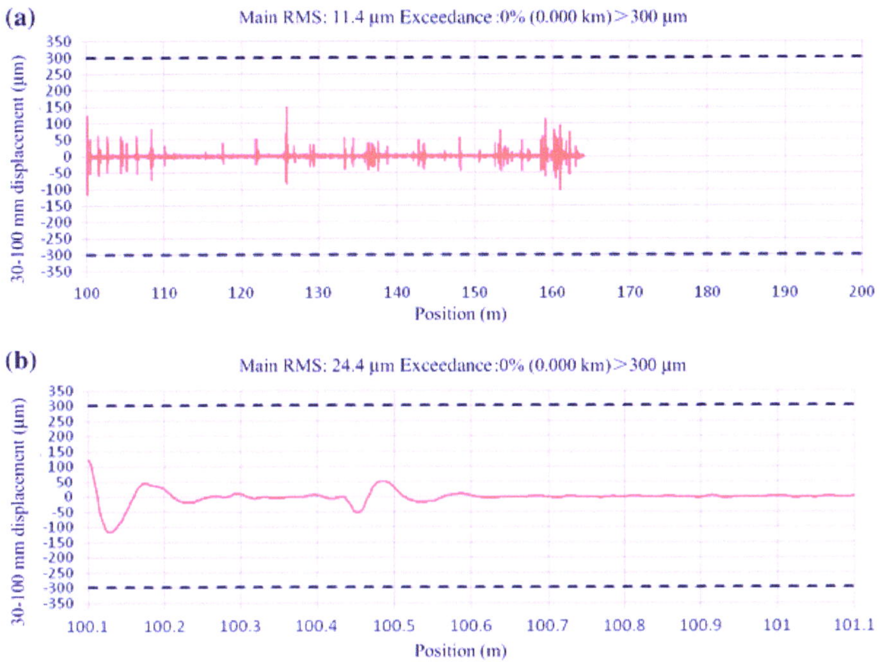

Figure 13: Profile recording filtered in the range of wavelength λ = 30–100 mm (**a**) and a section (1-m long) of the profile (**b**) (the dashed lines indicate the thresholds of ±300 μm).

CASE STUDY

As part of a Research Project some segments of tram network in Rome were detected, which are characterized by different levels of rail corrugation. The instrumentation used in measurement is the Corrugation Analysis Trolley and the profile measurements were conducted with three evaluation methods presented in the above sections. Figures 14, 15, and 16 show the main results about the profilometric data recorded on a corrugated rail in Via Gianicolense.

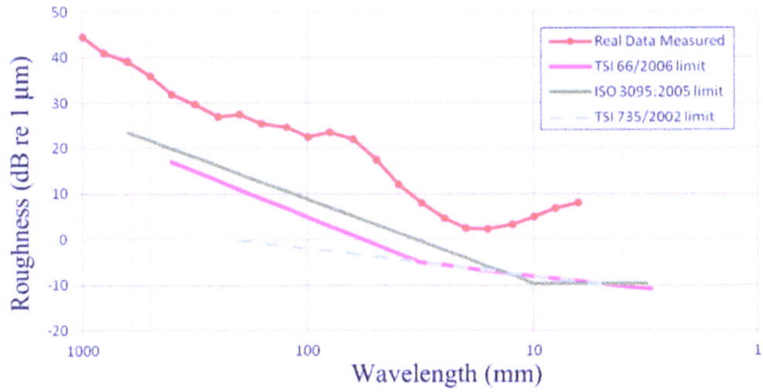

Figure 14: Acoustic roughness spectrum compared with standard limits curves.

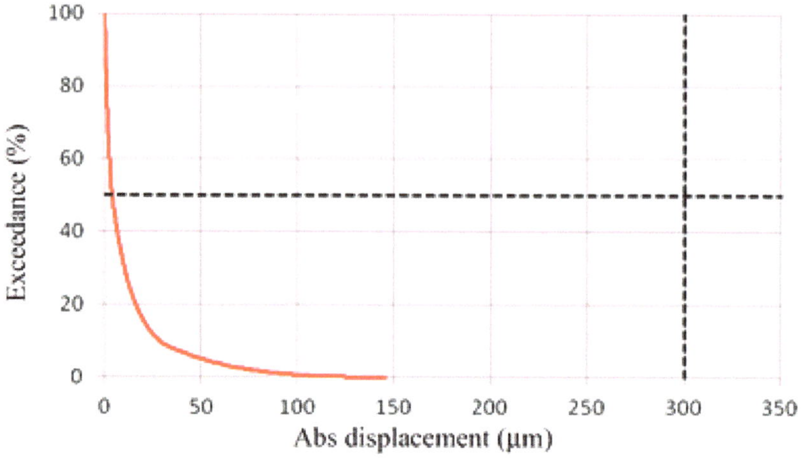

Figure 15: Percentage exceedance graph.

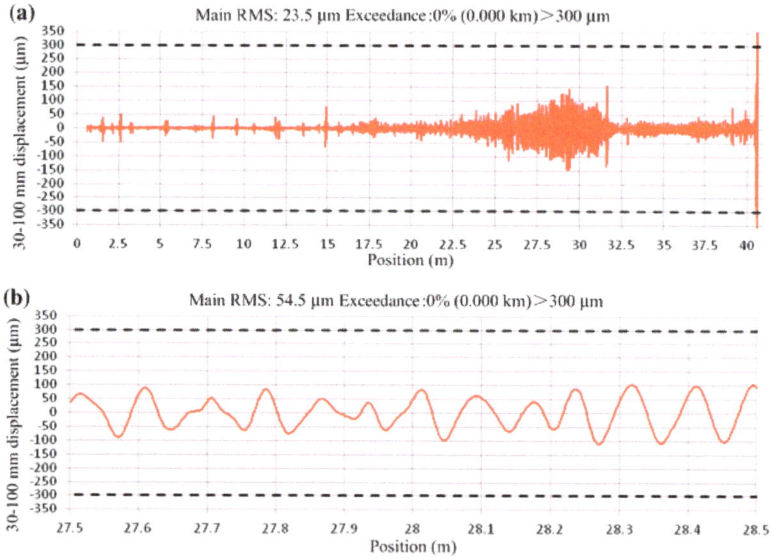

Figure 16: Profile recorded in via Gianicolense (**a**) and its selection of one-meter long (bottom) (**b**) in the wavelength range of $\lambda = 30–100$ mm (thresholds ±300 µm).

Figure 14 shows a comparison between the acoustic roughness spectrum and the TSI 2006/66/EC [31] limit curve in magenta (or the ISO 3095 [32] limits in green, the TSI 2002/735/EC [33] limits dashed line). Figure 15 shows the percentage exceedance graph, with the 50 % line highlighted, and Fig. 16 shows profilometric measure of 1 m in the wavelength range of 30–100 mm.

The examined rail exceeds limits if compared with limit curves supplied for railway field (due to the higher speed of the trains).

According to the second way to evaluate the condition of the rail, in the Fig. 15, it can be seen that 50 % of the length of the track shows an amplitude of irregularity with magnitudes less than 10 μm. The same percentage does not correspond any point of the exceedance curve: for example there is no correspondence at an amplitude of the irregularity with magnitude of 300 μm.

At last, using the third way to evaluate the profilometric data, the assessment of the measured section reveals that, for the range λ = 30–100 mm, the displacement does not exceed 300 μm in the overall recording.

DISCUSSION AND CONCLUSION

During the service life, each railhead is inevitably subjected to damaging. Within the several manifestations in which the damages can occur on the running surface, the most discussed in this paper are the periodic irregularities.

The particular attention to these types of defects is due to the acknowledged link existing between the corrugation of the rail surface and the generation of noise and vibrations (the corrugation interests also the surface of the wheels and naturally leads to an exacerbation of the phenomenon, but this is not discussed in this paper).

In the contact area, wheel and rail exchange interaction forces: the presence of irregularities on the running surfaces arises from dynamic loads, that, besides getting worse the damaged surface initiates vibrations of vehicle and track components (the formation of periodic surface defects is also accelerated by the speed-up of trains [36]).

Even an interaction between not-worn wheels and rails generates dynamic forces, due to the track configuration itself (e.g., the sleepers are supporting the rail with a specific spacing; they do not represent a uniform base for the vehicle riding).

Rail corrugation, a particular kind of periodic wear discussed in this paper, is one of the most serious problems in railway engineering. Its formation and development cause high level of vibrations between railway vehicle and track, noise and reduction of the useful life of the structural parts. Sometimes serious corrugation of rail leads to derailments [37].

The purpose of monitoring and diagnostics is to control the structural integrity, ensure no alteration of structural components, keep the performances of the railway/ tramway, and provide a useful pretext for the policy decisions of the interventions. Monitoring of rail irregularities represents, indeed, an investment to manage the evolution of the track damage; for example, it can be a chosen optimal to minimize the use of interventions in urgent cases, reducing the inefficiencies of the transport system only to planned interventions.

Dedicated equipment, and more in detail the manual corrugation measurement device CAT (specifically developed for the measurement of corrugation), have been described in this paper, together with three different methods to assess the profilometric data recorded.

The three evaluation methods for treating longitudinal irregularity recordings, all of them examined for the same section of tram rail, show contrasting results: the most rigorous method, which leads to gain the Acoustic Roughness Spectrum to compare with some reference levels provided by standards, is not very useful since the tram rails have always significantly accentuated longitudinal irregularities. Acoustic roughness spectrum calculated for a tram rail is always over the standard limit spectrum. Therefore, new limits should be established ad hoc for the tram rails, which are subject to operating conditions that are much more critical than the railway rails ones.

In this paper, the method for calculating the acoustic roughness spectrum has been developed starting from the profiles of the rails measured with the CAT. This methodology is entirely in line with the Standard UNI EN 15610:2009 [27] and it applies some algorithm and procedure used also to treat the road profiles.

The second method, which involves the determination of the curve Percentage Exceedance, seems to be very interesting: in an easy way can be read the whole range of irregularities found on the rail surface, highlighting the one with higher rate of presence. Though, unfortunately, the height thus obtained is not associated with its own wavelength (but the wavelength range of interest, which includes various profile components).

The third method, which assumes to provide the maximum amplitude of irregularities associated with a single wavelength, is the easiest method, and it is currently most used by managers of tram networks, to assess the status of the rails. This method, certainly expeditious, focuses only on one type of defect of the rail and can lead to erroneous evaluations and interpretations.

Acknowledgments

The Corrugation Analysis Trolley (CAT) mentioned in this paper, is owned by ATAC S.p.a. (ATAC is the Italian acronym that stands for Agenzia per la Mobilità del Comune di Roma, which is the agency that manages the Public transport in Rome) and it has been lent for temporary use to DICEA, to perform some collaborative research project.

REFERENCES

1. Kalker JJ (1980) Review of wheel-rail rolling contact theories, American Society of Mechanical Engineers. Appl Mech Div 40:77–91
2. Johnson KL (1987) Contact mechanics. Cambridge University Press, Cambridge
3. Lewis R, Olofsson U (2009) Wheel-rail interface handbook. CRC Press, Oxford
4. Oostermeijer KH (2006) Short pitch rail corrugation-cause and contributing factors. In: 7th World Congress on Railway Research (WCRR), Montréal
5. Popovici RI (2010) Friction in wheel-rail contacts. Dissertation, University of Twente, Enschede
6. Ringsberg JW (2001) Life prediction of rolling contact fatigue crack initiation. Int J Fatigue 23(7):575–586
7. Grassie SL (2012) Squats and squat-type defects in rails: the understanding to date. Proc Inst Mech Eng Part F 226(3):235–242

8. Steenbergen MJMM (2008) Quantification of dynamic wheel–rail contact forces at short rail irregularities and application to measured rail welds. J Sound Vib 312(4–5):606–629

9. Wu TX, Thompson DJ (2005) An investigation into rail corrugation due to micro-slip under multiple wheel/rail interactions. Wear 258(7–8):1115–1125

10. Iwnicki S (2006) Handbook of railway vehicle dynamics. CRC Press, Boca Raton

11. Thompson DJ (1996) On the relationship between wheel and rail surface roughness and rolling noise. J Sound Vib 193(1):149–160

12. Grassie SL (2012) Rail irregularities, corrugation and acoustic roughness: characteristics, significance and effects of reprofiling. Proc Inst Mech Eng Part F 226(5):542–557

13. Bonin G, Cantisani G, Carbonari M, Loprencipe G, Pancotto A (2007) Railway traffic vibrations: generation and propagation-theoretical aspects. In: 4th International SIIV Congress, Palermo

14. Zoccali P, Cantisani G, Loprencipe G (2015) Ground-vibrations induced by trains: filled trenches mitigation capacity and length influence. Constr Build Mater 74:1–8. doi:10.1016/j.conbuildmat.2014.09.083

15. Curcuruto S, Atzori D, Betti R, Marsico G, Mazzocchi E, Monaco E, Amoroso F, Limone V, Loprencipe G, De Felice F (2012) Propagation of vibration induced on track: Implementation of previsional models for low and high speed trains and comparison with experimental measurements. In: 19th International Congress on Sound and Vibration 2012, ICSV 2012, 4:3033–3040

16. Di Mascio P, Loprencipe G, Maggioni F (2014) Visco-elastic modeling for railway track structure layers [Modellazione del comportamento visco-elastico degli strati della sede ferroviaria]. Ingegneria Ferroviaria 69(3):207–222

17. Grassie SL (2005) Rail corrugation: advances in measurement, understanding and treatment. Wear 258(7):1224–1234

18. Nielsen JCO, Johansson A (2000) Out-of-round railway wheels-a literature survey. Proc Inst Mech Eng Part F 214(2):79–91

19. Grassie SL (2009) Rail corrugation: characteristics, causes, and treatments. Proc Inst Mech Eng Part F 223(6):581–596

20. Lichtberger B (2005) Track compendium- formation, permanent way, maintenance, economics. Eurail Press, Hamburg

21. Daniel WJT, Horwood RJ, Meehan PA, Wheatley N (2008) Analysis of rail corrugation in cornering. Wear 265(9–10):1183–1192

22. Brickle BJ, Elkins JA, Grassie SL, Handal SJ (1998) Rail corrugation mitigation in transit, Research Results Digest, Transit Cooperative Research Program, National Research Council

23. Grassie SL, Edwards JW (2008) Development of corrugation as a result of varying normal load. Wear 265(9–10):1150–1155

24. Nielsen JCO (2008) High-frequency vertical wheel-rail contact forces-validation

of a prediction model by field testing. Wear 265(9–10):1465–1471

25. Jones CJ, Létourneaux F, Fodiman P (2008) Testing a new rail roughness measurement standard. J Acoust Soc Am 123(5):3266
26. Bongini E, Grassie SL, Saxon MJ (2012) "Noise mapping" of a railway network: validation and use of a system based on measurement of axlebox vibration. In: Maeda T, Gautier PE, Hanson CE, Hemsworth B, Nelson JT, Schulte-Werning B et al (eds) Noise and vibration mitigation for rail transportation systems. Springer, Japan, pp 505–513
27. Ente Italiano di Normazione. UNI EN 15610:2009 Railway applications—Noise emission—Rail roughness measurement related to rolling noise generation
28. Loprencipe G, Cantisani G (2013) Unified analysis of road pavement profiles for evaluation of surface characteristics. Mod Appl Sci 7(8):1–14. doi:10.5539/mas.v7n8p1
29. nternational Organization for Standardization, Mechanical vibration—Road surface profiles—Reporting of measured data, ISO 8608:1995 (E), 09 Jan 1995
30. International Organization for Standardization—Technical specification, characterization of pavement texture by use of surface profiles—part 4: Spectral analysis of surface profiles, ISO/TS 13473-4:2008 (E), 05 Jan 2008
31. European Commission, Technical specification for interoperability relating to the subsystem 'rolling stock—noise' of the trans-European conventional rail system, TSI 2006/66/EC, 23 Dec 2005
32. European Committee for Standardization, Railway applications—Acoustics—Measurement of noise emitted by rail bound vehicles, EN ISO 3095:2005 (E), 31 Jan 2006
33. European Commission, Technical specification for interoperability relating to the rolling stock subsystem of the trans-European high-speed rail system referred to in Article 6(1) of Directive 96/48/EC, TSI 2002/735/EC, 30 May 2002
34. Hardy AEJ, Jones RRK (2004) Rail and wheel roughness-implications for noise mapping based on the calculation of railway noise procedure. AEA Technol Rail 1:46–52
35. European Committee for Standardization, Railway applications—Track—Acceptance of works—part 3: Acceptance of reprofiling rails in track, EN 13231-3:2012 (E), 29 Feb 2012 (accepted in the national standards bodies of Italy from the organization "Ente Nazionale Italiano di Unificazione" as UNI EN 13231-3:2012)
36. Sato Y, Matsumoto A, Knothe K (2002) Review on rail corrugation studies. Wear 253(1–2):130–139
37. Jin X, Wen Z, Wang K, Zhang W (2004) Effect of high-frequency vertical vibration of track on formation and evolution of corrugations. Tsinghua Sci Technol 9(3):274–280

18
CHAPTER

QUANTITATIVE EVALUATION OF CARBONATION IN CONCRETE USING NONLINEAR ULTRASOUND

Gun Kim, Jin-Yeon Kim, Kimberly E. Kurtis, Laurence J. Jacobs, Yann Le Pape, Maria Guimaraes

Keywords: NDE for cement-based materials, Nonlinear Rayleigh waves, Nonlinearity parameter, Carbonation

INTRODUCTION

Exposure of in-service concrete infrastructure to the natural environment causes continuous changes in microstructure and composition, often starting from their surfaces. It is well known that carbonation is a common occurrence in cement-based materials. During carbonation, calcium hydroxide ($Ca(OH)_2$) contained in hydrated cement paste reacts with carbon dioxide, present in air or water, forming calcium carbonate ($CaCO_3$) and water (H_2O). The gradual process of carbonation alters the surface properties and decreases the pH value of the pore solution from typical values (12–13.0) to a 9 or less [1–3] due to the virtually insolubility of the carbonation product ($CaCO_3$) relative to the slightly soluble reactant [4]. More recently, processes to intentionally carbonate ordinary Portland cement concrete and concrete produced with alternative cements are of growing interest as potential options for carbon sequestration [5, 6].

However, over time in reinforced concrete, as the relatively low-pH "carbonation front" reaches greater depths from the surface, depassivation of the reinforcement steel can occur and corrosion can initiate. Corrosion of reinforcement is considered to be a serious durability concern which ultimately limits the service life of reinforced concrete structures [7–9].

Citation: Gun Kim, Jin-Yeon Kim, Kimberly E. Kurtis, Laurence J. Jacobs, Yann Le Pape, Maria Guimaraes, Quantitative evaluation of carbonation in concrete using nonlinear ultrasound, DOI: 10.1617/s11527-014-0506-1.

Hence, for a number of reasons, it is critical to detect carbonation in concrete. Methods to do so non-destructively are preferred such that elements or structures can be monitored over time as carbonation progresses. Also, on large-scale structures, destructive testing is either not permitted or cost-prohibitive and is not capable of providing a representative statistical sampling of the structural condition. Automated monitoring using embarked nondestructive devices is the most promising approach to a full-extent degradation mapping of structures like atmospheric cooling towers or nuclear containment buildings.

Various evaluation methods have been used to detect carbonation in concrete. The use of an indicator solution, like phenolphthalein, is the most well-known method to measure the depth of the carbonation, and this must be done on a freshly fractured surface [3, 10]. Also, analytical methods like thermogravimetric analysis (TGA) [1–3, 11–13] and quantitative X-ray diffraction (QXRD) analysis [11, 14] can be used to assess the relative amount of $Ca(OH)_2$ and $CaCO_3$ at various depths in a sample obtained from a structure. Most recently, indentation methods, (i.e., micro [10] and nanoindentation [15]) have been used to characterize carbonation depths and the process of carbonation. Accordingly, it is well-established that carbonation significantly influences density, strength, elasticity moduli, pore size, and permeability among other properties of cement-based materials [16–18]. Research has also been directed toward increasing the durability against carbonation, which has been shown to be controlled by the use of SCMs [10, 19–23], water-to-cement ratio (w/c) [23–26], curing conditions and time [21, 27, 28], aggregate type [8], the different alkali contents [24, 26, 29], and environmental factors [23, 25]. However, in spite of this extensive research, little effort has been made to use nondestructive evaluation (NDE) methods to evaluate the carbonation process.

NDE techniques using ultrasound have been developed to characterize concrete microstructures [30–33]. Recent research [34–37] introduced nonlinear ultrasonic techniques that can characterize microstructural changes in cement-based materials, such as the formation of microcracks, with more sensitivity than conventional linear ultrasonic methods such as phase velocity or attenuation. For instance, Chen et al. [34] and Leśnicki et al. [35, 36] established the nonlinear impact resonance acoustic spectroscopy (NIRAS) technique based on the shift of the resonance frequency that is caused by material damage; the NIRAS method excites the fundamental resonance of the sample by impact hammer and detects the vibration signal using an accelerometer. They showed that NIRAS can quantify the nonlinear hysteresis of mortar bars with different aggregates, and assessed the influence of progressive alkali-silica reaction (ASR) damage on microstructure using the changes in the measured nonlinearity parameter. Bouchaala et al. [37] also observed a shift of resonance frequencies induced by the carbonation and demonstrated that the measured material nonlinearity appears to be decreased by the progress of carbonation. Nevertheless, these resonance-based techniques are limited to relatively small-sized specimens, so they are impractical for the interrogation of large or complicated components. Also, those resonance methods measure the changes in global damage state of the specimen; the measured nonlinearity in this case represents an averaged value for the entire specimen. On the other hand, the carbonation reaction largely progresses as a relatively uniform front due to the CO_2 penetration from the concrete surface; therefore its product or the changed microstructure is concentrated near the surface. The surface wave-based techniques are well-suited to evaluate material properties such as the density, Poisson's ratio, and the shear modulus at the surface, where carbonation affects these properties, since the wave's energy is concentrated near surface. Additionally, the fact that they only need access to one-side of the specimen shows their capability for the application to in situ interrogation of in-service structures.

It has been demonstrated that second harmonic generation (SHG) methods using nonlinear Rayleigh surface waves [38–44] have great potential. The experimentally obtained acoustic nonlinearity parameter, β, (defined in Eq. (3)) can be related to the microstructural changes caused by damage if a proper model is developed. Furthermore, the unwanted system nonlinearity from the instrumentation can be eliminated by varying the propagation distance. Most recently, Thiele et al. [42] developed an SHG technique by using an air-coupled receiver that allows for a more consistent measurement of the nonlinearity parameter. Kim et al. [38] proposed a similar SHG setup for cement-based materials and assessed the effect of shrinkage-reducing admixture (SRA) on reducing autogenous and drying shrinkage, showing that the measured acoustic nonlinearity parameter is decreased by the addition of SRA.

The objective of this paper is to quantitatively assess the carbonation effect on the microstructure in a prepared concrete specimen using the proposed SHG setup in [38]. Specifically, it is expected that microcracks in the specimen—which are inherent defects in cement-based composites [45]—are changed by the evolution of the carbonation which will in turn affect the nonlinearity parameter. A 45 kHz wedge transducer and a 100 kHz air-coupled (non-contact) transducer are implemented to generate and to receive the Rayleigh surface waves, respectively. In order to track the degree of carbonation, a phenolphthalein indicator solution is employed to measure carbonation depths on additional small concrete blocks as a function of days. Finally, the potential and advantages of applying the SHG technique for evaluating carbonation in cement-based materials are discussed.

NONLINEARITY PARAMETER: SECOND HARMONIC GENERATION (SHG)

Consider a plane Rayleigh surface wave propagating along the x axis in an isotropic, macroscopically homogeneous, and nonlinear elastic half space where the z axis refers to the depth of the material. First, the displacements in the x and z directions can be written in Eq. (1) and both displacement components indicate the shear and longitudinal wave contributions [46].

$$u_x(\omega) = A_1 \left(e^{b_1 z} - \frac{2b_1 b_2}{k_R^2 + b_2^2} e^{b_2 z} \right) e^{i\{k_R(x - c_R t)\}} \tag{1a}$$

$$u_z(\omega) = iA_1 \frac{b_1}{k_R} \left(e^{b_1 z} - \frac{2k_R^2}{k_R^2 + b_2^2} e^{b_2 z} \right) e^{i\{k_R(x - c_R t)\}}, \tag{1b}$$

where, C_R is the speed of Rayleigh surface waves, ω is the fundamental angular frequency ($=C_R k_R$), $b_1 = \sqrt{k_R^2 - k_l^2}$ and $b_2 = \sqrt{k_R^2 - k_s^2}$ with k_l, k_s, and k_R are the wave numbers of the longitudinal wave, the shear wave, and the Rayleigh waves.

In an isotropic material that has weak quadratic nonlinearity, the displacements of the second harmonic Rayleigh waves in the far-field can be expressed as [39, 47]:

$$u_x(2\omega) \approx A_2\left(e^{2b_1 z} - \frac{2b_1 b_2}{k_R^2 + b_2^2} e^{2b_2 z}\right) e^{i\{2k_R(x - c_R t)\}} \tag{2a}$$

$$u_z(2\omega) \approx iA_2 \frac{b_1}{k_R}\left(e^{2b_1 z} - \frac{2k_R^2}{k_R^2 + b_2^2} e^{2b_2 z}\right) e^{i\{2k_R(x - c_R t)\}} \tag{2b}$$

Hermann et al. [39] showed that the acoustic nonlinearity parameter can be expressed by the out-of-plane displacement components of the propagating Rayleigh waves at the surface ($z = 0$) as shown in Eq. (3)

$$\beta = \frac{u_z(2\omega)|_{z=0}}{u_z^2(\omega)|_{z=0}} \frac{8b_i i}{k_l^2 k_R x}\left(1 - \frac{2k_R^2}{k_R^2 + b_2^2}\right), \tag{3}$$

where, β is the acoustic nonlinearity parameter [39, 47–49]. The relationship between the fundamental and second harmonic components and the acoustic nonlinearity parameter can be written as [47]

$$\frac{A_2}{A_1^2} \propto \frac{\beta \omega^2 x}{C_R^2} \tag{4}$$

Since the wave speed of the Rayleigh surface wave is independent of the frequency of the wave, A_2/A_1^2 is directly proportional to β at a fixed distance as shown in Eq. (5). It is well established that the relationship between A_2/A_1^2 and the propagation distance is a good approximation for calculating the acoustic nonlinearity parameter for short propagation distances

$$\beta \propto \frac{A_2}{A_1^2 x}. \tag{5}$$

This paper denotes $A_2/A_1^2 x$ as the relative nonlinearity parameter, β_{re} as shown in Eq. (6)

$$\beta_{re} = \frac{A_2}{A_1^2 x}. \tag{6}$$

The amplitude of the fundamental (A_1) and second harmonic (A_2) can be experimentally measured using the setup described in the next section. Therefore, the changes in the elastic nonlinearity of a material can be assessed using a nondestructive, macroscopic measurement.

EXPERIMENTAL PROCEDURE

Specimen Preparation

Based on ACI 211.1 [50], one concrete specimen (21.5 × 56.5 × 17.5 cm) with Type I

Portland cement (ASTM C 150) [51] is prepared for both accelerated carbonation tests and nonlinear ultrasonic (NLU) measurements. River gravel coarse aggregate with a maximum size of aggregate (MSA) of 12.7 mm, and dry rodded unit weight (DRUW) of 1,646 kg/m^3 (103 pcf) and a natural sand with a fineness modulus (FM) of 3.04, absorption capacity of 1.93 %, and saturated surface dry (SSD) specific gravity of 2.65 are used as aggregate. A water-to-cement ratio (w/c) of 0.60 is selected to accelerate the rate and depth of carbonation [24]. Table 1shows further details on the mixture design of concrete specimen.

Table 1: Mixture design of concrete specimen, kg/m^3 (lb/yd^3)

	Concrete
Water	271 (365)
Cement (ASTM C150 type I)	361 (608)
Water-to-cement ratio (w/c)	0.60
Coarse aggregate	872 (1,470)
Fine aggregate	784 (1,322)

The freshly mixed concrete is cast, and cured for 14 days at ambient temperature while covered by polyethylene sheeting. The demolded specimen is then placed in an environmental chamber (23 °C and 95 % relative humidity) in relatively moist conditions to facilitate hydration until it reaches 28 days of age. Finally, the specimen is transferred into an environmental chamber (23 °C and 50 ± 2 % relative humidity) to dry for 14 days to bring it to an equilibrium moisture condition. Additionally, five concrete blocks, each sealed on 5 sides for one-dimensional penetration of CO_2, are prepared according to the same mix design and cured and conditioned in the same manner. An indicator solution, 1 % solution of phenolphthalein in deionized water with 10 % v/v ethanol, are prepared for tracking the carbonation depth on freshly cut concrete surfaces.

Accelerated Carbonation Experiments

The accelerated carbonation test on the prepared concrete specimen and blocks is carried out for 40 days in a carbonation chamber, with temperature, relative humidity, and CO_2concentration control. Note that the maximum carbonation rate occurs at approximately 55–65 % RH [2, 52]. In this experiment, the RH is fixed to 55 %. The temperature in the environmental chamber is maintained at 26 °C. During the accelerated carbonation test, a continuous supply of CO_2 gas is maintained by the pressure regulator mounted on the pressurized steel cylinder and 20 % CO_2 is continuously fed into the chamber [10]. In this condition, the increase in carbonation depth due to increases in temperature is negligible [2]; the rate of carbonation is governed by the diffusion of CO_2. The test is interrupted after 0, 7, 14, 25, and 31 days of exposure so as to estimate the level of carbonation penetration in each block and the phenolphthalein indicator solution is applied to a freshly cut surface.

Non-Contact, Nonlinear Ultrasonic Measurements

The non-contact (air-coupled) measurements of SHG in Rayleigh surface waves are performed with the setup shown in Fig. 1. Details on the experimental procedure to calculate the nonlinearity parameter in concrete are described in [38]. A narrow band longitudinal wave piezoelectric transducer with a diameter of 50 mm (Ultran GRD 50) is employed as a transmitter and is coupled to a Teflon wedge with vacuum grease. The longitudinal wave speed in Teflon is 1,450 m/s and the Rayleigh surface wave speed in the concrete specimen is 2,450 m/s. Therefore the wedge angle for generating the Rayleigh waves in the concrete specimen is 36.3° based on Snell's law. To maximize the detected wave amplitude and to have a minimum attenuation, the transmitter/wedge assembly is designed in such a way that the wedge's bottom surface is smooth enough and the assembly is tightly coupled to the concrete specimen using the same vacuum grease.

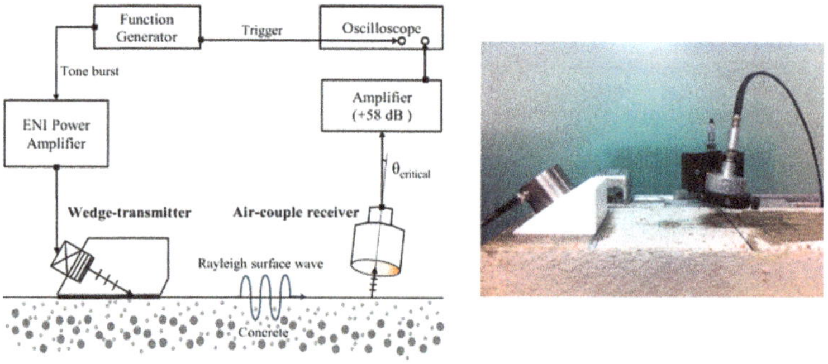

Figure 1: Air-coupled detection setup for Rayleigh surface waves.

A tone-burst signal of 16 cycles at 45 kHz is generated by a function generator (AGILENT 33250A) with a peak-to-peak voltage of 400 mV and is repeated every 200 ms. Then a power amplifier (ENI Power Amplifier) amplifies the signal as much as 50 dB and the amplified signal is fed to the wedge transmitter. To detect the propagating Rayleigh waves, a non-contact, air-coupled transducer with a diameter of 50 mm (Ultran GRD 100) is used. Here, the air-coupled receiver is tilted by 8° which is the critical angle for the Rayleigh wave at the air-concrete interface and it is slightly adjusted for small variations of wave speed during the carbonation. Then, the distance between the air-coupled receiver and the surface, called the lift-off distance, is fixed at 3.3 cm. At this distance, one can reduce the effects of attenuation and diffraction in the air [38, 42] while avoiding the multiple reflections of the acoustic signal between the concrete surface and transducer surface. To increase the signal-to-noise ratio (SNR), the received signal is amplified by 58 dB with a broadband preamplifier (Digital Wave Inc.). Additionally, the received time-domain signals are averaged 256 times with an oscilloscope (Tektronix TDS 5034B Digital); the calculated SNR is approximately 50 dB. Finally, the averaged signal is transferred to a computer for post-processing. The entire measurement is synchronized by a trigger signal from the function generator as shown in Fig. 1. The measurements are repeated 15–20 times.

During the accelerated carbonation process, the specimen is periodically removed from the chamber for the ultrasonic measurements, which are made after 0, 3, 7, 10, 14, 19, 25, 31, and 40 days of carbonation. Since the dimensions of the specimen limits the propagation distance, the propagation distance varies only from 250 to 295 mm by an incremental step of 5 mm. In the post-processing, a Hann window is applied to the first 5 cycles of the received time domain signals since it is expected that the boundary-reflected waves arrive at approximately 42 ms (the 6th cycle) and this reflected signal interferes with the directly propagating Rayleigh wave. It should be noted that it has been experimentally demonstrated that the first 5 cycles are not affected by these reflected waves [38]. Figure 2a shows the width of the Hann window (denoted by the blue squares) as well as the received time domain signal, and Fig. 2b shows the measured amplitudes of the fundamental (A_1) and second harmonic (A_2) from the frequency spectrum of the windowed signal obtained using the fast Fourier transform (FFT).

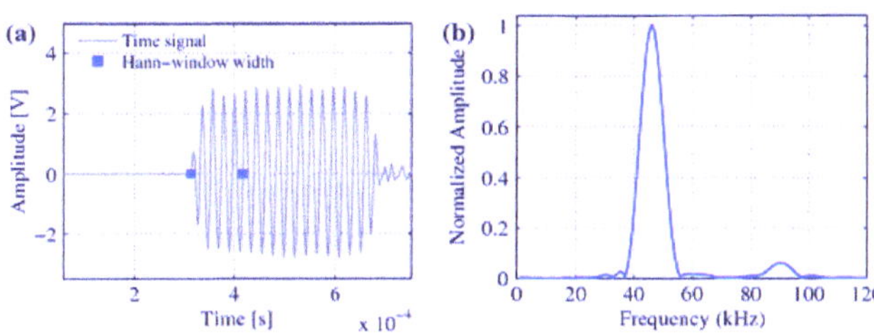

Figure 2: Time signal and Fourier spectrum for NLU measurement (45 and 90 kHz).

Figures 3 and 4 show an example of the measured A_1, A_2, and A_2/A_1^2 versus propagation distance on the concrete specimen cured for 31 days in the carbonation chamber. Obviously, the measured fundamental amplitude, A_1 tends to decrease with propagation distance mainly due to the beam spreading and attenuation effects in concrete. On the other hand, due to the cumulative nonlinear effects described in [53], the second harmonic amplitude, A_2 increases with propagation distance. Figure 4 shows that A_2/A_1^2 increases linearly with propagation distance, and the increase in A_2/A_1^2 is in agreement with previous experimental observations [38–44].

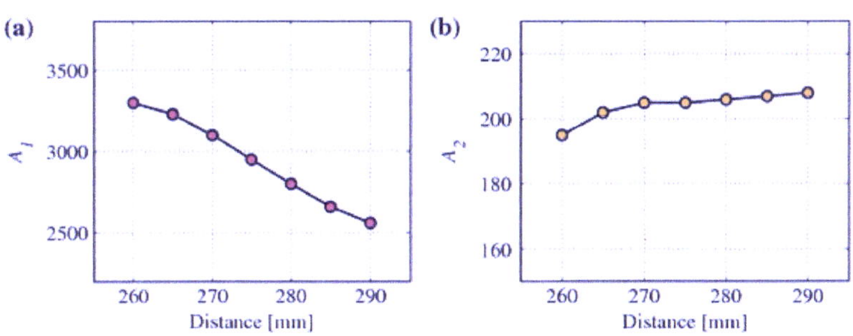

Figure 3: The measured fundamental and second harmonic amplitudes of nonlinear Rayleigh surface waves.

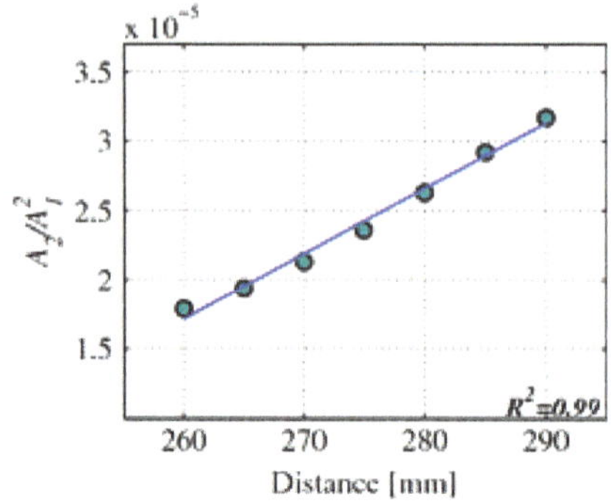

Figure 4: A_2/A_1^2 versus propagation distance.

RESULTS AND DISCUSSION

Carbonation Profile Using Phenolphthalein Indicator Solution

First, the destructive tests are performed to determine the carbonation depth. The indicator is sprayed uniformly on each freshly broken surface of the carbonated blocks after 0, 7, 14, 25, and 31 days of carbonation in the chamber. The results show that the uncarbonated area, of pH higher than approximately 9.2, appears as purple, while the carbonated area remains colorless. Figure 5 shows the photographs of each cross-section of the carbonated blocks. The depth of carbonation is then measured by averaging at 7 points perpendicular to one face of the split concrete blocks, and the measured depths are 1.7, 9, 14, 18, and 21 mm, respectively. As a result, it is determined that the carbonation depth is nonlinearly increased over time, as expected [10].

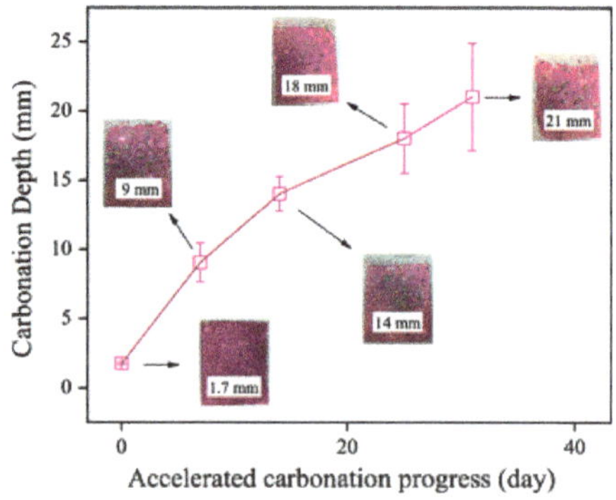

Figure 5: Averaged carbonation depth using phenolphthalein indicator solution.

Linear Ultrasonic Parameter (Phase Velocity) Results

The phase velocity is measured using the time domain signals received at two different propagation distances. The distance between the two measurement positions is divided by the difference in arrival times to obtain the phase velocity. The resulting phase velocities during the carbonation process are shown in Fig. 6 and Table 2. It is interesting to note that the measured phase velocity increases with the carbonation process. This increase suggests that the elastic moduli in fact increase more than the increase in density by the $CaCO_3$ formation. However, the results show only a maximum of 5.5 % change during the curing for 40 days. Furthermore, the changes in phase velocity are comparable to the measurement error bars. For these reasons, it is difficult to use the phase velocity as an indicator to detect microstructural changes due to the carbonation reaction.

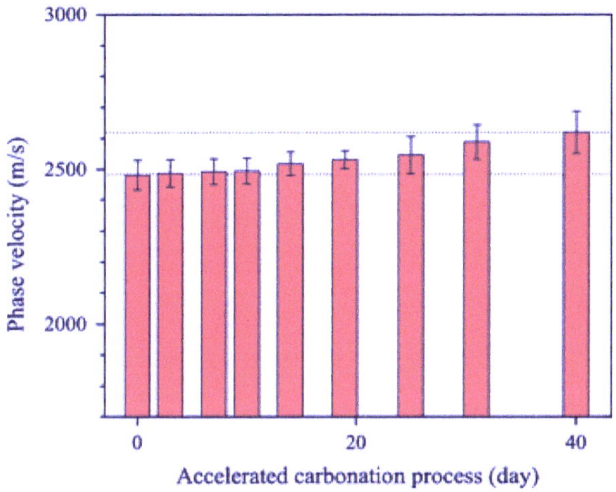

Figure 6: Trend of measured Rayleigh phase velocity with the accelerated carbonation tests.

Table 2: The averaged Rayleigh phase velocity

Exposure time (day)	0	3	7	10	14	19	25	31	40
Phase velocity (m/s)	2,481.84	2,487.17	2,492.71	2,495.57	2,518.57	2,531.37	2,546.04	2,588.22	2,618.94

Additionally, it is confirmed that the concrete specimen can be treated as a weakly dispersive medium, in agreement with the results of [38], since the phase velocity for the specimen at each carbonation state is almost constant within the effective frequency range (approximately 40–100 kHz). This is important since the carbonation layer could cause dispersion; a half-space with a finite layer of different material is in general, dispersive. However, our additional measurements on these carbonation specimens demonstrate that the measured velocity difference (velocities at the fundamental versus second harmonic frequencies) is quite small (maximum 4.9 % from the specimen with a 24 mm carbonation layer). Figure 3b also shows a cumulative second harmonic, so the effects of dispersion are negligible and the phase velocity matching of this second harmonic is approximated in this weakly dispersive system at these relatively small propagation distances (0.03 m).

Acoustic Nonlinearity Parameter Results

The measurement of the SHG in nonlinear Rayleigh surface waves is performed on the concrete specimen as it is exposed to the accelerated carbonation process for 40 days. This measurement uses 45 kHz for the fundamental frequency of the transmitted Rayleigh surface waves. The wavelength at the second harmonic frequency is approximately 27.2 mm, which is greater than the maximum size of the aggregate (MSA) of 12.7 mm so that the measured second harmonic amplitudes are minimally influenced by acoustic scattering. The relative nonlinearity parameter, β_{re} is measured at days 0, 3, 7, 10, 14, 19, 25, 31, and 40. Each data point in Fig. 7represents an average of the results from 15 to 20 repeated measurements on this specimen. The repeatability of air-coupled detection of the SHG is examined in [38], and the magnitude of the error bars in this figure is mainly due to variations in the contact condition between the wedge transducer (transmitter) and the specimen surface when the wedge transducer is removed and reattached to the specimen for repeated measurements.

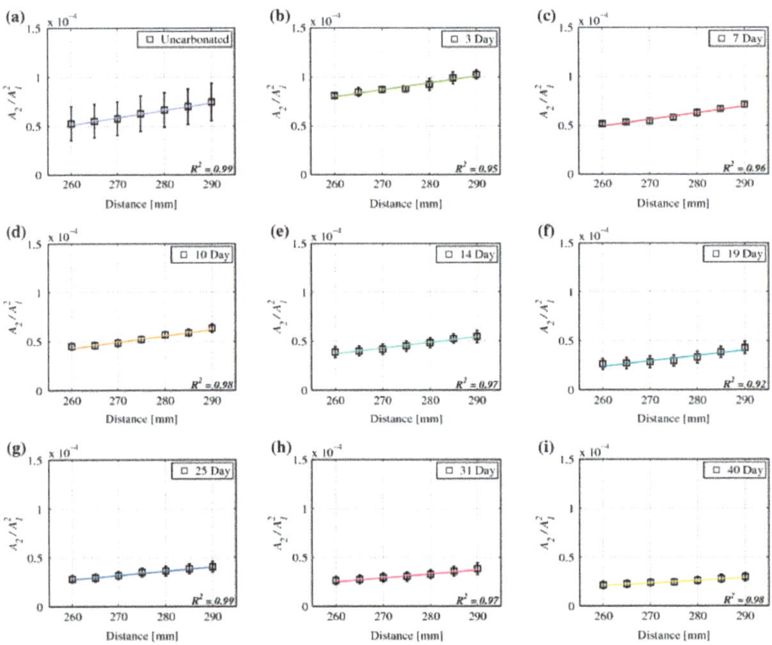

Figure 7: The changes in the slope of the linear fit by the carbonation process.

Figure 7 shows the slope of the linear fit of the A_2/A_1^2 versus the propagation distance, which represents β_{re}. Note that all of these β_{re} have R^2 correlation coefficients higher than 0.92. All measured relative nonlinearity parameters, β_{re} during the carbonation process are listed in Table 3; the measured β_{re} are shown to decrease by a maximum of 64.2 % from that of the uncarbonated concrete specimen. Comparing the results in Figs. 5 and 8, the measured nonlinearity parameters are shown to decrease in a monotonic fashion while the carbonation depth is increasing in an inversely proportional fashion to the nonlinearity parameter. It is interesting to note that the values of β_{re} tend to asymptotically approach a constant value after this maximum measurement period of 40 days since the carbonation depth is almost developed up to the wavelength (~27 mm) of the second harmonic Rayleigh surface waves, i.e. beyond this penetration depth, the Rayleigh waves at this frequency lose their sensitivity to the progression of carbonation.

Table 3: The measured relative nonlinearity parameters

Exposure time (day)	0	3	7	10	14	19	25	31	**40**
Relative nonlinearity parameter ($\times 10^{-7}$)	7.49	7.05	6.65	6.46	5.69	5.50	4.35	4.01	2.68

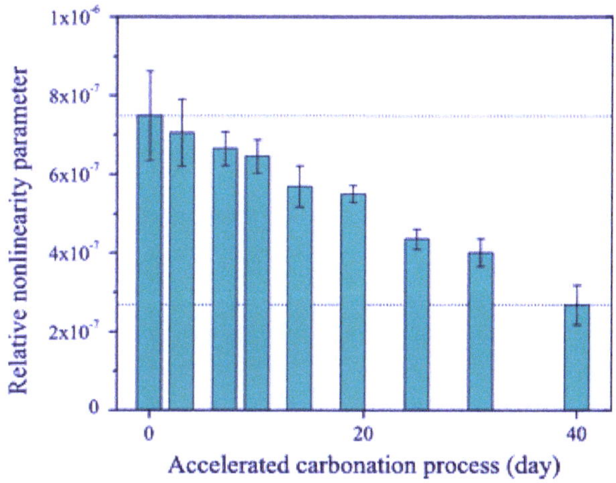

Figure 8: Trend of the measured relative nonlinearity parameters and the effect of carbonation on material nonlinearity.

Finally, Fig. 9 compares the relative changes in both the linear and nonlinear parameters, and shows that the sensitivity of the nonlinearity parameter is approximately 12 times higher than that of the Rayleigh phase velocity. More importantly, the relative nonlinearity parameter exhibits a sensitivity high enough to characterize the changes in microstructure, especially the decrease in porosity including filling of microcracks, induced by carbonation. Overall, this research verifies that the SHG technique using nonlinear Rayleigh surface waves provides a promising indicator capable of quantitatively evaluating carbonation in cement-based materials.

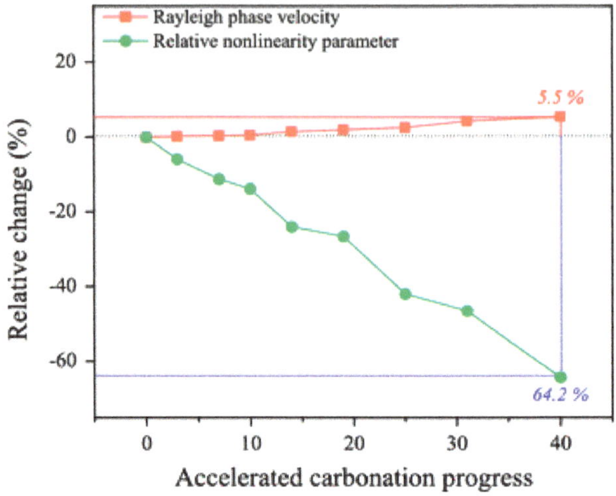

Figure 9: Sensitivity of nonlinear versus linear ultrasonic parameters.

Interpretation of Nonlinear Ultrasonic Measurements

The SHG results highlight that the measured relative nonlinearity parameter is significantly reduced by carbonation and show the sensitivity of the measured nonlinearity parameter to the microstructural changes in the carbonated layer. Note that the environmental humidity, CO_2concentration, and temperature along with the concrete w/c are optimized in this specimen to enhance the carbonation reaction. Based on the nonlinear ultrasonic measurements, it is shown that the carbonation of concrete leads to a reduction in nonlinearity. Most importantly, it is likely that the decrease in the measured relative nonlinearity parameter is mainly attributed to the fact that some pre-existing voids and perhaps some smaller microcracks become filled by $CaCO_3$, as described in [54, 55].

CONCLUSION

This paper applies the SHG technique in nonlinear Rayleigh surface waves to concrete to characterize the carbonation-induced microscale behavior of the concrete after exposure to a CO_2-rich environment, designed to rapidly develop a carbonation front in the concrete. In parallel with the nonlinear ultrasonic measurements which showed a decrease in nonlinearity with carbonation, the destructive characterization using a phenolphthalein indicator solution is performed to track the carbonation depth. The following conclusions are drawn from these experimental results:

- NDE of the microstructural behavior in the carbonated concrete is achieved by the proposed SHG technique. The relative nonlinearity parameter, β_{re} decreases by as much as 64.2 % after 40 days of curing in the carbonation chamber, and this decrease is attributed to the formation of carbonation product, $CaCO_3$ which fills pre-existing microcracks, reducing the relative nonlinearity parameter, β_{re}.

- An inverse relationship between the measured carbonation depths and relative nonlinearity parameters is observed.
- It is clearly demonstrated that the relative nonlinearity parameter, β_{re} shows a much higher sensitivity to the carbonation-induced microstructure changes than the measured Rayleigh phase velocity (64.2 vs. 5.5 %). It is therefore expected that the acoustic nonlinearity parameter can be used as a quantitative index for understanding the effect of carbonation if a model that can relate β_{re} to the microstructural changes is available.

Some challenges in the use of NDE to evaluate carbonation still remain to be addressed. These measurements used a relatively low frequency pair (i.e. 50–100 kHz) to reduce the effect of acoustic scattering due to the aggregate. However, more precise characterization could be possible using a range of higher and lower frequency pairs for carbonation penetration.

Acknowledgments

The authors gratefully acknowledge the research support from the Electric Power Research Institute (EPRI), Research Project No. 073289.

REFERENCES

1. Papadakis VG, Vayenas CG (1991) Experimental investigation and mathematical modeling of the concrete carbonation problem. Chem Eng Sci 46:1333–1338
2. Papadakis VG, Vayenas CG, Fardis MN (1991) Fundamental modeling and experimental investigation of concrete carbonation. ACI Mater J 88:363–373
3. Parrott LJ, Killoh DC (1989) Carbonation in a 36 year old, in situ concrete. Cem Concr Res 19:649–656
4. Dow C, Glasser FP (2003) Calcium carbonate efflorescence on Portland cement and building materials. Cem Concr Res 33:147–154
5. Bertos MF, Simons SJR, Hills CD, Carey PJ (2004) A review of accelerated carbonation technology in the treatment of cement-based materials and sequestration of CO_2. J Hazard Mater B112:193–205
6. Monteiro PJM, Clodic L, Battocchio F, Kanitpanyacharoen W, Chae SR, Ha J, Wenk H-K (2013) Incorporating carbon sequestration materials in civil infrastructure: a micro and nano-structural analysis. Cem Concr Compos 40:14–20
7. Basheer L, Kropp J, Cleland DJ (2001) Assessment of the durability of concrete from its permeation properties: a review. Constr Build Mater 15:93–103
8. Rozière E, Loukili A, Cussigh F (2009) A performance based approach for durability of concrete exposed to carbonation. Constr Build Mater 23:190–199
9. Koch GH, Brongers MPH, Thompson NG, Virmani YP, Payer, JH (2002) Corrosion cost and preventive strategies in the United States. Publication no. FHWA-RD-01-156. FHWA, Washington, DC

10. Holland RB (2012) Durability of precast prestressed concrete piles in marine environments. Dissertation, Georgia Institute of Technology
11. Kobayashi K, Suzuki K, Uno Y (1994) Carbonation of concrete structures and decomposition of C-S-H. Cem Concr Res 24:55–61
12. Villain G, Platret G (2006) Two experimental methods to determine carbonation profiles in concrete. ACI Mater J 103:265–271
13. Thiery M, Villain G, Dangla P, Platret G (2007) Investigation of the carbonation front shape on cementitious materials: effects of the chemical kinetics. Cem Concr Res 37:1047–1058
14. Moorehead DR (1986) Cementation by the carbonation of hydrated lime. Cem Concr Res 16:700–708
15. Vanzo J (2009) A nanochemomechanical investigation of carbonated cement paste. Master thesis, Massachusetts Institute of Technology
16. Johannesson B, Utgenannt P (2001) Microstructural changes caused by carbonation of cement mortar. Cem Concr Res 31:925–931
17. Jerga J (2004) Physico-mechanical properties of carbonated concrete. Constr Build Mater 18:645–652
18. Khan MI, Lynsdale CJ (2002) Strength, permeability, and carbonation of high-performance concrete. Cem Concr Res 32:123–131
19. Atiş CD (2003) Accelerated carbonation and testing of concrete made with fly ash. Constr Build Mater 17:147–152
20. Torgal FP, Miraldo S, Labrincha JA, Brito JD (2012) An overview on concrete carbonation in the context of eco-efficient construction: evaluation, use of SCMs and/or RAC. Constr Build Mater 36:141–150
21. Younsi A, Turcry P, Rozière E, Aït-Mokhtar A, Loukili A (2011) Performance-based design and carbonation of concrete with high fly ash content. Cem Concr Compos 33:993–1000
22. Papadakis VG (2000) Effect of supplementary cementing materials on concrete resistance against carbonation and chloride ingress. Cem Concr Res 30:291–299
23. Kamali S, Moranville M, Leclercq S (2008) Material and environmental parameter effects on the leaching of cement pastes: experiments and modelling. Cem Concr Res 38:575–585
24. Kobayashi K, Uno Y (1990) Influence of alkali on carbonation of concrete, part 2-influence of alkali in cement on rate of carbonation of concrete. Cem Concr Res 20:619–622
25. Papadakis VG, Fardis MN, Vayenas CG (1992) Effect of composition, environmental factors and cement-lime mortar coating on concrete carbonation. Mater Struct 25:293–304
26. De Schutter G, Audenaert K (2004) Evaluation of water absorption of concrete as a measure for resistance against carbonation and chloride migration. Mater Struct 37:591–596

27. Fattuhi NI (1988) Concrete carbonation as influenced by curing regime. Cem Concr Res 18:426–430
28. Rostami V, Shao Y, Boyd AJ, He Z (2012) Microstructure of cement paste subject to early carbonation curing. Cem Concr Res 42:186–193
29. Kobayashi K, Uno Y (1989) Influence of alkali on carbonation of concrete, part 1, preliminary tests with mortar specimen. Cem Concr Res 19:821–826
30. Popovics JS, Song WJ, Ghandehari M, Subramaniam KV, Achenbach JD, Shah SP (2000) Application of surface wave transmission measurements for crack depth determination in concrete. ACI Mater J 97:127–135
31. Popovics S, Rose JL, Popovics JS (1990) The behavior of ultrasonic pulses in concrete. Cem Concr Res 20:259–270
32. In C-W, Holland RB, Kim J-Y, Kurtis KE, Kahn LF, Jacobs LJ (2013) Monitoring and evaluation of self-healing in concrete using diffuse ultrasound. NDT&E Int 57:36–44
33. Selleck SF, Landis EN, Peterson ML, Shah SP, Achenbach JD (1998) Ultrasonic investigation of concrete with distributed damage. ACI Mater J 95:27–36
34. Chen J, Jayapalan AR, Kim J-Y, Kurtis KE, Jacobs LJ (2010) Rapid evolution of alkali-silica reactivity of aggregates using a nonlinear resonance spectroscopy technique. Cem Concr Res 40:914–923
35. Leśnicki KJ, Kim J-Y, Kurtis KE, Jacobs LJ (2011) Characterization of ASR damage in concrete using nonlinear impact resonance acoustic spectroscopy technique. NDT&E Int 44:721–727
36. Leśnicki KJ, Kim J-Y, Kurtis KE, Jacobs LJ (2013) Assessment of alkali-silica reaction damage through quantification of concrete nonlinearity. Mater Struct 46:497–509
37. Bouchaala F, Payan C, Garnier V, Balayssac JP (2011) Carbonation assessment in concrete by nonlinear ultrasound. Cem Concr Res 41:557–559
38. Kim G, In C-W, Kim J-Y, Kurtis KE, Jacobs LJ (2014) Air-coupled detection of nonlinear Rayleigh surface waves in concrete-application to microstructure characterization. NDT&E Int 67:64–70
39. Herrmann J, Kim J-Y, Jacobs LJ, Qu J, Littles JW, Savage MF (2006) Assessment of material damage in a nickel-base super alloy using nonlinear Rayleigh surface waves. J Appl Phys 99:124913
40. Walker SV, Kim J-Y, Qu J, Jacobs LJ (2012) Fatigue damage evaluation in A36 steel using nonlinear Rayleigh waves. NDT&E Int 48:10–15
41. Ruiz A, Ortiz N, Medina A, Kim J-Y, Jacobs LJ (2013) Application of ultrasonic methods for early detection of thermal damage in 2205 duplex stainless steel. NDT&E Int 54:19–26
42. Thiele S, Kim J-Y, Qu J, Jacobs LJ (2014) Air-coupled detection of nonlinear Rayleigh surface waves to assess material nonlinearity. Ultrasonics 54:1470–1475
43. Kim G, In C-W, Kim J-Y, Jacobs LJ, Kurtis KE (2014) Nondestructive detection

and characterization of carbonation in concrete. Rev Prog Quant Nondestr Eval 1581:805–813

44. Zeitvogel DT, Matlack KH, Kim J-Y, Jacobs LJ, Singh PM, Qu J (2014) Characterization of stress corrosion cracking in carbon steel using nonlinear Rayleigh surface waves. NDT&E Int 62:144–152
45. Mehta PK, Monteiro PJM (2006) Concrete: microstructure, properties, and materials, 3rd edn. McGraw-Hill, New York
46. Viktorov IA (1967) Rayleigh and Lamb waves. Plenum Press, New York
47. Matlack KH, Kim J-Y, Jacobs LJ, Qu J (2015) Review of second harmonic generation measurement techniques for material state determination. J Nondestruct Eval. doi:10.1007/s10921-014-0273-5
48. Huang CL (2013) Effect of beam diffraction on nonlinear Rayleigh surface wave measurements. Master thesis, Georgia Institute of Technology
49. Torello D, Thiele S, Matlack KH, Kim J-Y, Qu J, Jacobs LJ (2015) Diffraction, attenuation, and source corrections for nonlinear Rayleigh wave ultrasonic measurements. Ultrasonics 56:417–426
50. ACI Committee 211 (1991) Standard practice for selecting proportions for normal, heavyweight, and mass concrete (AU 211.1-91). ACI, Detroit, p 38
51. ASTM C 150-04 (2004) Standard specification for Portland cement. ASTM, PA
52. Bertolini L, Elsener B, Pedeferri P, Polder RB (2004) Corrosion of steel in concrete: prevention diagnosis, repair. Wiley-VCH, Weinheim
53. Kim J-Y, Jacobs LJ, Qu J, Littles JW (2006) Experimental characterization of fatigue damage in a nickel-base superalloy using nonlinear ultrasonic waves. J Acoust Soc Am 120:1266–1273
54. Hearn N (1998) Self-sealing, autogenous healing and continued hydration: what is the difference? Mater Struct 31:563–567
55. Edvardsen C (1999) Water permeability and autogenous healing of cracks in concrete. ACI Mater J 96:448–454

INDEX

M

N

O

P

Q

R

S

T

U

W

Z